Course   Student Solutions Manual for
           Introductory Algebra,
           Third Edition

           Julie Miller

http://create.mheducation.com

This McGraw-Hill Create text may include materials submitted to
McGraw-Hill for publication by the instructor of this course.
The instructor is solely responsible for the editorial content of such
materials. Instructors retain copyright of these additional materials.

ISBN-10: 0077343239   ISBN-13: 9780077343231

Contents

Credits

Chapter R

Section R.1 Practice Exercises

Answers will vary.

Section R.2 Practice Exercises

1. **a.** product
   b. factors
   c. numerator; b
   d. lowest
   e. 1;4
   f. reciprocals
   g. multiple
   h. least

3. Numerator 2; denominator 3; proper

5. Numerator 6; denominator 6; improper

7. Numerator 12; denominator 1; improper

9. $\dfrac{3}{4}$

11. $\dfrac{4}{3}$

13. $\dfrac{1}{6}$

15. $\dfrac{2}{2}$

17. $\dfrac{5}{2}$ or $2\dfrac{1}{2}$

19. $\dfrac{6}{2}$ or 3

21. The set of whole numbers includes the number 0 and the set of natural numbers does not.

23. Answers may vary. One example would be $\dfrac{2}{4}$.

25. Prime

27. Composite

29. Composite

31. Prime

33. $2 \times 2 \times 3 \times 3$

35. $2 \times 3 \times 7$

37. $2 \times 5 \times 11$

39. $3 \times 3 \times 3 \times 5$

41. $\dfrac{3}{15} = \dfrac{\not{3}}{\not{3} \times 5} = \dfrac{1}{5}$

43. $\dfrac{16}{6} = \dfrac{\not{2} \times 2 \times 2 \times 2}{\not{2} \times 3} = \dfrac{8}{3}$ or $2\dfrac{2}{3}$

45. $\dfrac{42}{48} = \dfrac{\not{2} \times \not{3} \times 7}{\not{2} \times 2 \times 2 \times 2 \times \not{3}} = \dfrac{7}{8}$

47. $\dfrac{48}{64} = \dfrac{\not{2} \times \not{2} \times \not{2} \times \not{2} \times 3}{\not{2} \times \not{2} \times \not{2} \times \not{2} \times 2 \times 2} = \dfrac{3}{4}$

49. $\dfrac{110}{176} = \dfrac{\not{2} \times 5 \times \not{11}}{\not{2} \times 2 \times 2 \times 2 \times \not{11}} = \dfrac{5}{8}$

51. $\dfrac{200}{150} = \dfrac{\not{2} \times 2 \times 2 \times \not{5} \times \not{5}}{\not{2} \times 3 \times \not{5} \times \not{5}} = \dfrac{4}{3}$ or $1\dfrac{1}{3}$

53. False: When adding/subtracting fractions, it is necessary to have a common denominator.

55. $\dfrac{10}{13} \times \dfrac{26}{15} = \dfrac{2 \times 2 \times \not{5} \times \not{13}}{3 \times \not{5} \times \not{13}} = \dfrac{4}{3}$ or $1\dfrac{1}{3}$

57. $\dfrac{3}{7} \div \dfrac{9}{14} = \dfrac{3}{7} \times \dfrac{14}{9} = \dfrac{2 \times \not{3} \times \not{7}}{3 \times \not{3} \times \not{7}} = \dfrac{2}{3}$

59. $\dfrac{9}{10} \times 5 = \dfrac{9}{10} \times \dfrac{5}{1} = \dfrac{3 \times 3 \times \not{5}}{2 \times \not{5}} = \dfrac{9}{2}$ or $4\dfrac{1}{2}$

61. $\dfrac{12}{5} \div 4 = \dfrac{12}{5} \div \dfrac{4}{1} = \dfrac{12}{5} \times \dfrac{1}{4} = \dfrac{\overset{3}{\cancel{12}} \times 1}{5 \times \cancel{4}} = \dfrac{3}{5}$

63. $\dfrac{5}{2} \times \dfrac{10}{21} \times \dfrac{7}{5} = \dfrac{\overset{1}{\cancel{5}} \times \overset{5}{\cancel{10}} \times \overset{1}{\cancel{7}}}{\underset{1}{\cancel{2}} \times \underset{3}{\cancel{21}} \times \underset{1}{\cancel{5}}} = \dfrac{5}{3} \text{ or } 1\dfrac{2}{3}$

65. $\dfrac{9}{100} \div \dfrac{13}{1000} = \dfrac{9}{100} \times \dfrac{1000}{13} = \dfrac{9 \times \overset{10}{\cancel{1000}}}{\underset{1}{\cancel{100}} \times 13} = \dfrac{90}{13}$

or $6\dfrac{12}{13}$

67. $\dfrac{1}{4}$ of $\$4200 = \dfrac{1}{4} \times \dfrac{4200}{1} = \dfrac{4200}{4} = \$1050$

69. The statement "five-sixths of the students passed with the first test" translates to " students who passed $= \dfrac{5}{6} \times 42$ "

$\dfrac{5}{6} \times 42 = \dfrac{5}{6} \times \dfrac{42}{1}$

$= \dfrac{5 \times \overset{7}{\cancel{42}}}{\underset{1}{\cancel{6}} \times 1}$

$= \dfrac{35}{1}$

$= 35$ passed the first test

71. $4 \text{ yd} \div \dfrac{1}{2} \text{ yd} = \dfrac{4}{1} \times \dfrac{2}{1} = \dfrac{8}{1} = 8, \ 8 \text{ pieces}$

73. $6 \text{ lb} \div \dfrac{3}{4} \text{ lb} = \dfrac{6}{1} \times \dfrac{4}{3} = \dfrac{24}{3} = 8, \ 8 \text{ jars}$

75. $\dfrac{5}{14} + \dfrac{1}{14} = \dfrac{6}{14} = \dfrac{\cancel{2} \times 3}{\cancel{2} \times 7} = \dfrac{3}{7}$

77. $\dfrac{17}{24} - \dfrac{5}{24} = \dfrac{12}{24} = \dfrac{1}{2}$

79. $6 = 2 \times 3$

$15 = 3 \times 5$

$2 \times 3 \times 5 = 30$

81. $20 = 2^2 \times 5$

$8 = 2^3$

$4 = 2^2$

$2^3 \times 5 = 40$

83. $\dfrac{1}{8} + \dfrac{3}{4} = \dfrac{1}{8} + \dfrac{6}{8} = \dfrac{7}{8}$

85. $\dfrac{11}{8} - \dfrac{3}{10} = \dfrac{55}{40} - \dfrac{12}{40} = \dfrac{43}{40} \text{ or } 1\dfrac{3}{40}$

87. $\dfrac{7}{26} - \dfrac{2}{13} = \dfrac{7}{26} - \dfrac{4}{26} = \dfrac{3}{26}$

89. $\dfrac{7}{18} + \dfrac{5}{12} = \dfrac{14}{36} + \dfrac{15}{36} = \dfrac{29}{36}$

91. $\dfrac{5}{4} - \dfrac{1}{20} = \dfrac{25}{20} - \dfrac{1}{20} = \dfrac{24}{20} = \dfrac{6}{5} \text{ or } 1\dfrac{1}{5}$

93. $\dfrac{5}{12} + \dfrac{5}{16} = \dfrac{20}{48} + \dfrac{15}{48} = \dfrac{35}{48}$

95. $\dfrac{1}{6} + \dfrac{3}{4} - \dfrac{5}{8} = \dfrac{4}{24} + \dfrac{18}{24} - \dfrac{15}{24} = \dfrac{7}{24}$

97. $\dfrac{4}{7} + \dfrac{1}{2} + \dfrac{3}{4} = \dfrac{16}{28} + \dfrac{14}{28} + \dfrac{21}{28} = \dfrac{51}{28} \text{ or } 1\dfrac{23}{28}$

99. $\dfrac{2}{3} + \dfrac{1}{4} = \dfrac{8}{12} + \dfrac{3}{12} = \dfrac{11}{12}$

He uses $\dfrac{11}{12}$ cup sugar.

101. $\dfrac{1}{2} - \dfrac{9}{25} = \dfrac{25}{50} + \dfrac{18}{50} = \dfrac{7}{50}$

The difference is $\dfrac{7}{50}$ of an inch.

103. $3\dfrac{1}{5} \times 2\dfrac{7}{8} = \dfrac{16}{5} \times \dfrac{23}{8} = \dfrac{\overset{2}{\cancel{16}} \times 23}{5 \times \cancel{8}} = \dfrac{46}{5} \text{ or } 9\dfrac{1}{5}$

105. $1\dfrac{2}{9} \div 7\dfrac{1}{3} = \dfrac{11}{9} \div \dfrac{22}{3} = \dfrac{\cancel{11}}{\cancel{9}_3} \times \dfrac{\cancel{3}}{\cancel{22}_2} = \dfrac{1}{6}$

107. $1\dfrac{2}{9} \div 6 = \dfrac{11}{9} \times \dfrac{1}{6} = \dfrac{11}{9 \times 6} = \dfrac{11}{54}$

109. $2\dfrac{1}{8} + 1\dfrac{3}{8} = \dfrac{17}{8} + \dfrac{11}{8}$

$= \dfrac{28}{8}$

$= \dfrac{\cancel{2} \times \cancel{2} \times 7}{\cancel{2} \times \cancel{2} \times 2}$

$= \dfrac{7}{2} \text{ or } 3\dfrac{1}{2}$

111. $3\dfrac{1}{2} - 1\dfrac{7}{8} = \dfrac{7}{2} - \dfrac{15}{8} = \dfrac{28}{8} - \dfrac{15}{8} = \dfrac{13}{8} \text{ or } 1\dfrac{5}{8}$

113. $1\dfrac{1}{6} + 3\dfrac{3}{4} = \dfrac{7}{6} + \dfrac{15}{4}$

$= \dfrac{14}{12} + \dfrac{45}{12}$

$= \dfrac{59}{12} \text{ or } 4\dfrac{11}{12}$

115. $1 - \dfrac{7}{8} = \dfrac{8}{8} - \dfrac{7}{8} = \dfrac{1}{8}$

117. $26\dfrac{3}{8} \div 3 = \dfrac{211}{8} \div \dfrac{3}{1}$

$= \dfrac{211}{8} \times \dfrac{1}{3}$

$= \dfrac{211}{24}$

$= 8\dfrac{19}{24} \text{ in.}$

119. $2\dfrac{3}{4} - 1\dfrac{1}{6} = \dfrac{11}{4} - \dfrac{7}{6} = \dfrac{33}{12} - \dfrac{14}{12}$

$= \dfrac{19}{12}$

$= 1\dfrac{7}{12} \text{ hr}$

121. $1\dfrac{1}{2} + \dfrac{3}{4} = \dfrac{3}{2} + \dfrac{3}{4} = \dfrac{6}{4} + \dfrac{3}{4} = \dfrac{9}{4} = 2\dfrac{1}{4} \text{ lb}$

123. $6\dfrac{1}{4} \times 4 = \dfrac{25}{4} \times \dfrac{4}{1} = \dfrac{25}{\cancel{4}} \times \dfrac{\cancel{4}}{1} = 25 \text{ in.}$

Calculator Exercises

1. 4⁄9   .444444444

3. 3⁄22   .1363636364

Section R.3 Practice Exercises

1. **(a)** tenths; hundredths; thousandths
   (b) percent
   (c) $\dfrac{1}{100}$; 0.01
   (d) 100%

3. Hundreds

5. Tenths

7. Hundredths

9. No; the symbols I, V, X, and so on each represent certain numerical values but the values are not dependent on the position of the symbol within the number.

11.
$$10\overline{)7.0} \quad \begin{array}{r} 0.7 \\ \hline 7.0 \\ 7\,0 \\ \hline \end{array}$$
Solution: 0.7

13.
$$25\overline{)9.00} \quad \begin{array}{r} 0.36 \\ \hline 9.00 \\ 7\,5 \\ \hline 1\,50 \\ 1\,50 \\ \hline \end{array}$$
Solution: 0.36

15.
$$9)\overline{11.00}^{1.22}$$
$$\underline{9}$$
$$2\,0$$
$$\underline{1\,8}$$
$$20$$
$$\underline{18}$$
$$2$$

Solution: $0.\overline{2}$

17.
$$33)\overline{7.0000}^{0.2121}$$
$$\underline{66}$$
$$40$$
$$\underline{33}$$
$$70$$
$$\underline{66}$$
$$4$$

Solution: $0.\overline{21}$

19. 214.1

21. 39.268

23. 40,000

25. 0.73

27. $0.45 = \dfrac{45}{100} = \dfrac{5 \cdot 9}{5 \cdot 20} = \dfrac{9}{20}$

29. $0.181 = \dfrac{181}{1000}$

31. $2.04 = \dfrac{204}{100} = \dfrac{4 \cdot 51}{4 \cdot 25} = \dfrac{51}{25}$ or $2\dfrac{1}{25}$

33. $13.007 = \dfrac{13,007}{1000}$ or $13\dfrac{7}{1000}$

35. $0.\overline{5} = \dfrac{5}{9}$

37. $1.\overline{1} = 1 + 0.\overline{1} = 1 + \dfrac{1}{9} = 1\dfrac{1}{9}$ or $\dfrac{10}{9}$

39. $30\% = 30 \times 0.01 = 0.30$

$$30\% = 30 \times \frac{1}{100} = \frac{3 \cdot 10}{10 \cdot 10} = \frac{3}{10}$$

41. $75\% = 75 \times 0.01 = 0.75$

$$75\% = 75 \times \frac{1}{100} = \frac{75}{100} = \frac{25 \cdot 3}{25 \cdot 4} = \frac{3}{4}$$

43. $3\dfrac{3}{4}\% = 3.75\% = 3.75 \times 0.01 = 0.0375$

$$3\frac{3}{4}\% = 3\frac{3}{4} \times \frac{1}{100}$$
$$= \frac{15}{4} \times \frac{1}{100}$$
$$= \frac{3 \cdot 5}{4} \times \frac{1}{5 \cdot 20}$$
$$= \frac{3}{4} \times \frac{1}{20}$$
$$= \frac{3}{80}$$

45. $15.7\% = 15.7 \times 0.01 = 0.157$

$$15.7\% = \frac{157}{10} \times \frac{1}{100} = \frac{157}{1000}$$

47. $270\% = 270 \times 0.01 = 2.70$

$$140\% = 270 \times \frac{1}{100} = \frac{270}{100} = \frac{27 \times \cancel{10}}{10 \times \cancel{10}} = \frac{27}{10}$$

49. Multiply by 100 and apply the % sign.

51. $0.05 = 0.05 \times 100\% = 5\%$

53. $0.90 = 0.90 \times 100\% = 90\%$

55. $1.2 = 1.2 \times 100\% = 120\%$

57. $7.5 = 7.5 \times 100\% = 750\%$

59. $0.135 = 0.135 \times 100\% = 13.5\%$

61. $0.003 = 0.003 \times 100\% = 0.3\%$

63.
$$50\overline{)3.00} \quad \begin{array}{r} 0.06 \\ \hline \end{array}$$
$$\underline{3\ 00}$$
$$0.06 = 0.06 \times 100\% = 6\%$$

65.
$$2\overline{)9.0} \quad \begin{array}{r} 4.5 \\ \hline \end{array}$$
$$\underline{8}$$
$$1\ 0$$
$$\underline{1\ 0}$$
$$4.5 = 4.5 \times 100\% = 450\%$$

67.
$$8\overline{)5.000} \quad \begin{array}{r} 0.625 \\ \hline \end{array}$$
$$\underline{4\ 8}$$
$$20$$
$$\underline{16}$$
$$40$$
$$\underline{40}$$
$$0.625 = 0.625 \times 100\% = 62.5\%$$

69.
$$16\overline{)5.0000} \quad \begin{array}{r} 0.3125 \\ \hline \end{array}$$
$$\underline{48}$$
$$20$$
$$\underline{16}$$
$$40$$
$$\underline{32}$$
$$80$$
$$\underline{80}$$
$$0.3125 = 0.3125 \times 100\% = 31.25\%$$

71.
$$6\overline{)5.000} \quad \begin{array}{r} 0.833 \\ \hline \end{array}$$
$$\underline{4\ 8}$$
$$20$$
$$\underline{18}$$
$$20$$
$$\underline{18}$$
$$2$$

$$0.83\overline{3} = 0.83\overline{3} \times 100\% = 83.\overline{3}\%$$

73.
$$15\overline{)14.000} \quad \begin{array}{r} 0.933 \\ \hline \end{array}$$
$$\underline{13\ 5}$$
$$50$$
$$\underline{45}$$
$$50$$
$$\underline{45}$$
$$5$$

$$0.93\overline{3} = 0.93\overline{3} \times 100\% = 93.\overline{3}\%$$

75. The discount is 30% of the original price of the suit. *of* translates as multiplication.
30%(140) = 0.30(140) = $42

77. Tom's taxes are 27% of his income. *of* translates as multiplication.
27%(12,500) = 0.27(12,500) = $3375

79. $\dfrac{5.95}{85} = 0.07 = 0.07 \times 100\% = 7\%$

81. 33% of the monthly income is spent on rent. *of* translates as multiplication.
33%(2400) = 0.33 × 2400 = $792

83. 8% of the monthly income is spent on utilities. *of* translates as multiplication.
8%(2400) = 0.08 × 2400 = $192

85. 75% of the mortgage has been paid. *of* translates as multiplication.
75%(90,000) = 0.75(90,000) = $67,500

Section R.4 Practice Exercises

1. a. perimeter; circumference
b. area;
c. $C = 2\pi r$; $A = \pi r^2$
d. *lwh*
e. acute; obtuse
f. complementary; 180°
g. 180
h. right

3. $P = 2w + 2l$
$= 2(6) + 2(10)$
$= 12 + 20$
$= 32$ m

5. $P = 4s$
$= 4(4.3)$
$= 17.2$ miles

7. $P = 2\frac{1}{3} + 5\frac{1}{6} + 4$
$= \frac{14}{6} + \frac{31}{6} + \frac{24}{6}$
$= \frac{69}{6}$
$= \frac{23}{2}$
$= 11\frac{1}{2}$ in.

9. $C = 2\pi r = 2(3.14)(5) = 31.4$ ft

11. a, f, g

13. $A = lw = (3)(11) = 33$ cm^2

15. $A = s^2$
$= (4.1)^2$
$= (4.1)(4.1)$
$= 16.81$ m^2

17. $A = bh = (14)(6) = 84$ in.^{2}

19. $A = \frac{1}{2}bh = 0.5(8.8)(2.3) = 10.12$ km^2

21. $A = \pi r^2 = (3.14)(2.1)^2 = 13.8474$ ft^2

23. $A = \frac{1}{2}(b_1 + b_2)h = \frac{1}{2}(14 + 8)6 = 66$ in.^{2}

25. $A = \frac{1}{2}bh = \frac{1}{2}(9)(7) = 31.5$ ft^2

27. c, d, h

29. $V = \pi r^2 h$
$= (3.14)(3.5)^2(8)$
$= 307.72$ cm^3

31. $V = lwh = (6.5)(1.5)(4) = 39$ in.^{3}

33. $V = \frac{4}{3}\pi r^3$
$= \frac{4}{3}(3.14)(3)^3$
$= \frac{4}{\cancel{3}}(3.14)(\cancel{27}^9)$
$= 113.04$ cm^3

35. $V = \frac{1}{3}\pi r^2 h$
$= \frac{1}{3}(3.14)(9)^2(20)$
$= \frac{1}{\cancel{3}}(3.14)(\cancel{81}^{27})(20)$
$= 1695.6$ cm^3

37. To find the amount of helium is to find the volume of the spherical balloon.
$V = \frac{4}{3}\pi r^3 = \frac{4}{3}(3.14)(9)^3 = 3052.08$ in.^{3}

39. The volume of the snow cone is
$V = \frac{1}{3}\pi r^2 h = \frac{1}{3}(3.14)(3)^2(12) = 113.04$ cm^3

41. The area of the wall is $(20)(8) = 160$ ft^2.

(a) To find the price per square foot, divide the cost by the area of the wall.
$\$40 \div 160$ ft$^2 \approx \$0.25/$ft^2

(b) The areas of the remaining walls are $(20)(8) = 160$ ft^2, $(16)(8) = 128$ ft^2, and $(16)(8) = 128$ ft^2. The total area is $160 + 128 + 128 = 416$ ft^2. The total price is $416(0.25) = \$104$.

43. Perimeter

45. To find the fence is to find the perimeter of the triangularly shaped garden.
$P = 12 + 22 + 20 = 54$ ft

47. (a) The area of the field is
$360(160) = 57600 \text{ ft}^2$

(b) The area of a piece of sod is $1(3) = 3$ ft^2
To find the number of pieces of sod, divide the area of the football field by the area of a piece of sod.
$57600 \div 3 = 19200$ pieces

49. (a) The area of the 8-in. diameter circular pizza is
$A = \pi r^2 = (3.14)(4) = 50.24 \text{ in.}^2$

(b) The area of the 12-in. diameter circular pizza is
$A = \pi r^2 = (3.14)(6) = 113.04 \text{ in.}^2$

(c) Two 8-inch pizzas have $2(50.24) = 100.48 \text{ in.}^2$ One 12-inch pizza has 113.04 in.^2 Therefore one 12-inch pizza has more.

51. The volume of a soup can in the shape of a right circular cylinder is
$V = \pi r^2 h = (3.14)(3.2)^2(9) = 289.3824 \text{ cm}^3$

53. True

55. True

57. True

59. Not possible; acute angles are less than $90°$.

61. Answers vary. One example: $100°, 80°$

63. (a) $\angle 1$ and $\angle 3$, $\angle 2$ and $\angle 4$

(b) $\angle 1$ and $\angle 2$, $\angle 2$ and $\angle 3$, $\angle 3$ and $\angle 4$, $\angle 1$ and $\angle 4$

(c) Since $\angle 1$ and $\angle 4$ are supplementary, $m(\angle 1) + m(\angle 4) = 180°$.
$m(\angle 1) + 80° = 180°$
$m(\angle 1) = 100°$
It follows that $m(\angle 3) = 100°$ and $m(\angle 2) = 80°$.

65. $90° - 33° = 57°$

67. $90° - 12° = 78°$

69. $180° - 33° = 147°$

71. $180° - 122° = 58°$

73. Vertical angles: $m(\angle 5) = m(\angle 7)$

75. Corresponding angles: $m(\angle 5) = m(\angle 1)$

77. Alternate exterior angles: $m(\angle 7) = m(\angle 1)$

79. Alternate interior angles: $m(\angle 3) = m(\angle 5)$

81. $m(\angle a) = 45°$ (Supplementary angles); $m(\angle b) = 135°$ (Vertical angles are equal); $m(\angle c) = 45°; m(\angle d) = 135°$ (Corresponding angles are equal); $m(\angle e) = 45°$; $m(\angle f) = 135°; m(\angle g) = 45°$

83. Scalene (no equal sides)

85. Isosceles (Two sides equal)

87. True

89. $45°$

91. No; a $90°$ angle plus an angle greater than $90°$ would make the sum of the angles greater than $180°$.

93. $40°$; the sum of angles of a triangle is $180°$.

95. $37°$; the angles of a right triangle are complementary.

97. $m(\angle a) = 80°$ (The sum of angles of a triangle is $180°$); $m(\angle b) = 80°$ (Vertical angles are equal); $m(\angle c) = 100°$ Supplementary angles); $m(\angle d) = 100°; m(\angle e) = 65°; m(\angle f) = 115°$; $m(\angle g) = 115°; m(\angle h) = 35°; m(\angle i) = 145°$; $m(\angle j) = 145°$

99. $m(\angle a) = 70°; m(\angle b) = 65°; m(\angle c) = 65°$; $m(\angle d) = 110°; m(\angle e) = 70°; m(\angle f) = 110°$; $m(\angle g) = 115°; m(\angle h) = 115°; m(\angle i) = 65°$; $m(\angle j) = 70°; m(\angle k) = 65°$

101. $P = 20 + 2(16) + 2(5) + 20 = 82$ ft

103. $\begin{pmatrix} \text{Area of} \\ \text{outer square} \end{pmatrix} - \begin{pmatrix} \text{Area of} \\ \text{inner square} \end{pmatrix} = 10^2 - 8^2$

$= 100 - 64$

$= 36 \text{ in.}^2$

105. $\begin{pmatrix} \text{Area of outer} \\ \text{rectangle} \end{pmatrix} - \begin{pmatrix} \text{Area of inner} \\ \text{circle} \end{pmatrix}$

$= (6.2)(4.1) - (3.14)(1.8)^2$

$= 25.42 - 10.1736$

$= 15.2464 \text{ cm}^2$

Chapter 1

Chapter 1 Opener

Across

1. $35(7) = 245$

3. $300 - 186 = 114$

5. 3550

Down

1. The least common denominator (LCD) is 24

5. $\dfrac{5}{2} + \dfrac{5}{8} + \dfrac{5}{4} = \dfrac{20}{8} + \dfrac{5}{8} + \dfrac{10}{8}$

   $= \dfrac{35}{8}$

   The numerator of the sum is 35.

Calculator Exercises

1. √(12)
   3.464101615

3. 4*π
   12.56637061

Section 1.1 Practice Exercises

1. (a) variable
   (b) constants
   (c) set
   (d) inequalities
   (e) a is less than b
   (f) c is greater than or equal to d
   (g) 5 is not equal to 6
   (h) opposites
   (i) $|a|, 0$

3. $3q = 3(5) = 15$

5. $8 + w = 8 + 12 = 20$

7. $\dfrac{6}{5}h = \dfrac{6}{5}(10) = 12$

9. $c - 2 - d = 15.4 - 2 - 8.1 = 5.3$

11. $abc = \left(\dfrac{1}{10}\right)\left(\dfrac{1}{4}\right)\left(\dfrac{1}{2}\right) = \dfrac{1}{80}$

13. (a) $1.29s = 1.29(3) = \$3.87$
    (b) $1.29s = 1.29(8) = \$10.32$
    (c) $1.29s = 1.29(10) = \$12.90$

15. (a) $850 - b = 850 - 475 = 375$ calories
    (b) $850 - b = 850 - 220 = 630$ calories
    (c) $850 - b = 850 - 580 = 270$ calories

17.

19. a. a terminating decimal; rational number

21. b. repeating decimal; rational number

23. a. a terminating decimal; rational number

25. c. a nonterminating, nonrepeating decimal; irrational number

27. a. a terminating decimal; rational number

29. a. a terminating decimal; rational number

31. b. a repeating decimal; rational number

33. c. a nonterminating, nonrepeating decimal; irrational number

35. Answers vary; for example: π, $-\sqrt{2}$, $\sqrt{3}$

37. Answers vary; for example: -4, -1, 0

39. Answers vary; for example: $-\dfrac{3}{4}$, $\dfrac{1}{2}$, 0.206

41. $-\dfrac{3}{2}$, -4, $0.\overline{6}$, 0, 1

43. 1

45. -4, 0, 1

47. (a) Since Kane's score is 0 and Pak's score is -8, $0 > -8$.
   (b) Since Scorenstam's score is 7 and Davies' score is -4, $7 > -4$.
   (c) Since Pak's score is -8 and McCurdy's score is 3, $-8 < 3$.
   (d) Since Kane's score is 0 and Davies's score is -4, $0 > -4$.

49. -18

51. 6.1

53. $\dfrac{5}{8}$

55. $-\dfrac{7}{3}$

57. 3

59. $-\dfrac{7}{3}$

61. 8

63. -72.1

65. 2

67. 1.5

69. -1.5

71. $\dfrac{3}{2}$

73. -10

75. $-\dfrac{1}{2}$

77. False; $|n|$ is never negative.

79. True; 5 is to the right of 2.

81. False; 6 is equal to 6.

83. True; -7 is equal to -7.

85. False; $\dfrac{3}{2}$ is to the right of $\dfrac{1}{6}$.

87. False; -5 is to the left of -2.

89. False; 8 is equal to 8.

91. True; 2 is to the right of 1.

93. True; $\dfrac{1}{9}$ is equal to $\dfrac{1}{9}$.

95. False; 7 is equal to 7.

97. True; -1 is to the left of 1.

99. True; 8 is equal to 8.

101. True; 2 is equal to 2.

103. For all $a < 0$ since $-a$ is the opposite of a.

Calculator Exercises

1. `(4+6)/(8-3)`   `2`

3. `100-2(5-3)^3`   `84`

5. `(12-6+1)²`

               49

7. `√(18-2)`

               4

9. `(20-3²)/(26-2²)`

               .5

Section 1.2 Practice Exercises

1. (a) quotient, product
   (b) sum, difference
   (c) base, exponent, power
   (d) 8^2
   (e) p^4
   (f) radical, square
   (g) order of operations

3. 56

5. -14

7. -19

9. $\dfrac{1}{6} \cdot \dfrac{1}{6} \cdot \dfrac{1}{6} \cdot \dfrac{1}{6} = \left(\dfrac{1}{6}\right)^4$

11. $a \cdot a \cdot a \cdot b \cdot b = a^3 b^2$

13. $(5c)^5$

15. (a) x

    (b) Yes, 1

17. $x^3 = x \cdot x \cdot x$

19. $(2b)^3 = 2b \cdot 2b \cdot 2b$

21. $10y^5 = 10 \cdot y \cdot y \cdot y \cdot y \cdot y$

23. $2wz^2 = 2 \cdot w \cdot z \cdot z$

25. $6^2 = 6 \cdot 6 = 36$

27. $\left(\dfrac{1}{7}\right)^2 = \dfrac{1}{7} \cdot \dfrac{1}{7} = \dfrac{1}{49}$

29. $(0.2)^3 = 0.2 \cdot 0.2 \cdot 0.2 = 0.008$

31. $2^6 = 2 \cdot 2 \cdot 2 \cdot 2 \cdot 2 \cdot 2 = 64$

33. $\sqrt{81} = 9$

35. $\sqrt{4} = 2$

37. $\sqrt{144} = 12$

39. $\sqrt{16} = 4$

41. $\sqrt{\dfrac{1}{9}} = \sqrt{\left(\dfrac{1}{3}\right)^2} = \dfrac{1}{3}$

43. $\sqrt{\dfrac{25}{81}} = \sqrt{\left(\dfrac{5}{9}\right)^2} = \dfrac{5}{9}$

45. $8 + 2 \cdot 6 = 8 + 12 = 20$

47. $(8 + 2)6 = 10 \cdot 6 = 60$

49. $4 + 2 \div 2 \cdot 3 + 1 = 4 + 3 + 1 = 8$

51. $81 - 4 \cdot 3 + 3^2 = 81 - 12 + 9 = 78$

53. $\dfrac{1}{4} \cdot \dfrac{2}{3} - \dfrac{1}{6} = \dfrac{1}{6} - \dfrac{1}{6} = 0$

55. $\left(\dfrac{11}{6} - \dfrac{3}{8}\right) \cdot \dfrac{4}{5} = \left(\dfrac{44}{24} - \dfrac{9}{24}\right) \cdot \dfrac{4}{5}$

$$= \dfrac{\overset{7}{\cancel{35}}}{\underset{6}{\cancel{24}}} \cdot \dfrac{\cancel{4}}{\cancel{5}}$$

$$= \dfrac{7}{6}$$

57. $3[5 + 2(8 - 3)] = 3[5 + 2(5)] = 3[15] = 45$

59. $10 + |-6| = 10 + 6 = 16$

61. $21 - |8 - 2| = 21 - 6 = 15$

63. $2^2 + \sqrt{9} \cdot 5 = 4 + 15 = 19$

11

65. $3 \cdot 5^2 = 3 \cdot 25 = 75$

67. $\sqrt{9+16} - 2 = \sqrt{25} - 2 = 5 - 2 = 3$

69. $[4^2 \cdot (6-4) \div 8] + [7 \cdot (8-3)]$
$= [16 \cdot 2 \div 8] + [7 \cdot 5]$
$= 4 + 35$
$= 39$

71. $48 - 13 \cdot 3 + [(50 - 7 \cdot 5) + 2]$
$= 48 - 39 + [15 + 2]$
$= 26$

73. $\dfrac{7 + 3(8-2)}{(7+3)(8-2)} = \dfrac{7+18}{(10)(6)} = \dfrac{25}{60} = \dfrac{5}{12}$

75. $\dfrac{15 - 5(3 \cdot 2 - 4)}{10 - 2(4 \cdot 5 - 16)} = \dfrac{15 - 5(2)}{10 - 2(4)} = \dfrac{5}{2}$

77. (a) Debt $= 52 + 20 + 65 + 43 = \$180$
Debt:Income $= 180 : 1200 = 12 : 100$
$= 0.12$

(b) Yes, Monica meets the criteria since $0.12 < 0.20$.

79. $A = lw = 360 \cdot 160 = 57,600 \text{ ft}^2$

81. $A = \frac{1}{2}(b_1 + b_2)h = \frac{1}{2}(6+8)3 = 21 \text{ ft}^2$

83. $3x$

85. $\dfrac{x}{7}$ or $x \div 7$

87. $2 - a$

89. $2y + x$

91. $4(x + 12)$

93. $3 - Q$

95. $2y^3 = 2(\ \ )^3 = 2(2)^3 = 2(8) = 16$

97. $|z - 8| = |(\ \ ) - 8| = |(10) - 8| = |2| = 2$

99. $5\sqrt{x} = 5\sqrt{(\ \ )} = 5\sqrt{(4)} = 5(2) = 10$

101. $yz - x = (\ \ )(\ \ ) - (\ \ )$
$= (2)(10) - (4)$
$= 20 - 4$
$= 16$

103. $\dfrac{\sqrt{\frac{1}{9}} + \frac{2}{3}}{\sqrt{\frac{4}{25}} + \frac{3}{5}} = \dfrac{\frac{1}{3} + \frac{2}{3}}{\frac{2}{5} + \frac{3}{5}} = \dfrac{\frac{3}{3}}{\frac{5}{5}} = \dfrac{1}{1} = 1$

105. $\dfrac{\left|-2\right|}{|-10| - |2|} = \dfrac{2}{10 - 2} = \dfrac{2}{8} = \dfrac{1}{4}$

107. (a) $36 \div 4 \cdot 3 = 9 \cdot 3 = 27$
Division must be performed before multiplication.

(b) $36 - 4 + 3 = 32 + 3 = 35$
Subtraction must be performed before addition.

109. This is acceptable, provided division and multiplication are performed in order from left to right, and subtraction and addition are performed in order from left to right.

Section 1.3 Practice Exercises

1. (a) Negative

(b) b

3. $\dfrac{9}{2} > \dfrac{3}{4}$

5. $0 > -\dfrac{5}{2}$

7. $\dfrac{3}{4} > -\dfrac{5}{2}$

9. $-2 + (-4) = -6$

11. $-7 + 10 = 3$

13. $6 + (-3) = 3$

15. $2 + (-5) = -3$

17. $-19 + 2 = -17$

19. $-4 + 11 = 7$

21. $-16 + (-3) = -19$

23. $-2 + (-21) = -23$

25. $0 + (-5) = -5$

27. $-3 + 0 = -3$

29. $-16 + 16 = 0$

31. $41 + (-41) = 0$

33. $4 + (-9) = -5$

35. $7 + (-2) + (-8) = -3$

37. $-17 + (-3) + 20 = -20 + 20 = 0$

39. $-3 + (-8) + (-12) = -11 + (-12) = -23$

41. $-42 + (3) + 45 + (-6) = -45 + 45 + (-6)$
$$= -6$$

43. $-5 + (-3) + (-7) + 4 + 8 = -8 + (-7) + 4 + 8$
$$= -3$$

45. $23.81 + (-2.51) = 21.3$

47. $-\dfrac{2}{7} + \dfrac{1}{14} = -\dfrac{4}{14} + \dfrac{1}{14} = -\dfrac{3}{14}$

49. $\dfrac{2}{3} + \left(-\dfrac{5}{6}\right) = \dfrac{4}{6} + \left(-\dfrac{5}{6}\right) = -\dfrac{1}{6}$

51. $-\dfrac{7}{8} + \left(-\dfrac{1}{16}\right) = -\dfrac{14}{16} + \left(-\dfrac{1}{16}\right) = -\dfrac{15}{16}$

53. $-\dfrac{1}{4} + \dfrac{3}{10} = -\dfrac{5}{20} + \dfrac{6}{20} = \dfrac{1}{20}$

55. $-2.1 + \left(-\dfrac{3}{10}\right) = -2.1 + -0.3 = -2.4$ or $-\dfrac{12}{5}$

57. $\dfrac{3}{4} + (-0.5) = 0.75 + (-0.5) = 0.25$ or $\dfrac{1}{4}$

59. $8.23 + (-8.23) = 0$

61. $-\dfrac{7}{8} + 0 = -\dfrac{7}{8}$

63. $-\dfrac{3}{2} + \left(-\dfrac{1}{3}\right) + \dfrac{5}{6} = -\dfrac{9}{6} + \left(-\dfrac{2}{6}\right) + \dfrac{5}{6} = -\dfrac{6}{6} = -1$

65. $-\dfrac{2}{3} + \left(-\dfrac{1}{9}\right) + 2 = -\dfrac{6}{9} + \left(-\dfrac{1}{9}\right) + \dfrac{18}{9} = \dfrac{11}{9}$

67. $-47.36 + 24.28 = -23.08$

69. $-0.000617 + (-0.0015) = -0.002117$

71. To add two numbers with different signs, subtract the smaller absolute value from the larger absolute value and apply the sign of the number with the larger absolute value.

73. $x + y + \sqrt{z} = -3 + (-2) + \sqrt{16} = -5 + 4 = -1$

75. $y + 3\sqrt{z} = -2 + 3\sqrt{16}$
$$= -2 + 3 \cdot 4$$
$$= -2 + 12$$
$$= 10$$

77. $|x| + |y| = |-3| + |-2| = 3 + 2 = 5$

79. $-x + y = -(-3) + (-2) = 3 + (-2) = 1$

81. $-6 + (-10)$; -16

83. $-3 + 8$; 5

85. $-21 + 17$; -4

87. $3(-14 + 20)$; 18

89. $(-7 + (-2)) + 5$; -4

91. $-5 + 13 + (-11)$; $-3°F$

93. $-8 + 1 + 2 + (-5)$; -10 pounds. Amara lost 10 pounds.

95. (a) $52.23 + (-52.95) = -\$0.72$

    (b) Yes

    (b) Yes

97. (a) $100 + 200 + (-500) + 300 + 100 + (-200)$

    (b) $\$0$

(b) −$48,000; a loss

Calculator Exercises

1. $-8+(-5)$

   -13

3. $627-(-84)$

   711

5. $-3.2-(-14.5)$

   11.3

7. $-12-9+4$

   -17

Section 1.4 Practice Exercises

1. **(a)** $a-b=a+-b$

   (b) positive

3. x^2

5. $-b+2$

7. $1+36 \div 9 \cdot 2 = 1 + 4 \cdot 2 = 1 + 8 = 9$

9. -3

11. -12

13. 4

15. $3-5=3+(-5)=-2$

17. $3-(-5)=3+5=8$

19. $-3-5=-3+(-5)=-8$

21. $-3-(-5)=-3+5=2$

23. $23-17=6$

25. $23-(-17)=23+17=40$

27. $-23-17=-23+(-17)=-40$

29. $-23-(-23)=0$

31. $-6-14=-6+(-14)=-20$

33. $-7-17=-7+(-17)=-24$

35. $13-(-12)=13+12=25$

37. $-14-(-9)=-14+9=-5$

39. $-\dfrac{6}{5}-\dfrac{3}{10}=-\dfrac{12}{10}+\left(-\dfrac{3}{10}\right)=-\dfrac{15}{10}=-\dfrac{3}{2}$

41. $\dfrac{3}{8}-\left(-\dfrac{4}{3}\right)=\dfrac{9}{24}+\dfrac{32}{24}=\dfrac{41}{24}$

43. $\dfrac{1}{2}-\dfrac{1}{10}=\dfrac{5}{10}-\dfrac{1}{10}=\dfrac{4}{10}=\dfrac{2}{5}$

45. $-\dfrac{11}{12}-\left(-\dfrac{1}{4}\right)=-\dfrac{11}{12}+\dfrac{3}{12}=-\dfrac{8}{12}=-\dfrac{2}{3}$

47. $6.8-(-2.4)=6.8+2.4=9.2$

49. $3.1-8.82=3.10+(-8.82)=-5.72$

51. $-4-3-2-1=-4+(-3)+(-2)+(-1)=-10$

53. $6-8-2-10=6+(-8)+(-2)+(-10)=-14$

55. $10+(-14)+6-22=-20$

57. $-112.846+(-13.03)-47.312=-173.188$

59. $0.085-(-3.14)+(0.018)=3.243$

61. $6-(-7); 13$

63. $3-18; -15$

65. $-5-(-11); 6$

67. $-1-(-13); 12$

69. $-32-20; -52$

71. $200+400+600+800-1000; \$1000$

73. $113°-(-39°)=152°F$

75. $8848-(-11,033 \text{ m})=19,881 \text{ m}$

77. $6+8-(-2)-4+1=14+2-4+1$
    $$=16-4+1$$
    $$=13$$

79. $-1-7+(-3)-8+10 = -8+(-3)-8+10$
$$= -9$$

81. $2-(-8)+7+3-15 = 2+8+7+3-15$
$$= 17+3-15$$
$$= 5$$

83. $-6+(-1)+(-8)+(-10) = -7+(-8)+(-10)$
$$= -25$$

85. $-4-\{11-[4-(-9)]\} = -4-\{11-[4+9]\}$
$$= -4-\{11-13\}$$
$$= -4-(-2)$$
$$= -2$$

87. $-\dfrac{13}{10}+\dfrac{8}{15}-\left(-\dfrac{2}{5}\right) = -\dfrac{39}{30}+\dfrac{16}{30}+\dfrac{12}{30} = -\dfrac{11}{30}$

89. $\left(\dfrac{2}{3}-\dfrac{5}{9}\right)-\left(\dfrac{4}{3}-(-2)\right)$
$$= \left(\dfrac{6}{9}-\dfrac{5}{9}\right)-\left(\dfrac{4}{3}+\dfrac{6}{3}\right)$$
$$= \dfrac{1}{9}-\dfrac{10}{3}$$
$$= \dfrac{1}{9}-\dfrac{30}{9}$$
$$= -\dfrac{29}{9}$$

91. $\sqrt{29+(-4)}-7 = \sqrt{25}-7 = 5-7 = -2$

93. $\big|10+(-3)\big|-\big|-12+(-6)\big| = |7|-|-18|$
$$= 7-18$$
$$= -11$$

95. $\dfrac{3-4+5}{4+(-2)} = \dfrac{4}{2} = 2$

97. $(a+b)-c = (-2+(-6))-(-1) = -8+1 = -7$

99. $a-(b+c) = -2-(-6+(-1))$
$$= -2-(-7)$$
$$= -2+7$$
$$= 5$$

101. $(a-b)-c = (-2-(-6))-(-1) = (4)+1 = 5$

103. $a-(b-c) = -2-(-6-(-1))$
$$= -2-(-5)$$
$$= -2+5$$
$$= 3$$

Problem Recognition Exercises

1. Add their absolute values and apply a negative sign.

3. (a) $14+(-8) = 6$
(b) $-14+8 = -6$
(c) $-14+(-8) = -22$
(d) $14-(-8) = 14+8 = 22$
(e) $-14-8 = -22$

5. (a) $-25+25 = 0$
(b) $25-25 = 0$
(c) $25-(-25) = 25+25 = 50$
(d) $-25-(-25) = 0$
(e) $-(25)+(-25) = -50$

7. (a) $3.5-7.1 = -3.6$
(b) $3.5-(-7.1) = 3.5+7.1 = 10.6$
(c) $-3.5+7.1 = 3.6$
(d) $-3.5-(-7.1) = -3.5+7.1 = 3.6$
(e) $-3.5+(-7.1) = -10.6$

9. (a) $-100-90-80 = -270$
(b) $-100-(90-80) = -100-10 = -110$
(c) $-100+(90-80) = -100+10 = -90$
(d) $-100-(90+80) = -100-170 = -270$

Calculator Exercises

1.

```
-6*5
              -30
```

3.

```
(-5)(-5)(-5)(-5)
             625
```

15

5.

```
-5^4
              -625
```

7.

```
(-2.4)²
         5.76
```

9.

```
-8.4/-2.1
              4
```

Section 1.5 Practice Exercises

1. **(a)** $\dfrac{1}{a}$

   (b) 0

   (c) 0

   (d) undefined

   (e) positive

   (f) negative

   (g) $-\dfrac{2}{3}\left(-\dfrac{3}{2}\right) = 1$

   (h) All of these.

3. True; $20 \le 20$

5. False; $6 \le 0$

7. -56

9. 143

11. -12.76

13. $\left(-\dfrac{2}{3}\right)\left(-\dfrac{9}{8}\right) = \dfrac{18}{24} = \dfrac{3}{4}$

15. $(-6)^2 = 36$

17. $-6^2 = -36$

19. $\left(-\dfrac{3}{5}\right)^3 = \left(-\dfrac{3}{5}\right)\left(-\dfrac{3}{5}\right)\left(-\dfrac{3}{5}\right) = -\dfrac{27}{125}$

21. $(-0.2)^4 = 0.0016$

23. $\dfrac{54}{-9} = -6$

25. $\dfrac{-100}{-10} = 10$

27. $\dfrac{-14}{-7} = 2$

29. $\dfrac{7}{-7} = -1$

31. $\dfrac{-8}{-8} = 1$

33. $\dfrac{13}{-65} = -\dfrac{1}{5}$

35. $(-2)(-7) = 14$

37. $-5 \cdot 0 = 0$

39. No number multiplied by 0 equals 6.

41. $(-6)(4) = -24$

43. $2 \cdot 3 = 6$

45. $2(-3) = -6$

47. $24 \div (-3) = -8$

49. $(-24) \div (-3) = 8$

51. $-6 \cdot 0 = 0$

53. Undefined

55. $0\left(-\dfrac{2}{5}\right) = 0$

57. $0 \div \left(-\dfrac{1}{10}\right) = 0$

59. $\dfrac{-9}{6} = -\dfrac{3}{2}$

61. $\dfrac{-30}{-100} = \dfrac{3}{10}$

63. $\dfrac{26}{-13} = -2$

65. $(1.72)(-4.6) = -7.912$

67. $-0.02(-4.6) = 0.092$

69. $\dfrac{14.4}{-2.4} = -6$

71. $\dfrac{-5.25}{-2.5} = 2.1$

73. $(-3)^2 = 9$

75. $-3^2 = -9$

77. $\left(-\dfrac{4}{3}\right)^3 = \left(-\dfrac{4}{3}\right)\left(-\dfrac{4}{3}\right)\left(-\dfrac{4}{3}\right) = -\dfrac{64}{27}$

79. $2.8(-5.1) = -14.28$

81. $(-6.8) \div (-0.02) = 340$

83. $\left(-\dfrac{2}{15}\right)\left(\dfrac{25}{3}\right) = -\dfrac{50}{45} = -\dfrac{\cancel{5}\cdot 10}{\cancel{5}\cdot 9} = -\dfrac{10}{9}$

85. $\left(-\dfrac{7}{8}\right) \div \left(-\dfrac{9}{16}\right) = \left(-\dfrac{7}{8}\right)\cdot\left(-\dfrac{16}{9}\right)$
$$= \dfrac{112}{72}$$
$$= \dfrac{\cancel{8}\cdot 14}{\cancel{8}\cdot 9}$$
$$= \dfrac{14}{9}$$

87. $(-2)(-5)(-3) = (10)(-3) = -30$

89. $(-8)(-4)(-1)(-3) = (32)(3) = 96$

91. $100 \div (-10) \div (-5) = (-10) \div (-5) = 2$

93. $-12 \div (-6) \div (-2) = 2 \div (-2) = -1$

95. $\dfrac{2}{5}\cdot\dfrac{1}{3}\cdot\left(-\dfrac{10}{11}\right) = \dfrac{2}{15}\cdot\left(-\dfrac{10}{11}\right) = -\dfrac{20}{165} = -\dfrac{4}{33}$

97. $\left(1\dfrac{1}{3}\right) \div 3 \div \left(-\dfrac{7}{9}\right) = \dfrac{4}{3}\cdot\dfrac{1}{3}\div\left(-\dfrac{7}{9}\right)$
$$= \dfrac{4}{9}\cdot\left(-\dfrac{9}{7}\right)$$
$$= -\dfrac{4}{7}$$

99. $12 \div (-2)(4) = (-6)(4) = -24$

101. $\left(-\dfrac{12}{5}\right) \div (-6) \cdot \left(-\dfrac{1}{8}\right) = \left(-\dfrac{12}{5}\right)\cdot\left(-\dfrac{1}{6}\right)\cdot\left(-\dfrac{1}{8}\right)$
$$= \dfrac{12}{30}\cdot\left(-\dfrac{1}{8}\right)$$
$$= \dfrac{2}{5}\cdot\left(-\dfrac{1}{8}\right)$$
$$= -\dfrac{2}{40}$$
$$= -\dfrac{1}{20}$$

103. $8 - 2^3 \cdot 5 + 3 - (-6) = 8 - 8\cdot 5 + 3 + 6$
$$= 8 - 40 + 3 + 6$$
$$= -23$$

105. $-(2-8)^2 \div (-6) \cdot 2 = -36 \div (-6) \cdot 2$
$$= 6\cdot 2$$
$$= 12$$

107. $\dfrac{6(-4) - 2(5-8)}{-6-3-5} = \dfrac{-24+6}{-14} = \dfrac{-18}{-14} = \dfrac{9}{7}$

109. $\dfrac{-4+5}{(-2)\cdot 5 + 10} = \dfrac{1}{-10+10} = \dfrac{1}{0}$ is undefined

111. $-4 - 3[2 - (-5+3)] - 8\cdot 2^2$
$$= -4 - 3[2 - (-2)] - 8\cdot 4$$
$$= -4 - 3[4] - 32$$
$$= -4 - 12 - 32$$
$$= -48$$

113. $-|-1| - |5| = -1 - 5 = -6$

115. $\dfrac{|2-9| - |5-7|}{10-15} = \dfrac{7-2}{-5} = \dfrac{5}{-5} = -1$

17

117. $\dfrac{6-3[2-(6-8)]^2}{-2|2-5|} = \dfrac{6-3[2-(-2)]^2}{-2\cdot 3}$

$\qquad = \dfrac{6-3\cdot 16}{-6}$

$\qquad = \dfrac{6-48}{-6}$

$\qquad = \dfrac{-42}{-6}$

$\qquad = 7$

119. $-x^2 = -(-2)^2 = -4$

121. $4(2x-z) = 4(2(-2)-6)$

$\qquad = 4(-4-6)$

$\qquad = 4(-10)$

$\qquad = -40$

123. $\dfrac{3x+2y}{y} = \dfrac{3(-2)+2(-4)}{-4}$

$\qquad = \dfrac{-6+(-8)}{-4}$

$\qquad = \dfrac{-14}{-4}$

$\qquad = \dfrac{7}{2}$

125. No, the first expression equals $10 \div (5x) = 2 \div x,$ and the second equals $10 \div 5 \cdot x = 2x$.

127. $-3.75(0.3) = -1.125$

129. $\left(\dfrac{16}{5}\right) \div \left(-\dfrac{8}{9}\right) = \dfrac{16}{5}\cdot\left(-\dfrac{9}{8}\right) = -\dfrac{18}{5}$

131. $-0.4 + 6(-0.42) = -2.92$

133. $-\dfrac{1}{4} - 6\left(-\dfrac{1}{3}\right) = -\dfrac{1}{4} + 2 = -\dfrac{1}{4} + \dfrac{8}{4} = \dfrac{7}{4}$

135. $-2(3) + 3 = -3;$ a loss of \$3

137. $2(5) + 3(-3) = 1;$ Lorna was 1 sale above quota for the week

139. $\dfrac{12+(-15)+4+(-9)+3}{5} = \dfrac{-5}{5} = -1$ The average loss was 1 oz.

141. **(a)** $-4-3-2-1 = -4+(-3)+(-2)+(-1)$

$\qquad\qquad = -10$

(b) $-4(-3)(-2)(-1) = 12(2) = 24$

(c) Part (a) is subtraction; part (b) is multiplication.

Problem Recognition Exercises

1. **(a)** $-8 - (-4) = -8 + 4 = -4$

(b) $-8(-4) = 32$

(c) $-8 + (-4) = -12$

(d) $-8 \div (-4) = 2$

3. **(a)** $-36 + 9 = -27$

(b) $-36(9) = -324$

(c) $-36 \div 9 = -4$

(d) $-36 - 9 = -45$

5. **(a)** $-5(-10) = 50$

(b) $-5 + (-10) = -15$

(c) $-5 \div (-10) = \dfrac{1}{2}$

(d) $-5 - (-10) = -5 + 10 = 5$

7. **(a)** $-4(-16) = 64$

(b) $-4 - (-16) = -4 + 16 = 12$

(c) $-4 \div (-16) = \dfrac{1}{4}$

(d) $-4 + (-16) = -20$

9. **(a)** $80(-5) = -400$

(b) $80 - (-5) = 80 + 5 = 85$

(c) $80 \div (-5) = -16$

(d) $80 + (-5) = 75$

11. **(a)** $|-6| + |2| = 6 + 2 = 8$

(b) $|-6 + 2| = |-4| = 4$

(c) $|-6| - |-2| = 6 - 2 = 4$

(d) $|-6 - 2| = |-8| = 8$

Section 1.6 Practice Exercises

1. (a) constant
   (b) coefficient
   (c) 1, 1
   (d) like

3. $(-2) + 9 = 7$

5. $-1 - (-19) = -1 + 19 = 18$

7. $-27 \div 5 = -\dfrac{27}{5} = -5.4$

9. $0(-15) = 0$

11. $\dfrac{25}{21} - \dfrac{6}{7} = \dfrac{25}{21} - \dfrac{18}{21} = \dfrac{7}{21} = \dfrac{1}{3}$

13. $\left(-\dfrac{11}{12}\right) \div \left(-\dfrac{5}{4}\right) = \left(-\dfrac{11}{12}\right) \cdot \left(-\dfrac{4}{5}\right) = \dfrac{44}{60} = \dfrac{11}{15}$

15. $-8 + 5$

17. $x + 8$

19. $4(5)$

21. $-12x$

23. $x + (-3); \ -3 + x$

25. $4p + (-9); \ -9 + 4p$

27. $(x + 4) + 9 = x + (4 + 9) = x + 13$

29. $-5(3x) = (-5 \cdot 3)x = -15x$

31. $\dfrac{6}{11}\left(\dfrac{11}{6}x\right) = \left(\dfrac{6}{11} \cdot \dfrac{11}{6}\right)x = x$

33. $-4\left(-\dfrac{1}{4}t\right) = \left(-4 \cdot -\dfrac{1}{4}\right)t = t$

35. Reciprocal

37. Zero

39. $6(5x + 1) = 6(5x) + 6(1) = 30x + 6$

41. $-2(a + 8) = -2a + (-2)(8) = -2a - 16$

43. $3(5c - d) = 3(5c) - 3d = 15c - 3d$

45. $-7(y - 2) = -7y - (-7)(2) = -7y + 14$

47. $-\dfrac{2}{3}(x - 6) = -\dfrac{2}{3}x - \left(-\dfrac{2}{3}\right)(6)$
    $$= -\dfrac{2}{3}x + \dfrac{12}{3}$$
    $$= -\dfrac{2}{3}x + 4$$

49. $\dfrac{1}{3}(m - 3) = \dfrac{1}{3}m - \dfrac{1}{3} \cdot 3 = \dfrac{1}{3}m - 1$

51. $-(2p + 10) = -2p - 10$

53. $6w + 10z - 16$

55. $4(x + 2y - z) = 4(x) + 4(2y) - 4(z)$
    $$= 4x + 8y - 4z$$

57. $-(-6w + x - 3y) = 6w - x + 3y$

59. $-9 + (16 + y) = 7 + y$

61. $2(3 + x) = 6 + 2x$

63. $4(6z) = 24z$

65. $20 + (30 + r) = 50 + r$

67. $20(30 + r) = 600 + 20r$

69. $20(30) = 600r$

71. b

73. i

75. g

77. d

79. h

81. Term: $2x$, coefficient 2;
    Term: $-y$, coefficient -1;
    Term: $18xy$, coefficient 18;
    Term: 5, coefficient 5.

83. Term: $-x$, coefficient -1;
    Term: $8y$, coefficient 8;

Term: $-9x^2y$, coefficient -9;

Term: -3, coefficient -3.

85. The variable factors are different

87. The variables are the same *and* raised to the same power.

89. Answers vary: $5y$, $-2x$, 6

91. $-4p - 2p + 8p - 15 + 3 = 2p - 12$

93. $2y^2 - 8y + y - 5y^2 - 3y^2$

$= 2y^2 - 5y^2 - 3y^2 - 8y + y$

$= -6y^2 - 7y$

95. $7x^3y + xy - 4$

97. $\dfrac{2}{5} + 2t - \dfrac{3}{5} + t - \dfrac{6}{5} = \dfrac{2}{5} - \dfrac{3}{5} - \dfrac{6}{5} + 2t + t$

$= -\dfrac{7}{5} + 3t$

99. $-3(2x - 4) + 10 = -6x + 12 + 10 = -6x + 22$

101. $4(w + 3) - 12 = 4w + 12 - 12 = 4w$

103. $5 - 3(x - 4) = 5 - 3x + 12 = 17 - 3x$

105. $-3(2t + 4w) + 8(2t - 4w)$

$= -6t - 12w + 16t - 32w$

$= -6t + 16t - 12w - 32w$

$= 10t - 44w$

107. $2(q - 5u) - (2q + 8u)$

$= 2q - 10u - 2q - 8u$

$= -18u$

109. $-\dfrac{1}{3}(6t + 9) + 10 = -2t - 3 + 10 = -2t + 7$

111. $10(5.1a - 3.1) + 4 = 51a - 31 + 4 = 51a - 27$

113. $-4m + 2(m - 3) + 2m = -4m + 2m - 6 + 2m$

$= -6$

115. $\dfrac{1}{2}(10q - 2) + \dfrac{1}{3}(2 - 3q) = 5q - 1 + \dfrac{2}{3} - q$

$= 4q - \dfrac{1}{3}$

117. $7n - 2(n - 3) - 6 + n = 7n - 2n + 6 - 6 + n$

$= 6n$

119. $6(x + 3) - 12 - 4(x - 3)$

$= 6x + 18 - 12 - 4x + 12$

$= 2x + 18$

121. $0.2(6c - 1.6) + c = 1.2c - 0.32 + c$

$= 2.2c - 0.32$

123. $6 + 2[-8 - 3(2x + 4)] + 10x$

$= 6 + 2[-8 - 6x - 12] + 10x$

$= 6 + 2[-6x - 20] + 10x$

$= 6 - 12x - 40 + 10x$

$= -2x - 34$

125. $1 - 3[2(z + 1) - 5(z - 2)]$

$= 1 - 3[2z + 2 - 5z + 10]$

$= 1 - 3[-3z + 12]$

$= 1 + 9z - 36$

$= 9z - 35$

127. Equivalent

129. Not equivalent. The terms are not *like* terms and cannot be combined.

131. Not equivalent. Subtraction is not commutative.

133. Equivalent

135. $14\dfrac{2}{7} + \left(2\dfrac{1}{3} + \dfrac{2}{3}\right)$ is easier.

137. (a) $10 + (1 + 9) + (2 + 8) + (3 + 7)$
$+ (4 + 6) + 5 = 55$

(b) $(1 + 19) + (2 + 18) + (3 + 17)$
$+ (4 + 16) + (5 + 15) + (6 + 14)$
$+ (7 + 13) \ \ + (8 + 12) + (9 + 11) + 10$
$= 210$

Group Activity

1. Substitute $C = 35$.

$$F = \frac{9}{5}C + 32$$
$$= \frac{9}{5}(35) + 32$$
$$= 63 + 32$$
$$= 95$$

3. Substitute $k = 0.05, L = 200, r = 0.5$.

$$R = k\left(\frac{L}{r^2}\right)$$
$$= 0.05\left(\frac{200}{0.5^2}\right)$$
$$= 0.05(800)$$
$$= 40$$

5. Substitute $\bar{x} = 69, \mu = 55, \sigma = 20, n = 25$

$$z = \frac{\bar{x} - \mu}{\frac{\sigma}{\sqrt{n}}}$$
$$= \frac{69 - 55}{\frac{20}{\sqrt{25}}}$$
$$= \frac{14}{\frac{20}{5}}$$
$$= \frac{14}{4}$$
$$= 3.5$$

7. Substitute $a = 2, b = -7, c = -15$

$$x = \frac{-b + \sqrt{b^2 - 4ac}}{2a}$$
$$= \frac{-(-7) + \sqrt{(-7)^2 - 4(2)(-15)}}{2(2)}$$
$$= \frac{7 + \sqrt{49 + 120}}{4}$$
$$= \frac{7 + \sqrt{169}}{4}$$
$$= \frac{7 + 13}{4}$$
$$= \frac{20}{4}$$
$$= 5$$

Chapter 1 Review Exercises

Section 1.1

1. (a) 7, 1

   (b) 7, −4, 0, 1

   (c) 7, 0, 1

   (d) 7, $\frac{1}{3}$, −4, 0, $-0.\overline{2}$, 1

   (e) $-\sqrt{3}$, π

   (f) 7, $\frac{1}{3}$, $-\sqrt{3}$, $-0.\overline{2}$, π, 1, −4, 0

3. $|-6| = 6$

5. $|0| = 0$

7. False

9. True

11. True

13. True

15. $x^2 - y = 8^2 - 4 = 64 - 4 = 60$

17. $\sqrt{x+2y} = \sqrt{8+2(4)}$

$\quad\quad\quad\quad = \sqrt{8+8}$

$\quad\quad\quad\quad = \sqrt{16}$

$\quad\quad\quad\quad = 4$

Section 1.2

19. $\dfrac{7}{y}$ or $7 \div y$

21. $a - 5$

23. $13z - 7$

25. $15^2 = 225$

27. $\dfrac{1}{\sqrt{100}} = \dfrac{1}{10}$

29. $\left(\dfrac{3}{2}\right)^3 = \dfrac{27}{8}$

31. $|-11| + |5| - (7-2) = 11 + 5 - 5 = 11$

33. $22 - 3(8 \div 4)^2 = 22 - 3(2)^2 = 22 - 12 = 10$

Section 1.3

35. $14 + (-10) = 4$

37. $-12 + (-5) = -17$

39. $-\dfrac{8}{11} + \dfrac{1}{2} = -\dfrac{16}{22} + \dfrac{11}{22} = -\dfrac{5}{22}$

41. $\left(-\dfrac{5}{2}\right) + \left(-\dfrac{1}{5}\right) = -\dfrac{25}{10} + \left(-\dfrac{2}{10}\right) = -\dfrac{27}{10}$

43. $2.9 + (-7.18) = -4.28$

45. $-5 + (-7) + 20 = -12 + 20 = 8$

47. When a and b are both negative or when a and b have different signs and the number with the larger absolute value is negative.

Section 1.4

49. $13 - 25 = -12$

51. $-8 - (-7) = -8 + 7 = -1$

53. $-\dfrac{7}{9} - \dfrac{5}{6} = -\dfrac{14}{18} - \dfrac{15}{18} = -\dfrac{29}{18}$

55. $7 - 8.2 = -1.2$

57. $-16.1 - (-5.9) = -16.1 + 5.9 = -10.2$

59. $\dfrac{11}{2} - \left(-\dfrac{1}{6}\right) - \dfrac{7}{3} = \dfrac{33}{6} + \dfrac{1}{6} - \dfrac{14}{6} = \dfrac{20}{6} = \dfrac{10}{3}$

61. $6 - 14 - (-1) - 10 - (-21) - 5$

$\quad = 6 - 14 + 1 - 10 + 21 - 5$

$\quad = -8 - 9 + 16$

$\quad = -17 + 16$

$\quad = -1$

63. $-7 - (-18);$

$\quad -7 - (-18) = 11$

65. $7 - 13;$

$\quad 7 - 13 = -6$

67. $(6 + (-12)) - 21;$

$\quad (6 + (-12)) - 21 = -6 - 21 = -27$

Section 1.5

69. $10(-17) = -170$

71. $(-52) \div 26 = -2$

73. $\dfrac{7}{4} \div \left(-\dfrac{21}{2}\right) = \dfrac{7}{4} \cdot \left(-\dfrac{2}{21}\right) = -\dfrac{14}{84} = -\dfrac{1}{6}$

75. $-\dfrac{21}{5} \cdot 0 = 0$

77. $0 \div (-14) = 0$

79. $-\dfrac{21}{14} = -\dfrac{3 \cdot 7}{2 \cdot 7} = -\dfrac{3}{2}$

81. $(5)(-2)(3) = (-10)(3) = -30$

83. $\left(-\dfrac{1}{2}\right)\left(\dfrac{7}{8}\right)\left(-\dfrac{4}{7}\right) = \left(-\dfrac{7}{16}\right)\left(-\dfrac{4}{7}\right) = \dfrac{7 \cdot 4}{16 \cdot 7} = \dfrac{1}{4}$

85. $40 \div 4 \div (-5) = 10 \div (-5) = -2$

87. $9 - 4[-2(4 - 8) - 5(3 - 1)]$
$= 9 - 4[-2(-4) - 5(2)]$
$= 9 - 4[8 - 10]$
$= 9 - 4[-2]$
$= 9 + 8$
$= 17$

89. $\dfrac{2}{3} - \left(\dfrac{3}{8} + \dfrac{5}{6}\right) \div \dfrac{5}{3} = \dfrac{2}{3} - \left(\dfrac{9}{24} + \dfrac{20}{24}\right) \cdot \dfrac{3}{5}$

$= \dfrac{16}{24} - \dfrac{29}{24} \cdot \dfrac{3}{5}$

$= \dfrac{16}{24} - \dfrac{29}{40}$

$= \dfrac{80}{120} - \dfrac{87}{120}$

$= -\dfrac{7}{120}$

91. $\dfrac{5 - [3 - (-4)^2]}{36 \div (-2)(3)}$

$= \dfrac{5 - [3 - 16]}{(-18)(3)}$

$= \dfrac{5 - [-13]}{-64}$

$= \dfrac{18}{-64}$

$= -\dfrac{1}{3}$

93. $3(x + 2) \div y = 3(4 + 2) \div (-9)$
$= 18 \div (-9)$
$= -2$

95. $-xy = -(4)(-9) = -(-36) = 36$

97. $x = \mu + z\sigma$
$x = (100) + (-1.96)(15)$
$x = 70.6$

99. False; any nonzero real number raised to an even power is positive.

101. True

103. True

Section 1.6

105. $2 + 3 = 3 + 2$

107. $5 + (-5) = 0$

109. $5 \cdot 2 = 2 \cdot 5$

111. $3 \cdot \dfrac{1}{3} = 1$

113. $5x - 2y = 5x + (-2y)$; then use commutative property of addition.

115. $3y, 10x, -12, xy$

117. $3a + 3b - 4b + 5a - 10$
$= 3a + 5a + 3b - 4b - 10$
$= 8a - b - 10$

119. $-2(4z + 9) = -8z - 18$

121. $2p - (p + 5w) + 3w = 2p - p - 5w + 3w$
$= p - 2w$

123. $\dfrac{1}{2}(-6z) + q - 4\left(3q + \dfrac{1}{4}\right) = -3q + q - 12q - 1$
$= -14q - 1$

125. $-4[2(x + 1) - (3x + 8)] = -4[2x + 2 - 3x - 8]$
$= -4[-x - 6]$
$= 4x + 24$

Chapter 1 Test

1. Rational; all repeating decimals are rational numbers.

3. (a) False

(b) True

(c) True

(d) True

(b) $4x^3 = 4 \cdot x \cdot x \cdot x$

5. (a) Twice the difference of a and b

(b) The difference twice a and b

7. $18 + (-12) = 6$

9. $-15 - (-3) = -15 + 3 = -12$

11. $-\dfrac{1}{8} + \left(-\dfrac{3}{4}\right) = -\dfrac{1}{8} + \left(-\dfrac{6}{8}\right) = -\dfrac{7}{8}$

13. $-14 + (-2) - 16 = -14 + (-18) = -32$

15. $38 \div 0$ is undefined.

17. $-22 \cdot 0 = 0$

19. $\begin{aligned}
\dfrac{2}{5} \div \left(-\dfrac{7}{10}\right) \cdot \left(-\dfrac{7}{6}\right) &= \dfrac{2}{5} \cdot \left(-\dfrac{10}{7}\right) \cdot \left(-\dfrac{7}{6}\right) \\
&= -\dfrac{4}{7} \cdot \left(-\dfrac{7}{6}\right) \\
&= \dfrac{28}{42} \\
&= \dfrac{2}{3}
\end{aligned}$

21. $\begin{aligned}
8 - [(2-4) - (8-9)] &= 8 - [(-2) - (-1)] \\
&= 8 - [-1] \\
&= 8 + 1 \\
&= 9
\end{aligned}$

23. $\dfrac{|4-10|}{2-3(5-1)} = \dfrac{|-6|}{2-3(4)} = \dfrac{6}{-10} = -\dfrac{3}{5}$

25. $4500 - (-1750) = \$6250$
Hector's profit was $6250 more in February.

27. $\begin{aligned}
-5x - 4y + 3 - 7x + 6y - 7 \\
= -5x - 7x - 4y + 6y + 3 - 7 \\
= -12x + 2y - 4
\end{aligned}$

29. $3k - 20 + (-9k) + 12 = -6k - 8$

31. $\begin{aligned}
\dfrac{1}{2}(12p - 4) + \dfrac{1}{3}(2 - 6p) &= 6p - 2 + \dfrac{2}{3} - 2p \\
&= 4p - \dfrac{4}{3}
\end{aligned}$

33. $\begin{aligned}
3x - 2y &= 3(4) - 2(-3) \\
&= 12 - (-6) \\
&= 12 + 6 \\
&= 18
\end{aligned}$

35. $\begin{aligned}
-y^2 - 4x + z &= -(-3)^2 - 4(4) + (-7) \\
&= -9 - 4(4) + (-7) \\
&= -9 - 16 + (-7) \\
&= -9 + (-16) + (-7) \\
&= -32
\end{aligned}$

37. $6 - 8;$
$\quad 6 - 8 = -2$

Chapter 2

Chapter 2 Opener

−10	+	2	+	−3	=	−11
+		·		·		−
12	÷	−2	·	3	=	−18
+		+		+		+
6	−	4	−	10	=	−8
=		=		=		=
8	·	0	−	1	=	−1

Section 2.1 Practice Exercises

1. **(a)** equation

   (b) solution

   (c) linear

   (d) solution set

3. Expression

5. Equation

7. Substitute the value into the equation and determine if the right-hand side is equal to the left-hand side.

9. No; $4 - 1 \neq 5$

   $3 \neq 5$

11. Yes; $5(-2) = -10$

    $-10 = -10$

13. Yes; $3(-2) + 9 = 3$

    $-6 + 9 = 3$

    $3 = 3$

15. $$x + 6 = 5$$
    $$x + 6 + (-6) = 5 + (-6)$$
    $$x = -1$$

17. $$q - 14 = 6$$
    $$q - 14 + 14 = 6 + 14$$
    $$q = 20$$

19. $$2 + m = -15$$
    $$-2 + 2 + m = -15 - 2$$
    $$m = -17$$

21. $$-23 = y - 7$$
    $$-23 + 7 = y - 7 + 7$$
    $$-16 = y \text{ or } y = -16$$

23. $$4 + c = 4$$
    $$-4 + 4 + c = 4 - 4$$
    $$c = 0$$

25. $$4.1 = 2.8 + a$$
    $$4.1 - 2.8 = -2.8 + 2.8 + a$$
    $$a = 1.3$$

27. $$5 = z - \frac{1}{2}$$
    $$5 + \frac{1}{2} = z - \frac{1}{2} + \frac{1}{2}$$
    $$\frac{11}{2} = z \text{ or } z = \frac{11}{2}$$
    $$\text{or } z = 5\frac{1}{2}$$

29. $$x + \frac{5}{2} = \frac{1}{2}$$
    $$x + \frac{5}{2} - \frac{5}{2} = \frac{1}{2} - \frac{5}{2}$$
    $$x = -\frac{4}{2} = -2$$

31. $$-6.02 + c = -8.15$$
    $$6.02 - 6.02 + c = -8.15 + 6.02$$
    $$c = -2.13$$

33. $$3.245 + t = -0.0225$$
    $$3.245 - 3.245 + t = -0.0225 + 3.245$$
    $$t = -3.2675$$

35. $6x = 54$

$$\frac{6x}{6} = \frac{54}{6}$$

$$x = 9$$

37. $12 = -3p$

$$\frac{12}{-3} = \frac{-3p}{-3}$$

$$-4 = p \text{ or } p = -4$$

39. $-5y = 0$

$$\frac{-5y}{-5} = \frac{0}{5}$$

$$y = 0$$

41. $\qquad -\dfrac{y}{5} = 3$

$$-\frac{y}{5} \cdot (-5) = 3(-5)$$

$$y = -15$$

43. $\qquad \dfrac{4}{5} = -t$

$$\frac{4}{5}(-1) = -t(-1)$$

$$t = -\frac{4}{5}$$

45. $\qquad \dfrac{2}{5}a = -4$

$$\frac{5}{2} \cdot \frac{2}{5}a = \frac{5}{2}(-4) = -\frac{20}{2}$$

$$a = -10$$

47. $\qquad -\dfrac{1}{5}b = -\dfrac{4}{5}$

$$(-5)\left(-\frac{1}{5}b\right) = (-5)\left(-\frac{4}{5}\right)$$

$$b = 4$$

49. $\qquad -41 = -x$

$$(-1)(-41) = (-1)(-x)$$

$$x = 41$$

51. $\qquad 3.81 = -0.03p$

$$\frac{3.81}{-0.03} = \frac{-0.03p}{-0.03}$$

$$p = -127$$

53. $5.82y = -15.132$

$$\frac{5.82y}{5.82} = \frac{-15.132}{5.82}$$

$$y = -2.6$$

55. Let $x =$ the number. $-8 + x = 42$;

$$-8 + x = 42$$

$$8 - 8 + x = 42 + 8$$

$$x = 50$$

57. Let $x =$ the number. $x - (-6) = 18$;

$$x - (-6) = 18$$

$$x + 6 - 6 = 18 - 6$$

$$x = 12$$

59. $x \cdot 7 = -63$ or $7x = -63$

$$\frac{7x}{7} = \frac{-63}{7}$$

$$x = -9$$

61. $\qquad \dfrac{x}{12} = \dfrac{1}{3}$

$$12 \cdot \frac{x}{12} = 12 \cdot \frac{1}{3}$$

$$x = 4$$

63. Let $x =$ the number. $x + \dfrac{5}{8} = \dfrac{13}{8}$;

$$x + \frac{5}{8} = \frac{13}{8}$$

$$x + \frac{5}{8} - \frac{5}{8} = \frac{13}{8} - \frac{5}{8}$$

$$x = \frac{8}{8} = 1$$

65. (a) $\qquad a - 9 = 1$

$$a - 9 + 9 = 1 + 9$$

$$a = 10$$

(b) $-9a = 1$

$$\frac{-9a}{-9} = \frac{1}{-9}$$

$$a = -\frac{1}{9}$$

79. $-\frac{1}{3}d = 12$

$$(-3)\left(-\frac{1}{3}d\right) = (-3)(12)$$

$$d = -36$$

67. (a) $-\frac{2}{3}h = 8$

$$\left(-\frac{3}{2}\right)\left(-\frac{2}{3}h\right) = \left(-\frac{3}{2}\right)(8)$$

$$h = -12$$

81. $4 = \frac{1}{2} + z$

$$4 - \frac{1}{2} = \frac{1}{2} - \frac{1}{2} + z$$

$$z = \frac{7}{2} \text{ or } 3\frac{1}{2}$$

(b) $\frac{2}{3} + h = 8$

$$-\frac{2}{3} + \frac{2}{3} + h = 8 - \frac{2}{3}$$

$$h = 7\frac{1}{3} = \frac{22}{3}$$

83. $1.2y = 4.8$

$$\frac{1.2y}{1.2} = \frac{4.8}{1.2}$$

$$y = 4$$

69. $\frac{r}{3} = -12$

$$3 \cdot \frac{r}{3} = 3(-12)$$

$$r = -36$$

85. $4.8 = 1.2 + y$

$$4.8 - 1.2 = 1.2 - 1.2 + y$$

$$y = 3.6$$

71. $k + 16 = 32$

$$k + 16 - 16 = 32 - 16$$

$$k = 16$$

87. $0.0034 = y - 0.405$

$$0.0034 + 0.405 = y - 0.405 + 0.405$$

$$y = 0.4084$$

89. Yes

91. No

73. $16k = 32$

$$\frac{16k}{16} = \frac{32}{16}$$

$$k = 2$$

93. Yes

95. Yes

97. For example: $y + 9 = 15$

75. $7 = -4q$

$$\frac{7}{-4} = \frac{-4q}{-4}$$

$$q = -\frac{7}{4} \text{ or } -1\frac{3}{4}$$

99. For example: $2p = -8$

101. For example: $5a + 5 = 5$

103. $5x - 4x + 7 = 8 - 2$

$$x + 7 = 6$$

$$x = -1$$

77. $-4 + q = 7$

$$-4 + 4 + q = 4 + 7$$

$$q = 11$$

105. $6p - 3p = 15 + 6$

$3p = 21$

$p = \dfrac{21}{3} = 7$

Section 2.2 Practice Exercises

1. (a) conditional
   (b) contradiction
   (c) identity

3. $10 - 4w + 7w - 2 + w = 10 + 3w + w - 2$
$ = 4w + 8$

5. $8y - (2y + 3) - 19 = 8y - 2y - 3 - 19$
$ = 6y - 22$

7. $7 = p - 12$

$12 + 7 = p - 12 + 12$

$19 = p$

9. $-7y = 21$

$\dfrac{-7y}{-7} = \dfrac{21}{-7}$

$y = -3$

11. $z - 23 = -28$

$z - 23 + 23 = -28 + 23$

$z = -5$

13. $6z + 1 = 13$

$6z + 1 - 1 = 13 - 1$

$\dfrac{6z}{6} = \dfrac{12}{6}$

$z = 2$

15. $3y - 4 = 14$

$3y - 4 + 4 = 4 + 14$

$\dfrac{3y}{3} = \dfrac{18}{3}$

$y = 6$

17. $-2p + 8 = 3$

$-2p + 8 - 8 = 3 - 8$

$\dfrac{-2p}{-2} = \dfrac{-5}{-2}$

$p = \dfrac{5}{2}$ or $2\dfrac{1}{2}$

19. $0.2x + 3.1 = -5.3$

$0.2x + 3.1 - 3.1 = -5.3 - 3.1$

$\dfrac{0.2x}{0.2} = \dfrac{-8.4}{0.2}$

$x = -42$

21. $\dfrac{5}{8} = \dfrac{1}{4} - \dfrac{1}{2}p$

$\dfrac{5}{8} - \dfrac{1}{4} = -\dfrac{1}{4} + \dfrac{1}{4} - \dfrac{1}{2}p$

$(-2)\left(\dfrac{3}{8}\right) = (-2)\left(-\dfrac{1}{2}p\right)$

$p = -\dfrac{6}{8} = -\dfrac{3}{4}$

23. $7w - 6w + 1 = 10 - 4$

$w + 1 = 6$

$w + 1 - 1 = 6 - 1$

$w = 5$

25. $11h - 8 - 9h = -16$

$2h - 8 + 8 = -16 + 8$

$\dfrac{2h}{2} = \dfrac{-8}{2}$

$h = -4$

27. $3a + 7 = 2a - 19$

$3a - 2a + 7 - 7 = 2a - 2a - 19 - 7$

$a = -26$

29. $-4r - 28 = -58 - r$

$-4r + r - 28 + 28 = -58 + 28 - r + r$

$\dfrac{-3r}{-3} = \dfrac{-30}{-3}$

$r = \dfrac{30}{3} = 10$

31.
$$-2z - 8 = -z$$
$$-2z + 2z - 8 = -z + 2z$$
$$-8 = z \text{ or } z = -8$$

33.
$$\frac{5}{6}x + \frac{2}{3} = -\frac{1}{6}x - \frac{5}{3}$$
$$\frac{5}{6}x + \frac{1}{6}x + \frac{2}{3} - \frac{2}{3} = -\frac{1}{6}x + \frac{1}{6}x - \frac{5}{3} - \frac{2}{3}$$
$$x = -\frac{7}{3} \text{ or } -2\frac{1}{3}$$

35.
$$3y - 2 = 5y - 2$$
$$3y - 5y - 2 + 2 = 5y - 5y - 2 + 2$$
$$\frac{-2y}{-2} = \frac{0}{-2}$$
$$y = 0$$

37.
$$4q + 14 = 2$$
$$4q + 14 - 14 = 2 - 14$$
$$\frac{4q}{4} = \frac{-12}{4}$$
$$q = -3$$

39.
$$-9 = 4n - 1$$
$$-9 + 1 = 4n - 1 + 1$$
$$\frac{-8}{4} = \frac{4n}{4}$$
$$n = -2$$

41.
$$3(2p - 4) = 15$$
$$6p - 12 = 15$$
$$6p - 12 + 12 = 15 + 12$$
$$\frac{6p}{6} = \frac{27}{6}$$
$$p = \frac{9}{2} \text{ or } 4\frac{1}{2}$$

43.
$$6(3x + 2) - 10 = -4$$
$$18x + 12 - 10 = -4$$
$$18x + 2 = -4$$
$$\frac{18x}{18} = \frac{-6}{18}$$
$$x = -\frac{1}{3}$$

45.
$$3.4x - 2.5 = 2.8x + 3.5$$
$$3.4x - 2.8x - 2.5 = 2.8x - 2.8x + 3.5$$
$$0.6x - 2.5 + 2.5 = 3.5 + 2.5$$
$$\frac{0.6x}{0.6} = \frac{6}{0.6}$$
$$x = 10$$

47.
$$17(s + 3) = 4(s - 10) + 13$$
$$17s + 51 = 4s - 40 + 13$$
$$17s - 4s = -27 - 51$$
$$\frac{13s}{13} = \frac{-78}{13}$$
$$s = -6$$

49.
$$6(3t - 4) + 10 = 5(t - 2) - (3t + 4)$$
$$18t - 24 + 10 = 5t - 10 - 3t - 4$$
$$18t - 14 = 2t - 14$$
$$16t = 0$$
$$t = 0$$

51.
$$5 - 3(x + 2) = 5$$
$$5 - 3x - 6 = 5$$
$$-3x - 1 + 1 = 5 + 1$$
$$-3x = 6$$
$$x = -2$$

53.
$$3(2z - 6) - 4(3z + 1) = 5 - 2(z + 1)$$
$$6z - 18 - 12z - 4 = 5 - 2z - 2$$
$$-6z - 22 = -2z + 3$$
$$-4z = 25$$
$$z = -\frac{25}{4}$$

55.
$$-2[(4p + 1) - (3p - 1)] = 5(3 - p) - 9$$
$$-2[4p + 1 - 3p + 1] = 15 - 5p - 9$$
$$-8p - 2 + 6p - 2 = 6 - 5p$$
$$-2p - 4 = 6 - 5p$$
$$3p = 10$$
$$p = \frac{10}{3} \text{ or } 3\frac{1}{3}$$

57. $3(-0.9n + 0.5) = -3.5n + 1.3$
$-2.7n + 1.5 = -3.5n + 1.3$
$0.8n = -0.2$
$n = -0.25$

59. $2(k - 7) = 2k - 13$
$2k - 14 = 2k - 13$
$-14 = -13$
Contradiction; no solution

61. Conditional equation; $7x + 3 = 6x - 12$
$x = -15$

63. $3 - 5.2p = -5.2p + 3$
$3 = 3$
Identity; all real numbers

65. One solution

67. Infinitely many solutions

69. $4p - 6 = 8 + 2p$
$4p - 2p - 6 = 8 + 2p - 2p$
$2p - 6 + 6 = 8 + 6$
$2p = 14$
$\dfrac{2p}{2} = \dfrac{14}{2}$
$p = 7$

71. $2k - 9 = -8$
$2k - 9 + 9 = -8 + 9$
$2k = 1$
$\dfrac{2k}{2} = \dfrac{1}{2}$
$k = \dfrac{1}{2}$

73. $7(w - 2) = -14 - 3w$
$7w - 14 = -14 - 3w$
$10w = 0$
$w = 0$

75. $2(x + 2) - 3 = 2x + 1$
$2x + 4 - 3 = 2x + 1$
$2x + 1 = 2x + 1$
$2x - 2x + 1 = 2x - 2x + 1$
$1 = 1$
All real numbers are solutions.

77. $0.5b = -23$
$\dfrac{0.5b}{0.5} = \dfrac{-23}{0.5}$
$b = -46$

79. $8 - 2q = 4$
$8 - 8 - 2q = 4 - 8$
$-2q = -4$
$\dfrac{-2q}{-2} = \dfrac{-4}{-2}$
$q = 2$

81. $2 - 4(y - 5) = -4$
$2 - 4y + 20 = -4$
$-4y + 22 = -4$
$-4y = -26$
$y = \dfrac{-26}{-4}$
$y = \dfrac{13}{2}$

83. $0.4(a + 20) = 6$
$0.4 + 8 = 6$
$0.4a + 8 - 8 = 6 - 8$
$0.4a = -2$
$\dfrac{0.4a}{0.4} = \dfrac{-2}{0.4}$
$a = -5$

85. $10(2n + 1) - 6 = 20(n - 1) + 12$
$20n + 10 - 6 = 20n - 20 + 12$
$20n + 4 = 20n - 8$
$20n - 20n + 4 = 20n - 20n - 8$
$4 \neq -8$
No solution

87.
$$c + 0.123 = 2.328$$
$$c + 0.123 - 0.123 = 2.328 - 0.123$$
$$c = 2.205$$

89.
$$\frac{4}{5}t - 1 = \frac{1}{5}t + 5$$
$$5\left(\frac{4}{5}t - 1\right) = 5\left(\frac{1}{5}t + 5\right)$$
$$4t - 5 = t + 25$$
$$4t - t - 5 = t - t + 25$$
$$3t - 5 + 5 = 25 + 5$$
$$3t = 30$$
$$\frac{3t}{3} = \frac{30}{3}$$
$$t = 10$$

91.
$$8 - (3q + 4) = 6 - q$$
$$8 - 3q - 4 = 6 - q$$
$$4 - 3q = 6 - q$$
$$-2q = 2$$
$$q = -1$$

93.
$$x + a = 10$$
$$-5 + a = 10$$
$$a = 15$$

95.
$$ax = 12$$
$$a(3) = 12$$
$$a = 4$$

97. For example: $5x + 2 = 2 + 5x$

Section 2.3 Practice Exercises

1. (a) clearing fractions

(b) clearing decimals

3.
$$5(x + 2) - 3 = 4x + 5$$
$$5x + 10 - 3 = 4x + 5$$
$$5x + 7 = 4x + 5$$
$$x + 7 = 5$$
$$x = -2$$

5.
$$3(2y + 3) - 4(-y + 1) = 7y - 10$$
$$6y + 9 + 4y - 4 = 7y - 10$$
$$10y + 5 = 7y - 10$$
$$3y + 5 = -10$$
$$3y = -15$$
$$y = -5$$

7.
$$7x + 2 = 7(x - 12)$$
$$7x + 2 = 7x - 84$$
$$2 \neq -84$$

No solution

All real numbers

9. 18, 36

11. 100; 1000; 10,000

13. 30, 60

15.
$$\frac{1}{2}x + 3 = 5$$
$$2\left(\frac{1}{2}x + 3\right) = 2(5)$$
$$x + 6 = 10$$
$$x = 4$$

17.
$$\frac{1}{6}y + 2 = \frac{5}{12}$$
$$12\left(\frac{1}{6}y + 2\right) = 12\left(\frac{5}{12}\right)$$
$$2y + 24 = 5$$
$$\frac{2y}{2} = -\frac{19}{2}$$
$$y = -\frac{19}{2}$$

19.

$$\frac{1}{3}q + \frac{3}{5} = \frac{1}{15}q - \frac{2}{5}$$

$$15\left(\frac{1}{3}q + \frac{3}{5}\right) = 15\left(\frac{1}{15}q - \frac{2}{5}\right)$$

$$5q + 9 = q - 6$$

$$4q = -15$$

$$q = -\frac{15}{4}$$

21.

$$\frac{12}{5}w + 7 = 31 - \frac{3}{5}w$$

$$5\left(\frac{12}{5}w + 7\right) = 5\left(31 - \frac{3}{5}w\right)$$

$$12w + 35 = 155 - 3w$$

$$15w = 120$$

$$w = 8$$

23.

$$\frac{1}{4}(3m - 4) - \frac{1}{5} = \frac{1}{4}m + \frac{3}{10}$$

$$20\left[\frac{1}{4}(3m - 4) - \frac{1}{5}\right] = 20\left(\frac{1}{4}m + \frac{3}{10}\right)$$

$$5(3m - 4) - 4 = 5m + 6$$

$$15m - 20 - 4 = 5m + 6$$

$$15m - 24 = 5m + 6$$

$$10m = 30$$

$$m = 3$$

25.

$$\frac{1}{6}(5s + 3) = \frac{1}{2}(s + 11)$$

$$6\left[\frac{1}{6}(5s + 3)\right] = 6\left[\frac{1}{2}(s + 11)\right]$$

$$5s + 3 = 3(s + 11)$$

$$5s + 3 = 3s + 33$$

$$2s = 30$$

$$s = 15$$

27. $\frac{2}{3}x + 4 = \frac{2}{3}x - 6$

$$4 \neq -6$$

No solution

No solution

29.

$$\frac{1}{6}(2c - 1) = \frac{1}{3}c - \frac{1}{6}$$

$$6\left(\frac{1}{6}(2c - 1)\right) = 6\left(\frac{1}{3}c - \frac{1}{6}\right)$$

$$2c - 1 = 2c - 1$$

$$-1 = -1$$

All real numbers

All real numbers

31.

$$\frac{2x + 1}{3} + \frac{x - 1}{3} = 5$$

$$3\left(\frac{2x + 1}{3} + \frac{x - 1}{3}\right) = 3(5)$$

$$2x + 1 + x - 1 = 15$$

$$3x = 15$$

$$x = 5$$

33.

$$\frac{3w - 2}{6} = 1 - \frac{w - 1}{3}$$

$$6\left(\frac{3w - 2}{6}\right) = 6\left(1 - \frac{w - 1}{3}\right)$$

$$3w - 2 = 6 - 2(w - 1)$$

$$3w - 2 = 8 - 2w$$

$$5w = 10$$

$$w = 2$$

35.

$$\frac{x + 3}{3} - \frac{x - 1}{2} = 4$$

$$6\left(\frac{x + 3}{3}\right) - 6\left(\frac{x - 1}{2}\right) = 6(4)$$

$$2(x + 3) - 3(x - 1) = 24$$

$$2x + 6 - 3x + 3 = 24$$

$$-x + 9 = 24$$

$$-x = 15$$

$$x = -15$$

37. $9.2y - 4.3 = 50.9$

$$9.2y = 55.2$$

$$y = \frac{55.2}{9.2} = 6$$

39. $0.05z + 0.2 = 0.15z - 10.5$

$\qquad -0.10z = -10.7$

$\qquad z = \dfrac{-10.7}{-0.10} = 107$

41. $0.2p - 1.4 = 0.2(p - 7)$

$\qquad 0.2p - 1.4 = 0.2p - 1.4$

$\qquad 0 = 0$

All real numbers

All real numbers

43. $0.20x + 53.60 = x$

$\qquad 0.80x = 53.60$

$\qquad x = 67$

45. $0.15(90) + 0.05p = 0.10(90 + p)$

$\qquad 13.5 + 0.05p = 9 + 0.10p$

$\qquad -0.05p = -4.5$

$\qquad p = 90$

47. $0.40(y + 10) - 0.60(y + 2) = 2$

$\qquad 0.40y + 4 - 0.60y - 1.2 = 2$

$\qquad -0.2y + 2.8 = 2$

$\qquad -0.2y = -0.8$

$\qquad y = 4$

49. $\qquad 0.12x + 3 - 0.8x = 0.22x - 0.6$

$\qquad 100(0.12x + 3 - 0.8x) = 100(0.22x - 0.6)$

$\qquad 12x + 300 - 80x = 22x - 60$

$\qquad -90x + 300 = -60$

$\qquad -90x = -360$

$\qquad x = 4$

51. $0.06(x - 0.5) = 0.06x + 0.01$

$\qquad 0.06x - 0.03 = 0.06x + 0.01$

$\qquad -0.03 \neq 0.01$

No solution

No solution

53. $\qquad -3.5x + 1.3 = -0.3(9x - 5)$

$\qquad 10(-3.5x + 1.3) = 10(-0.3(9x - 5))$

$\qquad -35x + 13 = -3(9x - 5)$

$\qquad -35x + 13 = -27x + 15$

$\qquad -8x + 13 = 15$

$\qquad -8x = 2$

$\qquad x = -0.25$

55. $0.2x - 1.8 = -3$

$\qquad 0.2x = -1.2$

$\qquad x = -6$

57. $\qquad \dfrac{1}{4}(x + 4) = \dfrac{1}{5}(2x + 3)$

$\qquad 20\left[\dfrac{1}{4}(x + 4)\right] = 20\left[\dfrac{1}{5}(2x + 3)\right]$

$\qquad 5(x + 4) = 4(2x + 3)$

$\qquad 5x + 20 = 8x + 12$

$\qquad -3x = -8$

$\qquad x = \dfrac{8}{3}$

59. $0.05(2t - 1) - 0.03(4t - 1) = 0.2$

$\qquad 0.1t - 0.05 - 0.12t + 0.03 = 0.2$

$\qquad -0.02t - 0.02 = 0.2$

$\qquad -0.02t = 0.22$

$\qquad t = -11$

61. $\qquad \dfrac{2k + 5}{4} = 2 - \dfrac{k + 2}{3}$

$\qquad 12\left(\dfrac{2k + 5}{4}\right) = 12\left(2 - \dfrac{k + 2}{3}\right)$

$\qquad 3(2k + 5) = 24 - 4(k + 2)$

$\qquad 6k + 15 = 24 - 4k - 8$

$\qquad 6k + 15 = -4k + 16$

$\qquad 10k = 1$

$\qquad k = \dfrac{1}{10}$

63.
$$\frac{1}{8}v + \frac{2}{3} = \frac{1}{6}v + \frac{3}{4}$$
$$24\left(\frac{1}{8}v + \frac{2}{3}\right) = 24\left(\frac{1}{6}v + \frac{3}{4}\right)$$
$$3v + 16 = 4v + 18$$
$$-v = 2$$
$$v = -2$$

65.
$$\frac{1}{2}a + 0.4 = -0.7 - \frac{3}{5}a$$
$$10\left(\frac{1}{2}a + 0.4\right) = 10\left(-0.7 - \frac{3}{5}a\right)$$
$$5a + 4 = -7 - 6a$$
$$11a = -11$$
$$a = -1$$

67.
$$0.8 + \frac{7}{10}b = \frac{3}{2}b - 0.8$$
$$10\left(0.8 + \frac{7}{10}b\right) = 10\left(\frac{3}{2}b - 0.8\right)$$
$$8 + 7b = 15b - 8$$
$$-8b = -16$$
$$b = \frac{-16}{-8} = 2$$

Problem Recognition Exercises

1. Expression: $-4b + 18$

3. Equation: $\dfrac{y}{4} = -2$
$$4 \cdot \frac{y}{4} = 4(-2)$$
$$y = -8$$

5. Equation:
$$3(4h - 2) - (5h - 8) = 8 - (2h + 3)$$
$$12h - 6 - 5h + 8 = 8 - 2h - 3$$
$$7h + 8 = -2h + 5$$
$$9h = -3$$
$$h = \frac{1}{3}$$

7. Expression:
$$3(8z - 1) + 10 - 6(5 + 3z)$$
$$= 24z - 3 + 10 - 30 - 18z$$
$$= 6z - 23$$

9. Equation: $6c + 3(c + 1) = 10$
$$6c + 3c + 3 = 10$$
$$9c = 7$$
$$c = \frac{7}{9}$$

11. Equation:
$$0.5(2a - 3) - 0.1 = 0.4(6 + 2a)$$
$$a - 1.5 - 0.1 = 2.4 + 0.8a$$
$$0.2a = 4$$
$$a = 20$$

13. Equation:
$$-\frac{5}{9}w + \frac{11}{12} = \frac{23}{36}$$
$$36\left(-\frac{5}{9}w + \frac{11}{12}\right) = 36\left(\frac{23}{36}\right)$$
$$-20w + 33 = 23$$
$$-20w = -10$$
$$w = \frac{-10}{-20} = \frac{1}{2}$$

15. Expression:
$$\frac{3}{4}x + \frac{1}{2} - \frac{1}{8}x + \frac{5}{4}$$
$$= \frac{6}{8}x - \frac{1}{8}x + \frac{2}{4} + \frac{5}{4}$$
$$= \frac{5}{8}x + \frac{7}{4}$$

17. Equation: no solution

19. Equation:

$$\frac{2x-1}{4}+\frac{3x+2}{6}=2$$

$$12\left(\frac{2x-1}{4}+\frac{3x+2}{6}\right)=12(2)$$

$$3(2x-1)+2(3x+2)=24$$

$$6x-3+6x+4=24$$

$$12x+1=24$$

$$12x=23$$

$$x=\frac{23}{12}$$

21. Equation:

$$4b-8-b=-3b+2(3b-4)$$

$$3b-8=-3b+6b-8$$

$$3b-8=3b-8$$

all real numbers

23. Equation: $\dfrac{4}{3}(6y-3)=0$

$$8y-4=0$$

$$8y=4$$

$$y=\frac{1}{2}$$

25. Expression:

$$3(x+6)-7(x+2)-4(1-x)$$

$$=3x+18-7x-14-4+4x$$

$$=0$$

27. Expression: $3-2[4a-5(a+1)]$

$$=3-2[4a-5a-5]$$

$$=3-2[-a-5]$$

$$=3+2a+10$$

$$=2a+13$$

29. Equation:

$$4+2[8-(6+x)]=-2(x-1)-4+x$$

$$4+2[8-6-x]=-2x+2-4+x$$

$$4+2[2-x]=-x-2$$

$$4+4-2x=-x-2$$

$$-2x+8=-x-2$$

$$-x=-10$$

$$x=10$$

31. Expression:

$$\frac{1}{6}y+y-\frac{1}{3}(4y-1)$$

$$=\frac{1}{6}y+y-\frac{4}{3}y+\frac{1}{3}$$

$$=\frac{1}{6}y+\frac{6}{6}y-\frac{8}{6}y+\frac{1}{3}$$

$$=-\frac{1}{6}y+\frac{1}{3}$$

Section 2.4 Practice Exercises

1. **(a)** consecutive

　(b) even

　(c) odd

　(d) 1

　(e) 2

　(f) 2

3. $x-5{,}682{,}080$

5. $10x$

7. $3x-20$

9. Let x represent the number.
　6 less than $x=-10$

$$x-6=-10$$

$$x=-4$$

The number is -4.

11. Let x represent the unknown number.
Twice(sum of x and 7) = 8
$$2(x+7) = 8$$
$$2x+14 = 8$$
$$2x = -6$$
$$x = -3$$
The number is -3.

13. Let x represent the unknown number.
(x added to 5) = Twice x
$$x+5 = 2x$$
$$x = 5$$
The number is 5.

15. Let x represent the unknown number.
(Sum of $6x$ and 10) = (difference of x and 15)
$$6x+10 = x-15$$
$$5x = -25$$
$$x = -5$$
The number is -5.

17. Let x represent the unknown number.
(3 times the difference of x and 4) = (6 less than x)
$$3(x-4) = x-6$$
$$3x-12 = x-6$$
$$3x = x+6$$
$$2x = 6$$
$$x = 3$$
The number is 3.

19. **(a)** Let x represent the smallest of three consecutive integers. The next two consecutive integers are $x + 1$ and $x + 2$.

(b) Let x represent the largest of three consecutive integers. The next two consecutive integers are $x - 1$ and $x - 2$.

21. Let x = first integer. Then, $x + 1$ is the next integer.
(1st integer) + (2nd integer) = -67
$$x+x+1 = -67$$
$$2x+1 = -67$$
$$2x = -68$$
$$x = -34$$
The integers are -34 and -33.

23. Let x = first odd integer. Then, $x + 2$ is the next odd integer.
(1st integer) + (2nd integer) = 28
$$x+x+2 = 28$$
$$2x+2 = 28$$
$$2x = 26$$
$$x = 13$$
The integers are 13 and 15.

25. Let $x, x + 1, x + 2, x + 3, x + 4$ be the lengths of the sides of the pentagon.
$$\text{Perimeter} = \begin{pmatrix} \text{sum of the} \\ \text{lengths of the} \\ \text{five sides} \end{pmatrix}$$
$$80 = x+x+1+x+2+x+3+x+4$$
$$80 = 5x+10$$
$$5x = 70$$
$$x = 14$$
The sides are 14 in., 15 in., 16 in., 17 in., and 18 in.

27. Let x = first even integer. Then, $x + 2$ and $x + 4$ are the next two even integers.
(sum of three integers) = 2(smallest integer) $- 48$
$$x+x+2+x+4 = 2x-48$$
$$3x+6 = 2x-48$$
$$3x = 2x-54$$
$$x = -54$$
The integers are -54, -52, and -50.

The integers are -87, -85, and -83.

29. Let x = first odd integer. Then, $x + 2$ and $x + 4$ are the next two odd integers.
$$8\begin{pmatrix} \text{sum of} \\ \text{three integers} \end{pmatrix} = 210+10\begin{pmatrix} \text{middle} \\ \text{integer} \end{pmatrix}$$
$$8(x+x+2+x+4) = 210+10(x+2)$$
$$8(3x+6) = 210+10x+20$$
$$24x+48 = 10x+230$$
$$14x = 182$$
$$x = 13$$
The integers are 13, 15, and 17.

31. Let x = the length of the first piece.
Then, the length of the second piece is
$x + 20$.

$$\begin{pmatrix} \text{length of} \\ \text{the 1st piece} \end{pmatrix} + \begin{pmatrix} \text{length of} \\ \text{2nd piece} \end{pmatrix} = 86 \text{ cm}$$

$$x + (x + 20) = 86$$
$$2x = 66$$
$$x = 33$$

The lengths of the pieces are 33 cm and
53 cm.

33. Let x = the number of playlists on
Clarann's iPod. Then, $x - 12$ is the number
of playlists on Karen's iPod.
(Clarann's playlists) + (Karen's playlists) = 58

$$x + x - 12 = 58$$
$$2x - 12 = 58$$
$$2x = 70$$
$$x = 35$$

Karen's iPod has 23 playlists and
Clarann's Ipod has 35 playlists.

35. Let x = Democrats. Then, $x - 31$ is
Republicans.
(Democrats) + (Republicans) = 433

$$x + x - 31 = 433$$
$$2x - 31 = 433$$
$$2x = 464$$
$$x = 232$$
$$x - 31 = 201$$

There were 201 Republicans and 232
Democrats.

37. Let x = number of people who watch
Criminal Minds. Then, $x - 1.06$ is the
number who watched *CSI*.
Total viewers = 27.40

$$x + x - 1.06 = 27.40$$
$$2x - 1.06 = 27.40$$
$$2x = 28.46$$
$$x = 14.23$$

CSI had 13.17 million viewers and
Criminal Minds had 14.23 million
viewers.

39. Let x = length of the Congo River.
Then, the length of the Nile is $x + 2455$.

$$\begin{pmatrix} \text{length of} \\ \text{Congo} \end{pmatrix} + \begin{pmatrix} \text{length of} \\ \text{Nile} \end{pmatrix} = 11{,}195 \text{ km}$$

$$x + (x + 2455) = 11195$$
$$2x = 8740$$
$$x = 4370$$

The length of the Congo is 4370 km; the
length of the Nile is 6825 km.

41. Let x = the land area of Africa. Then, the
land area of Asia is $x + 14{,}514{,}000$.

$$\begin{pmatrix} \text{area of} \\ \text{Africa} \end{pmatrix} + \begin{pmatrix} \text{area of} \\ \text{Asia} \end{pmatrix} = 74{,}644{,}000 \text{ km}^2$$

$$x + x + 14{,}514{,}000 = 74{,}644{,}000$$
$$2x = 60{,}130{,}000$$
$$x = 30{,}065{,}000$$

The area of Africa is 30,065,000 km^2. The
area of Asia is 44,579,000 km^2.

43. Let x = First day's distance. Then, $x - 4.1$
is Second day's distance.
(First day's distance) + (Second day's) = 20.5

$$x + x - 4.1 = 20.5$$
$$2x - 4.1 = 20.5$$
$$2x = 24.6$$
$$x = 12.3$$
$$x - 4.1 = 8.2$$

They walked 12.3 mi on the first day and
8.2 mi on the second day.

45. Let x = shortest piece of PVC piece. Then,
$(3x - 5)$ is length of the longest piece and
$(x + 8)$ is the length of the middle piece.
(Middle pc) + (Small pc) + (Long pc) = 48

$$(x + 8) + x + (3x - 5) = 48$$
$$5x + 3 = 48$$
$$5x = 45$$
$$x = 9$$
$$(3x - 5) = 27 - 5$$
$$= 22$$
$$x + 8 = 9 + 8$$
$$= 17$$

The pieces are 9 in., 17 in., and 22 in.

47. Let x = smallest integer. Then, $x + 1$ is the next integer and $x + 2$ is the largest.

$$(3 \text{ times the largest}) = \left(\begin{array}{c} 47 + \text{the sum} \\ \text{of the two smaller} \end{array}\right)$$

$$3(x + 2) = 47 + x + x + 1$$
$$3x + 6 = 2x + 48$$
$$x = 42$$

The integers are 42, 43, and 44.

49. Let x = earnings for Carrie Underwood. Then, $0.5x - 2$ = earnings for Jennifer Hudson.

$$\left(\begin{array}{c} \text{Carrie} \\ \text{Underwood} \end{array}\right) + \left(\begin{array}{c} \text{Jennifer} \\ \text{Hudson} \end{array}\right) = 19$$

$$x + 0.5x - 2 = 19$$
$$1.5x - 2 = 19$$
$$1.5x = 21$$
$$x = 14$$

$$0.5x - 2 = 0.5(14) - 2$$
$$= 5$$

Carrie Underwood earned \$14 million and Jennifer Hudson earned \$5 million.

51. Let x represent the unknown number.
5(the difference of x and 3) = 4 less than $4 \cdot x$

$$5(x - 3) = 4x - 4$$
$$5x - 15 = 4x - 4$$
$$x = 11$$

The number is 11.

The number is 20.

53. Let x = the first page number and $x + 1$ be the next page number.

$$\left(\begin{array}{c} \text{1st page} \\ \text{number} \end{array}\right) + \left(\begin{array}{c} \text{2nd page} \\ \text{number} \end{array}\right) = 941$$

$$x + x + 1 = 941$$
$$2x = 940$$
$$x = 470$$

The page numbers are 470 and 471.

55. Let x represent the unknown number.
(3 added to $5x$) = (43 more than x)

$$3 + 5x = 43 + x$$
$$4x = 40$$
$$x = 10$$

The number is 10.

57. Let x = the deepest point in the Arctic Ocean. Then, the deepest point in the Pacific Ocean is $2x + 676$.

$$\left(\begin{array}{c} \text{Deepest point} \\ \text{in Pacific} \end{array}\right) = 10,920 \text{ m}$$

$$2x + 676 = 10,920$$
$$2x = 10,244$$
$$x = 5122$$

The deepest point in the Arctic Ocean is 5122 m.

59. Let x = number.

$$\left(\begin{array}{c} \text{twice} \\ \text{a number} \end{array}\right) + \frac{3}{4} = \left(\begin{array}{c} \text{four times} \\ \text{the number} \end{array}\right) - \frac{1}{8}$$

$$2x + \frac{3}{4} = 4x - \frac{1}{8}$$
$$-2x = -\frac{1}{8} - \frac{3}{4}$$
$$-2x = -\frac{7}{8}$$
$$x = \frac{7}{16}$$

The number is $\frac{7}{16}$.

61. Let x = number.

$$\left(\begin{array}{c} \text{product of} \\ \text{a number} \\ \text{and 3.86} \end{array}\right) = \left(\begin{array}{c} 7.15 \\ \text{more than} \\ \text{the number} \end{array}\right)$$

$$3.86x = 7.15 + x$$
$$2.86x = 7.15$$
$$x = 2.5$$

The number is 2.5.

Section 2.5 Practice Exercises

1. **(a)** simple

   (b) 100

3. Let x = first integer. Then, $x + 1$ is the next integer.
$$\begin{pmatrix} 3 \text{ times the} \\ \text{larger} \end{pmatrix} = \begin{pmatrix} 45 \text{ more than} \\ \text{the smaller} \end{pmatrix}$$
$$3(x+1) = x + 45$$
$$3x + 3 = x + 45$$
$$2x + 3 = 45$$
$$2x = 42$$
$$x = 21$$

   The numbers are 21 and 22.

5. Let x = the percent.
$$45 = x(360)$$
$$360x = 45$$
$$x = 0.125$$
$$x = 12.5\%$$

7. Let x = the percent.
$$544 = x(640)$$
$$640x = 544$$
$$x = 0.85$$
$$x = 85\%$$

9. Let x = the number.
$$0.5\% \text{ of } 150 = x$$
$$0.005(150) = x$$
$$x = 0.75$$

11. Let x = the number.
$$142\% \text{ of } 740 = x$$
$$1.42(740) = x$$
$$x = 1050.8$$

13. Let x = the number.
$$177 = 20\% \text{ of } x$$
$$177 = 0.20x$$
$$x = \frac{177}{0.20} = 885$$

15. Let x = the number.
$$275 = 12.5\% \text{ of } x$$
$$2200 = 0.125x$$
$$x = \frac{275}{0.125} = 2200$$

17. Let x = the amount of tax.
$$\begin{pmatrix} \text{Sales} \\ \text{tax} \end{pmatrix} = \begin{pmatrix} \text{tax} \\ \text{rate} \end{pmatrix} \begin{pmatrix} \text{price} \\ \text{of the drill} \end{pmatrix}$$
$$x = 7\% \text{ of } 99.99$$
$$x = 0.07(99.99)$$
$$x = 6.9993$$
$$\begin{pmatrix} \text{total} \\ \text{price} \end{pmatrix} = \begin{pmatrix} \text{sale} \\ \text{price} \end{pmatrix} + \begin{pmatrix} \text{sales} \\ \text{tax} \end{pmatrix}$$
$$= 99.99 + 6.9993$$
$$= 106.9893$$
$$\approx 106.99$$

   Molly will have to pay \$106.99.

19. Let x = number of cases of prostate cancer.
$$x = 33\% \text{ of } 700000$$
$$= 0.33(70000)$$
$$= 231000$$

   There are approximately 231,000 cases of prostate cancer.

21. Let x = percent of cases of pancreas cancer.
$$x = \frac{14000}{700000} = 0.02 = 2\%$$

   2% of cancer cases were diagnosed as pancreas cancer.

23. Let x = Javon's taxtable income.
$$\begin{pmatrix} \text{federal} \\ \text{income} \\ \text{tax} \end{pmatrix} = \begin{pmatrix} \text{tax} \\ \text{rate} \end{pmatrix} \begin{pmatrix} \text{taxable} \\ \text{income} \end{pmatrix}$$
$$23520 = 28\% \text{ of } x$$
$$23520 = 0.28x$$
$$x = \frac{23520}{0.28}$$
$$x = 84000$$

   Javon's taxable income was \$84,000.

25. $I = Prt$
$$I = (\$3000)(3.5\%)(4 \text{ years})$$
$$I = (105)(4) = \$420$$

   Pam will earn \$420.

27. Let P = amount borrowed.

Total paid = $P + Prt$

$$\$1260 = P + P(5\%)(1)$$
$$1260 = P + 0.05P$$
$$1.05P = 1260$$
$$P = \$1200$$

Bob borrowed $1200.

29. Interest = $1950 - $1500 = $450

$$I = Prt$$
$$450 = 1500(r)(5)$$
$$450 = 7500r$$
$$r = 0.06 = 6\%$$

The rate is 6%.

31. Let P = amount invested.

Total saved = $P + Prt$

$$3500 = P + P(3\%)(2)$$
$$3500 = P + 0.06P$$
$$3500 = 1.06P$$
$$P \approx 3302$$

Perry needs to invest $3302.

33. **(a)** Let x = discount on the CD/MP3 player.

$$\text{Discount} = \left(\begin{array}{c}\text{Percent}\\\text{off}\end{array}\right)(\text{cost})$$
$$x = 12\% \text{ of } 170$$
$$x = 0.12(170)$$
$$x = 20.4$$

The discount on the CD/MP 3 player is $20.40.

(b) sale price = cost − discount

$$= 170 - 20.40$$
$$= 149.60$$

The sale price is $149.60.

35. Let x = original price of the digital camera.

$$\left(\begin{array}{c}\text{Original}\\\text{price}\end{array}\right) - \left(\text{Discount}\right) = \left(\begin{array}{c}\text{Sale}\\\text{price}\end{array}\right)$$
$$x - 15\% \text{ of } x = 400.00$$
$$x - 0.15x = 400.00$$
$$0.85x = 400.00$$
$$x = 470.5882353$$
$$x \approx 470.59$$

The original price of the digital camera is approximately $470.59.

37. Let x = percent discount.

$$\left(\begin{array}{c}\text{Original}\\\text{price}\end{array}\right) - \left(\text{Discount}\right) = \left(\begin{array}{c}\text{Sale}\\\text{price}\end{array}\right)$$
$$250 - x(250) = 220$$
$$-250x = -30$$
$$x = 0.12$$
$$x = 12\%$$

The percent discount is 12%.

39. Let x = original dosage.

$$\left(\begin{array}{c}\text{Original}\\\text{dosage}\end{array}\right) + \left(\text{increase}\right) = \left(\begin{array}{c}\text{New}\\\text{dosage}\end{array}\right)$$
$$x + 20\% \text{ of } x = 18$$
$$x + 0.20x = 18$$
$$1.20x = 18$$
$$x = 15$$

The original dosage was 15 cc.

41. Let x = tax rate.

$$(\text{price}) + (\text{tax rate of price}) = \$1890$$
$$1800 + x(1800) = 1890$$
$$1800x = 90$$
$$x = \frac{90}{1800}$$
$$= 0.05$$

The tax rate is 5%.

43. Let x = original ticket price before taxes.
(price) + (11% of price) = 74.37

$$x + 0.11x = 74.37$$
$$1.11x = 74.37$$
$$x = 67$$

The original ticket price before taxes was $67.

45. Let x = original purchase price.
(price) + (24% of price) = 260400

$$x + 0.24x = 260400$$
$$1.24x = 260400$$
$$x = 210000$$

The original purchase price was $210,000.

47. Let x = Alina's salary.
salary = 1600 + (12% commission)

$$x = 1600 + 0.12(25000)$$
$$x = 4600$$

Alina made $4600 that month.

49. Let x = amount sold over $200.

$$\left(\begin{array}{c} 4\% \text{ on amount} \\ \text{sold over } \$200 \end{array}\right) = \$25.80$$
$$0.04x = 25.80$$
$$x = \frac{25.80}{0.04} = 645$$

Diane sold $645 over $200 worth of merchandise.

Calculator Exercises

1. 880/(2π)
140.0563499

3. 20/(5π)
1.273239545

Section 2.6 Practice Exercises

1. $3(2y+3) - 4(-y+1) = 7y - 10$
$$6y + 9 + 4y - 4 = 7y - 10$$
$$10y + 5 = 7y - 10$$
$$3y = -15$$
$$y = -5$$

3. $\frac{1}{2}(x-3) + \frac{3}{4} = 3x - \frac{3}{4}$
$$(4)\left[\frac{1}{2}(x-3) + \frac{3}{4}\right] = (4)\left[3x - \frac{3}{4}\right]$$
$$2(x-3) + 3 = 12x - 3$$
$$2x - 3 = 12x - 3$$
$$-10x = 0$$
$$x = 0$$

5. $0.5(y+2) - 0.3 = 0.4y + 0.5$
$$0.5y + 1 - 0.3 = 0.4y + 0.5$$
$$0.1y = -0.2$$
$$y = -2$$

7. $8b + 6(7 - 2b) = -4(b+1)$
$$8b + 42 - 12b = -4b - 4$$
$$-4b + 42 = -4b - 4$$
$$-4b = -4b - 46$$
$$0 \neq -46$$

No solution

9. $P = a + b + c$
$$P - b - c = a + b + c - b - c$$
$$a = P - b - c$$

11. $x = y - z$
$$y = x + z$$

13. $p = 250 + q$
$$p - 250 = 250 - 250 + q$$
$$q = p - 250$$

15. $A = bh$

$$\frac{A}{h} = \frac{bh}{h}$$

$$b = \frac{A}{h}$$

17. $PV = nrt$

$$\frac{PV}{nr} = t$$

19. $x - y = 5$

$$x = 5 + y$$

21. $3x + y = -19$

$$y = -3x - 19$$

23. $2x + 3y = 6$

$$3y = -2x + 6$$

$$y = \frac{-2x + 6}{3} = -\frac{2}{3}x + 2$$

25. $-2x - y = 9$

$$-2x = y + 9$$

$$x = \frac{y + 9}{-2} = -\frac{1}{2}y - \frac{9}{2}$$

27. $4x - 3y = 12$

$$-3y = -4x + 12$$

$$y = \frac{-4x + 12}{-3} = \frac{4}{3}x - 4$$

29. $ax + by = c$

$$by = -ax + c$$

$$y = \frac{-ax + c}{b} \text{ or } y = -\frac{a}{b}x + \frac{c}{b}$$

31. $A = P(1 + rt)$

$$A = P + Prt$$

$$A - P = Prt$$

$$t = \frac{A - P}{Pr} = \frac{A}{Pr} - \frac{1}{r}$$

33. $a = 2(b + c)$

$$a = 2b + 2c$$

$$a - 2b = 2c$$

$$c = \frac{a - 2b}{2} \text{ or } c = \frac{a}{2} - b$$

35. $Q = \frac{x + y}{2}$

$$2Q = x + y$$

$$y = 2Q - x$$

37. $M = \frac{a}{S}$

$$a = MS$$

39. $P = I^2 R$

$$R = \frac{P}{I^2}$$

41. Let x = width. Then, length equals $x + 2$.

$P = 2w + 2l$

$24 = 2x + 2(x + 2)$

$24 = 2x + 2x + 4$

$4x = 20$

$x = 5$

The width is 5 feet; the length is 7 feet.

43. Let x = width. Then, the length is $4x$.

$P = 2w + 2l$

$300 = 2x + 2(4x)$

$300 = 2x + 8x$

$10x = 300$

$x = 30$

The width is 30 yd. and the length is 120 yd.

45. Let x = width. Then, the length is $2x - 5$.
$$P = 2w + 2l$$
$$590 = 2x + 2(2x - 5)$$
$$590 = 2x + 4x - 10$$
$$6x = 600$$
$$x = 100$$

The width is 100 m; length is 195 m.

47. Let x = length of the two sides that are the same. Then, the length of the third side is $x + 5$.
$$P = a + b + c$$
$$71 = x + x + x + 5$$
$$71 = 3x + 5$$
$$3x = 66$$
$$x = 22$$

The sides are 22 m, 22 m, and 27 m.

49. Adjacent supplementary angles form a straight angle. The words *supplementary* and *straight* both begin with the same letter.

51. Let x = one angle. Then, $x - 20°$ is the other angle.
(sum of the two angles) $= 90°$
$$x + (x - 20) = 90$$
$$2x = 110$$
$$x = 55$$

The angles are 55° and 35°.

53. Let x = one angle. Then, $3x$ is the other angle.
(sum of the angles) $= 180°$
$$x + 3x = 180$$
$$4x = 180$$
$$x = 45°$$

The angles are 45° and 135°.

55. Vertical angles are equal.
$$x + 17 = 2x - 3$$
$$x = 20$$

The angles are $20 + 17 = 37°$ and $2(20) - 3 = 37°$.

57. Let x = smallest angle. Then the middle angle is $2x$ and the largest angle is $3x$.
(sum of angles of a triangle) $= 180°$
$$x + 2x + 3x = 180$$
$$6x = 180$$
$$x = 30°$$

The measures of the angles are 30°, 60°, and 90°.

59. Let x = largest angle. Then, the middle angle is $x - 30°$, and the smallest angle is $\frac{1}{2}$ of $x = \frac{1}{2}x$.
(sum of angles of a triangle) $= 180°$
$$x + x - 30 + \frac{1}{2}x = 180$$
$$\frac{5}{2}x = 210$$
$$x = 210 \cdot \frac{2}{5} = 84°$$

The measures of the angles are 84°, 84° − 30° = 54°, and $\frac{1}{2}(84°) = 42°$.

61. The sum of complementary angles is 90°.
$$(3x + 5) + (2x) = 90$$
$$5x = 85$$
$$x = 17$$

The measures of the angles are $3(17) + 5 = 56°$, $2(17) = 34°$.

63. (a) $A = lw$

(b) $A = lw$
$$w = \frac{A}{l}$$

(c) $w = \frac{A}{l}$
$$w = \frac{1740.5}{59} = 29.5 \text{ feet}$$

43

65. (a) $P = l + l + w + w = 2l + 2w$

(b)
$$P = 2l + 2w$$
$$P - 2w = 2l$$
$$\frac{P - 2w}{2} = \frac{2l}{2}$$
$$l = \frac{P - 2w}{2}$$

(c) $l = \frac{P - 2w}{2}$
$$l = \frac{338 - 2(66)}{2} = \frac{206}{2}$$
$$l = 103$$

The length is 103 m.

67. (a) $C = 2\pi r$

(b) $C = 2\pi r$
$$\frac{C}{2\pi} = \frac{2\pi r}{2\pi}$$
$$r = \frac{C}{2\pi}$$

(c) $r = \frac{C}{2\pi}$
$$r = \frac{880}{2(3.14)} = \frac{880}{6.28}$$
$$r \approx 140$$
The radius is approximately 140 ft.

69. (a) $A = \pi r^2$
$$A = \pi (11.5)^2$$
$$A = 132.25\pi$$
$$A \approx 415.48 \text{ m}^2$$

(b) $V = \pi r^2 h$
$$V = \pi (11.5)^2 (25)$$
$$V = 3306.25\pi$$
$$V \approx 10,386.89 \text{ m}^3$$

Section 2.7 Practice Exercises

1. $ax - by = c$
$$ax = c + by$$
$$x = \frac{c + by}{a}$$

3. $7x + xy = 18$
$$xy = 18 - 7x$$
$$y = \frac{18 - 7x}{x}$$

5. $3(2y + 5) - 8(y - 1) = 3y + 3$
$$6y + 15 - 8y + 8 = 3y + 3$$
$$-2y + 23 = 3y + 3$$
$$-5y = -20$$
$$y = 4$$

7. $200 - t$

9. $100 - x$

11. $3000 - y$

13. Let x = number of \$3 tickets. Then, the number of \$2 tickets is $81 - x$.

$$3x + 2(81 - x) = 215$$
$$3x + 162 - 2x = 215$$
$$x + 162 = 215$$
$$x = 53$$
$$81 - x = 81 - 53 = 28$$
The church sold 53 tickets at \$3 and 28 tickets at \$2.

15. Let x = the number of songs costing \$0.90 each. Then, the number of songs costing \$1.50 each is $25 - x$.

$$\begin{pmatrix} \text{cost} \\ \text{of songs} \\ \text{at \$0.90} \\ \text{each} \end{pmatrix} + \begin{pmatrix} \text{cost} \\ \text{of songs} \\ \text{at \$1.50} \\ \text{each} \end{pmatrix} = \begin{pmatrix} \text{Total} \\ \text{cost} \end{pmatrix}$$

$$0.90x + 1.50(25 - x) = 27.30$$
$$0.90x + 37.50 - 1.50x = 27.30$$
$$-0.60x = -10.2$$
$$x = 17$$

$25 - x = 25 - 17 = 8$

Josh downloaded 17 songs for $0.90 each and 8 songs at $1.50 each.

17. Let x = the number of bottles of soda costing $1.60 each. Then, the number of bottles of flavored water costing $2.00 each is $45 - x$.

$$\begin{pmatrix} \text{cost of} \\ \text{bottles} \\ \text{of soda} \\ \text{at \$1.60} \\ \text{each} \end{pmatrix} + \begin{pmatrix} \text{cost of} \\ \text{bottles of} \\ \text{flavored} \\ \text{water at} \\ \text{\$2.00} \\ \text{each} \end{pmatrix} = \begin{pmatrix} \text{Total} \\ \text{cost} \end{pmatrix}$$

$$1.60x + 2.00(45 - x) = 80$$
$$1.60x + 90 - 2.00x = 80$$
$$-0.40x = -10$$
$$x = 25$$

$45 - x = 45 - 25 = 20$

Angelina bought 25 bottles of sodas and 20 bottles of flavored water.

19. $x + 7$ or $7 + x$

21. $d + 2000$

23. Let x = number of gallons of 5% ethanol. Then, the number of gallons of 9% ethanol is $x + 2000$.

	5% Ethanol	10% Ethanol	Final Mixture: 9% Ethanol
Number of gallons of fuel mixture	x gallons	2000 gallons	$x + 2000$ gallons
Number of gallons of pure ethanol	$0.05x$	0.10 (2000)	$0.09(x + 2000)$

$$0.05x + 0.10(2000) = 0.09(x + 2000)$$
$$0.05x + 200 = 0.09x + 180$$
$$-0.04x = -20$$
$$x = 500$$

500 gallons of 5% ethanol fuel mixture

25. Let x = number of mL of the 1% saline solution. Then, the number of mL of the 9% of saline solution is $x + 24$.

$$\begin{pmatrix} \text{Number} \\ \text{of mL of} \\ \text{pure} \\ \text{saline} \\ \text{in 1\%} \\ \text{solution} \end{pmatrix} + \begin{pmatrix} \text{Number} \\ \text{of mL of} \\ \text{pure} \\ \text{saline} \\ \text{in 16\%} \\ \text{solution} \end{pmatrix} = \begin{pmatrix} \text{Number} \\ \text{of mL of} \\ \text{pure} \\ \text{saline} \\ \text{in 9\%} \\ \text{solution} \end{pmatrix}$$

$$0.01x + 0.16(24) = 0.09(x + 24)$$
$$0.01x + 3.84 = 0.09x + 2.16$$
$$-0.08x = -1.68$$
$$x = 21$$

The pharmacist needs to use 21 mL of the 1% saline solution.

27. Let x = number of ounces of 50% acid solution. Then, the number of ounces of 30% of acid solution is $x + 15$.

$$\begin{pmatrix} \text{Number} \\ \text{of ounces} \\ \text{of pure} \\ \text{acid} \\ \text{in 50\%} \\ \text{solution} \end{pmatrix} + \begin{pmatrix} \text{Number} \\ \text{of ounces} \\ \text{of pure} \\ \text{acid} \\ \text{in 21\%} \\ \text{solution} \end{pmatrix} = \begin{pmatrix} \text{Number} \\ \text{of ounces} \\ \text{of acid} \\ \text{in 30\%} \\ \text{solution} \end{pmatrix}$$

$$0.50x + 0.21(15) = 0.30(x + 15)$$
$$0.50x + 3.15 = 0.30x + 4.5$$
$$0.20x = 1.35$$
$$x = 6.75$$

The contractor needs to mix 6.75 ounces of 50% acid solution.

29. (a) $d = rt = (60)(5) = 300$ mi
    (b) $d = rt = 5x$
    (c) $d = 5(x + 12) = 5x + 60$

31. Let x mph = speed walking down to the lake. Then, the speed walking uphill is $x - 2$.

	Distance	Rate	Time
Downhill to the lake	$2x$ mi	x mph	2 hours
Uphill from the lake	$4(x-2)$ mi	$x - 2$ mph	4 hours

$$2x = 4(x-2)$$
$$2x = 4x - 8$$
$$-2x = -8$$
$$x = 4$$

She walks 4 mph to the lake.

33. Let x mph = speed hiking uphill. Then, speed hiking downhill is $x + 1$ mph.

$$\begin{pmatrix} \text{Distance} \\ \text{hiking} \\ \text{uphill} \end{pmatrix} = \begin{pmatrix} \text{Distance} \\ \text{hiking} \\ \text{downhill} \end{pmatrix}$$

$$3x = 2(x+1)$$
$$3x = 2x + 2$$
$$x = 2$$
$$d = rt = 2(3) = 6$$

Bryan hiked 6 mi up the canyon.

35. Let x mph = speed of the plane in still air. Then, the speed of the plane with the wind is $x + 40$ mph and the speed of the plane against the wind is $x - 40$ mph.

$$\begin{pmatrix} \text{Distance} \\ \text{traveled} \\ \text{with the wind} \end{pmatrix} = \begin{pmatrix} \text{Distance} \\ \text{traveled} \\ \text{against} \\ \text{the wind} \end{pmatrix}$$

$$3.5(x+40) = 4(x-40)$$
$$3.5x + 140 = 4x - 160$$
$$-0.5x = -300$$
$$x = 600$$

The plane travels 600 mph in still air.

The speed of the boat is 15 mph in still water.

37. Let x mph = speed of slower car. Then, the speed of the faster car is $x + 4$ mph.

$$\begin{pmatrix} \text{Distance} \\ \text{traveled} \\ \text{by slower} \\ \text{car} \end{pmatrix} + \begin{pmatrix} \text{Distance} \\ \text{traveled} \\ \text{by faster} \\ \text{car} \end{pmatrix} = \begin{pmatrix} \text{total} \\ \text{distance} \end{pmatrix}$$

$$2x + 2(x+4) = 200$$
$$2x + 2x + 8 = 200$$
$$4x + 8 = 200$$
$$4x = 192$$
$$x = 48$$
$$x + 4 = 48 + 4 = 52$$

The slower car travels 48 mph and the faster car travels 52 mph.

39. Let x mph = speed of the 1st vehicle. Then, the speed of the 2nd vehicle is $x + 10$ mph. Time taken by the 1st vehicle is 4 hours and by the 2nd vehicle is 3 hours.

$$\begin{pmatrix} \text{Distance} \\ \text{traveled} \\ \text{by the } 1^{st} \\ \text{vehicle} \end{pmatrix} - \begin{pmatrix} \text{Distance} \\ \text{traveled} \\ \text{by the } 2^{nd} \\ \text{vehicel} \end{pmatrix} = \begin{pmatrix} \text{distance} \\ \text{between} \\ \text{them} \end{pmatrix}$$

$$4x - 3(x+10) = 10$$
$$4x - 3x - 30 = 10$$
$$x - 30 = 10$$
$$x = 40$$
$$x + 10 = 40 + 10 = 50$$

The speeds of the vehicles are 40 mph and 50 mph.

41. Let x mph = Sarah's speed. Then, the speed of Jeanette is $2x$ mph. Time taken by Sarah and Jeanette is 2 hours.

$$\begin{pmatrix} \text{Distance} \\ \text{traveled by} \\ \text{Jeanette} \end{pmatrix} - \begin{pmatrix} \text{Distance} \\ \text{traveled} \\ \text{by Sarah} \end{pmatrix} = \begin{pmatrix} \text{Distance} \\ \text{between} \\ \text{them} \end{pmatrix}$$

$$2(2x) - 2(x) = 7$$
$$4x - 2x = 7$$
$$2x = 7$$
$$x = 3.5$$
$$2x = 2(3.5) = 7$$

The speed of Sarah is 3.5 mph and Jeanette runs 7 mph.

43. **(a)** 10% of peanuts = 0.10(20) = 2
There are 2 lb of peanuts.

(b) $0.10x$

(c) $0.10(x + 3) = 0.10x + 0.30$

45. Let x = number of pounds of the $12 coffee. Then, the number of pounds of the $8 coffee is $50 - x$.

	$12 coffee	$8 coffee	Total
Number of pounds	x	$50 - x$	50
Value of coffee	$12x$	$8(50 - x)$	$8.80(50)

$$12x + 8(50 - x) = 8.80(50)$$
$$12x + 400 - 8x = 440$$
$$4x = 40$$
$$x = 10$$
$$50 - x = 50 - 10 = 40$$

10 lb of coffee sold at $12 per pound and 40 lb of coffee sold at $8 per pound

47. Let x hours = time that it will take for the boats to reach each other.

$$\begin{pmatrix} \text{Distance} \\ \text{traveled} \\ \text{by the} \\ \text{distress} \\ \text{boat} \end{pmatrix} + \begin{pmatrix} \text{Distance} \\ \text{traveled} \\ \text{by the} \\ \text{guard} \\ \text{cruiser} \end{pmatrix} = \begin{pmatrix} \text{total} \\ \text{distance} \end{pmatrix}$$

$$3x + 25x = 21$$
$$28x = 21$$
$$x = \frac{21}{28}$$
$$x = \frac{3}{4}$$

$$x = \frac{3}{4} \text{ hr} = \frac{3}{4}(60) = 45 \text{ min}$$

The boats will meet in $\frac{3}{4}$ hr or 45 min.

49. Let x = number of packages of wax. Then, the number of bottles of sunscreen is $21 - x$.

$$\begin{pmatrix} \text{Cost} \\ \text{of} \\ \text{packages} \\ \text{of} \\ \text{wax} \end{pmatrix} + \begin{pmatrix} \text{Cost} \\ \text{of} \\ \text{bottles} \\ \text{of} \\ \text{sunscreen} \end{pmatrix} = \begin{pmatrix} \text{Total} \\ \text{cost} \end{pmatrix}$$

$$3.00x + 8.00(21 - x) = 88.00$$
$$3.00x + 168 - 8.00x = 88.00$$
$$-5.00x = -80$$
$$x = 16$$
$$21 - x = 21 - 16 = 5$$

Sam purchased 16 packages of wax and 5 bottles of sunscreen.

51. Let x = number of quarts of 85% chlorine solution. Then, the number of quarts of 45% chlorine solution is $x + 5$.

$$\begin{pmatrix} \text{Number} \\ \text{of quarts} \\ \text{of pure} \\ \text{chlorine} \\ \text{in 85\%} \\ \text{solution} \end{pmatrix} + \begin{pmatrix} \text{Number} \\ \text{of quarts} \\ \text{of pure} \\ \text{chlorine} \\ \text{in 25\%} \\ \text{solution} \end{pmatrix} = \begin{pmatrix} \text{Number} \\ \text{of quarts} \\ \text{of pure} \\ \text{chlorine} \\ \text{in 45\%} \\ \text{solution} \end{pmatrix}$$

$$0.85x + 0.25(5) = 0.45(x + 5)$$
$$0.85x + 1.25 = 0.45x + 2.25$$
$$0.40x = 1$$
$$x = 2.5$$

2.5 quarts of 85% chlorine solution

53. Let x = number of L of pure water. Then, the number of L of 15% alcohol solution is $x + 12$.

$$\begin{pmatrix} \text{Number} \\ \text{of L of} \\ \text{alcohol} \\ \text{in pure} \\ \text{water} \end{pmatrix} + \begin{pmatrix} \text{Number} \\ \text{of L} \\ \text{of pure} \\ \text{alcohol} \\ \text{in 40\%} \\ \text{solution} \end{pmatrix} = \begin{pmatrix} \text{Number} \\ \text{of L} \\ \text{of pure} \\ \text{alcohol} \\ \text{in 15\%} \\ \text{solution} \end{pmatrix}$$

$$0x + 0.40(12) = 0.15(x + 12)$$
$$4.8 = 0.15x + 1.8$$
$$3.0 = 0.15x$$
$$x = 20$$

20 L of pure water

55. Let x km/hr = speed of the Acela Express. Then, the speed of the Japanese bullet train is $x + 60$ km/r.

$$\begin{pmatrix} \text{Distance} \\ \text{traveled} \\ \text{by the} \\ \text{Acela} \\ \text{Express} \end{pmatrix} = \begin{pmatrix} \text{Distance} \\ \text{traveled} \\ \text{by the} \\ \text{Bullet} \\ \text{train} \end{pmatrix}$$

$$3.375x = 2.7(x + 60)$$
$$3.375x = 2.7x + 162$$
$$0.675x = 162$$
$$x = 240$$
$$x + 60 = 240 + 60 = 300$$

The Acela Express travels 240 km/hr and the Japanese bullet train travels 300 km/hr.

Section 2.8 Practice Exercises

1. (a) linear inequality

   (b) inequality

   (c) set-builder, interval

3. $3(x + 2) - (2x - 7) = -(5x - 1) - 2(x + 6)$
$$x + 13 = -7x - 11$$
$$8x = -24$$
$$x = -3$$

5.

7.

9.

11.

13.

15.

17. $[6, \infty)$

19. $(-\infty, 2.1]$

21. $(-2, 7]$

23. $\left\{ x \middle| x > \dfrac{3}{4} \right\}$

$$\left(\dfrac{3}{4}, \infty \right)$$

25. $\{x | -1 < x < 8\}$
$(-1, 8)$

27. $\{x | x < -14\}$
$(-\infty, -14)$

29. $[18, \infty)$

31. $(-\infty, -0.6)$

33. $[-3.5, 7.1)$

35. (a) $x + 3 - 3 = 6 - 3$
$$x = 3$$

   (b) $x + 3 - 3 > 6 - 3$
$$x > 3$$

$\{x | x > 3\}; (3, \infty)$

48

37. (a) $p - 4 + 4 = 9 + 4$

$p = 13$

(b) $p - 4 + 4 \leq 9 + 4$

$p \leq 13$

$\{p \mid p \leq 13\}; (-\infty, 13]$

39. (a) $\dfrac{4c}{4} = \dfrac{-12}{4}$

$c = -3$

(b) $\dfrac{4c}{4} < \dfrac{-12}{4}$

$c < -3$

$\{c \mid c < -3\}; (-\infty, -3)$

41. (a) $\dfrac{-10z}{-10} = \dfrac{15}{-10}$

$z = -\dfrac{15}{10} = -\dfrac{3}{2}$

(b) $-10z \leq 15$

$\dfrac{-10z}{-10} \leq \dfrac{15}{-10}$

$z \geq -\dfrac{3}{2}$

$\left\{z \mid z \geq -\dfrac{3}{2}\right\}; \left[-\dfrac{3}{2}, \infty\right)$

43. $(-1, 4]$

45. $0 < x + 3 < 8$

$0 - 3 < x + 3 - 3 < 8 - 3$

$-3 < x < 5$

$(-3, 5)$

47. $8 \leq 4x \leq 24$

$\dfrac{8}{4} \leq \dfrac{4x}{4} \leq \dfrac{24}{4}$

$2 \leq x \leq 6$

$[2, 6]$

49. $x + 5 - 5 \leq 6 - 5$

$x \leq 1$

(a) $\{x \mid x \leq 1\}$

(b) $(-\infty, 1]$

51. $3q - 7 < 2q + 3$

$3q - 7 + 7 < 2q + 3 + 7$

$3q - 2q < 2q - 2q + 10$

$q < 10$

(a) $\{q \mid q > 10\}$

(b) $(10, \infty)$

53. $4 < 1 + x$

$1 - 1 + x > 4 - 1$

$x > 3$

(a) $\{x \mid x \geq 3\}$

(b) $(3, \infty)$

55. $3c > 6$

$\dfrac{3c}{3} > \dfrac{6}{3}$

$c > 2$

(a) $\{c \mid c > 2\}$

(b) $(2, \infty)$

57. $-3c > 6$

$$\frac{-3c}{-3} > \frac{6}{-3}$$

$$c < -2$$

(a) $\{c \mid c < -2\}$

(b) $(-\infty, -2)$

59. $-h \le -14$

$$(-1)(-h) \le (-1)(-14)$$

$$h \ge 14$$

(a) $\{h \mid h \ge 14\}$

(b) $[14, \infty)$

61. $12 \ge -\dfrac{x}{2}$

$$(-2)(12) \ge (-2)\left(-\frac{x}{2}\right)$$

$$-24 \le x \text{ or } x \ge -24$$

(a) $\{x \mid x \ge -24\}$

(b) $[-24, \infty)$

63. $-2 \le p + 1 < 4$

$$-2 - 1 \le p + 1 - 1 < 4 - 1$$

$$-3 \le p < 3$$

(a) $\{p \mid -3 \le p < 3\}$

(b) $[-3, 3)$

65. $-3 < 6h - 3 < 12$

$$0 < 6h < 15$$

$$0 < h < \frac{15}{6}$$

$$0 < h < \frac{5}{2}$$

(a) $\left\{h \mid 0 < h < \dfrac{5}{2}\right\}$

(b) $\left(0, \dfrac{5}{2}\right)$

67. $-24 < -2x < -20$

$$\frac{-24}{-2} < \frac{-2x}{-2} < \frac{-20}{-2}$$

$$10 < x < 12$$

(a) $\{x \mid 10 < x < 12\}$

(b) $(10, 12)$

69. $-3 + \le \dfrac{1}{4}x - 1 < 5$

$$-3 + 1 \le \frac{1}{4}x - 1 + 1 < 5 + 1$$

$$(4)(-2) \le (4)\frac{1}{4}x < (4)6$$

$$-8 \le x < 24$$

(a) $\{x \mid -8 \le x < 24\}$

(b) $[-8, 24)$

71. $-\dfrac{2}{3}y < 6$

$$\left(-\frac{3}{2}\right)\left(-\frac{2}{3}y\right) < \left(-\frac{3}{2}\right)(6)$$

$$y > -9$$

(a) $\{y \mid y > -9\}$

(b) $(-9, \infty)$

73. $-2x - 4 \le 11$

$\qquad -2x \le 15$

$\qquad \dfrac{-2x}{-2} \le \dfrac{15}{-2}$

$\qquad x \ge -\dfrac{15}{2}$

(a) $\left\{ x \mid x \ge -\dfrac{15}{2} \right\}$

(b) $\left[-\dfrac{15}{2}, \infty \right)$

75. $-12 > 7x + 9$

$\qquad -21 > 7x$

$\qquad \dfrac{7x}{7} < \dfrac{-21}{7}$

$\qquad x < -3$

(a) $\{x \mid x < -3\}$

(b) $(-\infty, -3)$

77. $-7b - 3 \le 2b$

$\qquad -9b \le 3$

$\qquad b \ge -\dfrac{1}{3}$

(a) $\left\{ b \mid b \ge -\dfrac{1}{3} \right\}$

(b) $\left[-\dfrac{1}{3}, \infty \right)$

79. $4n + 2 < 6n + 8$

$\qquad -2n < 6$

$\qquad n > -3$

(a) $\{n \mid n > -3\}$

(b) $(-3, \infty)$

81. $8 - 6(x - 3) > -4x + 12$

$\qquad 8 - 6x + 18 > -4x + 12$

$\qquad -2x > -14$

$\qquad x < 7$

(a) $\{x \mid x < 7\}$

(b) $(-\infty, 7)$

83. $3(x + 1) - 2 \le \dfrac{1}{2}(4x - 8)$

$\qquad 3x + 3 - 2 \le 2x - 4$

$\qquad 3x + 1 \le 2x - 4$

$\qquad x + 1 \le -4$

$\qquad x \le -5$

(a) $\{x \mid x \le -5\}$

(b) $(-\infty, -5]$

85. $4(z - 1) - 6 \ge 6(2z + 3) - 12$

$\qquad 4z - 4 - 6 \ge 12z + 18 - 12$

$\qquad 4z - 10 + 10 \ge 12z + 6 + 10$

$\qquad 4z - 12z \ge 12z - 12z + 16$

$\qquad -8z \ge 16$

$\qquad z \le -2$

(a) $\{x \mid x \le -2\}$

(b) $(-\infty, -2]$

51

87.
$$2a + 3(a + 5) > -4a - (3a - 1) + 6$$
$$2a + 3a + 15 > -4a - 3a + 1 + 6$$
$$5a + 15 - 15 > -7a + 7 - 15$$
$$5a + 7a > -7a + 7a - 8$$
$$\frac{12a}{12} > \frac{-8}{12}$$
$$a > -\frac{2}{3}$$

(a) $\left\{ a \mid a > -\frac{2}{3} \right\}$

(b) $\left(-\frac{2}{3}, \infty \right)$

89.
$$\frac{7}{6}p + \frac{4}{3} \ge \frac{11}{6}p - \frac{7}{6}$$
$$6\left(\frac{7}{6}p + \frac{4}{3} \right) \ge 6\left(\frac{11}{6}p - \frac{7}{6} \right)$$
$$7p + 8 \ge 11p - 7$$
$$-4p \ge -15$$
$$p \le \frac{15}{4}$$

(a) $\left\{ p \mid p \le \frac{15}{4} \right\}$

(b) $\left(-\infty, \frac{15}{4} \right]$

91. $\dfrac{y - 6}{3} > y + 4$
$$y - 6 > 3y + 12$$
$$-2y > 18$$
$$y < -9$$

(a) $\{ y \mid y < -9 \}$

(b) $(-\infty, -9)$

93. $-1.2a - 0.4 < -0.4a + 2$
$$-0.8a < 2.4$$
$$a > -3$$

(a) $\{ a \mid a > -3 \}$

(b) $(-3, \infty)$

95. $-2x + 5 \ge -x + 5$
$$-x \ge 0$$
$$x \le 0$$

(a) $\{ x \mid x \le 0 \}$

(b) $(-\infty, 0]$

97. $-2(-2) + 5 < 4$
$$9 < 4; \text{ No}$$

99. $4(1 + 7) - 1 > 2 + 1$
$$10 > 3; \quad \text{Yes}$$

101. $L \ge 10$

103. $w > 75$

105. $t \le 72$

107. $L \ge 8$

109. $2 < h < 5$

111. Let x = August rainfall.
$$\left(\begin{array}{c} \text{Average summer} \\ \text{rainfall} \end{array} \right) \ge (7.4 \text{ inches})$$
$$\left(\frac{5.9 + 6.1 + x}{3} \right) \ge 7.4$$
$$12 + x \ge 22.2$$
$$x \ge 10.2$$
More than 10.2 inches of rain is needed.

113. Let x = fifth test score.

$$\left(\begin{array}{c}\text{average test}\\\text{scores}\end{array}\right) \geq (80 \text{ percent})$$

$$\left(\frac{85+75+72+82+x}{5}\right) \geq 80$$

$$314 + x \geq 400$$

$$x \geq 86$$

Trevor needs at least 86 to get a B in the course.

115. (a) Let x = number of birdhouses ordered.

$$(\text{original price}) - \left(\begin{array}{c}\text{percent}\\\text{discount}\end{array}\right) = \left(\begin{array}{c}\text{total}\\\text{cost}\end{array}\right)$$

$$\left(\begin{array}{c}\$9 \text{ of}\\\text{number}\\\text{ordered}\end{array}\right) - \left(\begin{array}{c}10\% \text{ of}\\\text{original cost}\end{array}\right) = \left(\begin{array}{c}\text{total}\\\text{cost}\end{array}\right)$$

$$9x - 0.10(9x) = \left(\begin{array}{c}\text{total}\\\text{cost}\end{array}\right)$$

$$9(190) - 0.10(9)(190) = 1539$$

The total cost is $1539.

(b) Total cost for 200 birdhouses is
$9(200) - 0.20(9)(200) = 1440$, or
$1440.
The cost of 190 is $1539. Therefore, it is cheaper to purchase 200 birdhouses because the discount is greater.

117. Let x = number of text messages.

$$\left(\begin{array}{c}\text{cell phone plan}\\\text{with texting}\end{array}\right) \geq (\$18.00)$$

$$4.95 + 0.09x \geq 18$$

$$0.09x \geq 12.95$$

$$x \geq 145$$

If there were more than 145 text messages, the unlimited plan would be the better deal.

119. Let x = number of babysitting hours.

$$(\text{savings} + \text{babysitting hours}) \geq (\text{expenses})$$

$$700 + 10x \geq 1095$$

$$10x \geq 395$$

$$x \geq 39.5$$

Madison must babysit a minimum of 39.5 hours.

Group Activity

1. $\text{BMI} = \dfrac{703W}{h^2}$

$$= \frac{703(160 \text{ lb})}{(5'4'')^2}$$

$$= \frac{703(160 \text{ lb})}{(64'')^2}$$

$$= \frac{112480}{4096}$$

$$= 27.4609375$$

$$\approx 27.5$$

No, the person is considered overweight.

3.
$$18.5 \leq \frac{703W}{72^2} \leq 24.9$$

$$72^2(18.5) \leq 703W \leq 72^2(24.9)$$

$$\frac{72^2(18.5)}{703} \leq W \leq \frac{72^2(24.9)}{703}$$

$$136.4 \leq W \leq 183.6$$

The person's ideal weight range is $136.4 \leq W \leq 183.6$.

Chapter 2 Review Exercises

Section 2.1

1. (a) Equation
(b) Expression
(c) Equation
(d) Equation

3. (a) No
(b) Yes
(c) No
(d) Yes

5.
$$a + 6 = -2$$
$$a + 6 - 6 = -2 - 6$$
$$a = -8$$

7. $-\dfrac{3}{4} + k = \dfrac{9}{2}$
$$k = \frac{9}{2} + \frac{3}{4} = \frac{18}{4} + \frac{3}{4} = \frac{21}{4}$$

9. $-5x = 21$
$$x = \frac{21}{-5} = -\frac{21}{5}$$

11. $-\dfrac{2}{5}k = \dfrac{4}{7}$
$$-\frac{5}{2}\left(-\frac{2}{5}k\right) = -\frac{5}{2} \cdot \frac{4}{7}$$
$$k = -\frac{20}{14} = -\frac{10}{7}$$

13. Let x = the number.
$$\begin{pmatrix} \text{quotient of} \\ x \text{ and } -6 \end{pmatrix} = -10$$
$$\frac{x}{-6} = -10$$
$$x = (-10)(-6) = 60$$

The number is 60.

15. Let x = the number.
$$x - 4 = -12$$
$$x = -8$$
The number is -8.

Section 2.2

17. $4d + 2 = 6$
$$4d = 4$$
$$d = 1$$

19. $-7c = -3c - 9$
$$-4c = -9$$
$$c = \frac{-9}{-4} = \frac{9}{4}$$

21. $\dfrac{b}{3} + 1 = 0$
$$\frac{b}{3} = -1$$
$$b = (-1)(3) = -3$$

23. $-3p + 7 = 5p + 1$
$$-8p = -6$$
$$p = \frac{-6}{-8} = \frac{3}{4}$$

25. $4a - 9 = 3(a - 3)$
$$4a - 9 = 3a - 9$$
$$a = 0$$

27. $7b + 3(b - 1) + 3 = 2(b + 8)$
$$7b + 3b - 3 + 3 = 2b + 16$$
$$10b = 2b + 16$$
$$8b = 16$$
$$b = 2$$

29. A contradiction has no solution and an identity is true for all real numbers.

31. $3x - 19 = 2x + 1$
$$x = 20; \text{ conditional}$$

33. $2x - 8 = 2(x - 4)$
$$0 = 0; \text{ identity}$$

35. $4x - 4 = 3x - 2$
$$x = 2; \text{ conditional}$$

Section 2.3

37. $\dfrac{y}{15} - \dfrac{2}{3} = \dfrac{4}{5}$
$$15\left(\frac{y}{15} - \frac{2}{3}\right) = 15\left(\frac{4}{5}\right)$$
$$y - 10 = 12$$
$$y = 22$$

39.
$$\frac{x-6}{3} - \frac{2x+8}{2} = 12$$
$$6\left(\frac{x-6}{3} - \frac{2x+8}{2}\right) = 12(6)$$
$$2(x-6) - 3(2x+8) = 72$$
$$2x - 12 - 6x - 24 = 72$$
$$-4x - 36 = 72$$
$$-4x = 108$$
$$x = -27$$

41.
$$\frac{1}{4}y - \frac{3}{4} = \frac{1}{2}y + 1$$
$$4\left(\frac{1}{4}y - \frac{3}{4}\right) = 4\left(\frac{1}{2}y + 1\right)$$
$$y - 3 = 2y + 4$$
$$-y = 7$$
$$y = -7$$

43.
$$\frac{2}{7}(w+4) = \frac{1}{2}$$
$$14 \cdot \frac{2}{7}(w+4) = 14 \cdot \frac{1}{2}$$
$$4(w+4) = 7$$
$$4w = -9$$
$$w = -\frac{9}{4}$$

45. $4.9z + 4.6 = 3.2z - 2.2$
$$1.7z = -6.8$$
$$z = -4$$

47. $62.84t - 123.66 = 4(2.36 + 2.4t)$
$$62.84t - 123.66 = 9.44 + 9.6t$$
$$53.24t = 133.10$$
$$t = 2.5$$

49. $0.20(x+4) + 0.65x = 0.20(854)$
$$0.20x + 0.8 + 0.65x = 170.8$$
$$0.85x = 170$$
$$x = 200$$

51. $3 - (x+4) + 5 = 3x + 10 - 4x$
$$3 - x - 4 + 5 = -x + 10$$
$$4 \neq 10$$
No solution

53. $9 - 6(2z+1) = -3(4z-1)$
$$9 - 12z - 6 = -12z + 3$$
$$-12z + 3 = -12z + 3$$
$$0 = 0$$
Identity; all real numbers

Section 2.4

55. Let x = the number.
$$20 + (x+6) = 37$$
$$26 + x = 37$$
$$x = 11$$
The number is 11.

57. Let x = the number.
$$5x - 8 = x - 48$$
$$4x = -40$$
$$x = -10$$
The number is -10.

59. Let $x = 1^{\text{st}}$ integer. Then, $x + 1$ is the next and $x + 2$ is the largest.

$$\begin{pmatrix} 10 \text{ times} \\ \text{the smallest} \end{pmatrix} = 213 + \begin{pmatrix} \text{sum of} \\ \text{other two} \\ \text{integers} \end{pmatrix}$$
$$10x = 213 + (x+1) + (x+2)$$
$$10x = 216 + 2x$$
$$8x = 216$$
$$x = 27$$
The numbers are 27, 28, and 29.

61. Let $x = 1^{st}$ integer. Then, the next four integers are $x + 1$, $x + 2$, $x + 3$, $x + 4$.

$P = $ (sum of the sides)

$190 = x + x + 1 + x + 2 + x + 3 + x + 4$

$190 = 5x + 10$

$180 = 5x$

$x = 36$

Sides are 36 cm, 37 cm, 38 cm, 39 cm, and
40 cm.

63. Let $x = $ population of Kentucky (in millions of people). Then, $x + 2.1$ is the population of Indiana.

$\left(\begin{array}{c}\text{population} \\ \text{of Indiana}\end{array}\right) + \left(\begin{array}{c}\text{population} \\ \text{of Kentucky}\end{array}\right) = 10.3$

$x + x + 2.1 = 10.3$

$2x = 8.2$

$x = 4.1$

Kentucky has 4.1 million people; Indiana has 6.2 million people.

Section 2.5

65. Let $x = $ the number.

$x = (4\% \text{ of } 720)$

$x = 0.04(720)$

$x = 28.8$

67. Let $x = $ the percent.

$68.4 = x\% \text{ of } 72$

$68.4 = x(72)$

$x = \dfrac{68.4}{72} = 0.95 = 95\%$

69. Let $x = $ a number.

$8.75 = 0.5\% \text{ of } x$

$8.75 = 0.005x$

$x = \dfrac{8.75}{0.005} = 1750$

71. Let $x = $ price before tax and tip.

$\left(\begin{array}{c}\text{price} \\ \text{of} \\ \text{dinner}\end{array}\right) = \left(\begin{array}{c}\text{price} \\ \text{before} \\ \text{tax} \\ \text{and tip}\end{array}\right) + \left(\begin{array}{c}20\% \\ \text{tip}\end{array}\right) + \left(\begin{array}{c}6\% \\ \text{tax}\end{array}\right)$

$50.40 = x + 0.20x + 0.06x$

$50.40 = 1.26x$

$x = 40$

The dinner was $40 before tax and tip.

73. Amount $= P + I$

$A = P + Prt$

$14,400 = P + P(0.04)(5)$

$14,400 = 1.2P$

$P = \$12,000$

Eduardo originally invested $12,000.

Section 2.6

75. $K = C + 273$

$K - 273 = C$ or $C = K - 273$

77. $P = 3s$

$\dfrac{P}{3} = s$ or $s = \dfrac{P}{3}$

79. $a + bx = c$

$bx = c - a$

$x = \dfrac{c - a}{b}$

81. $4(a + b) = Q$

$4a + 4b = Q$

$4b = Q - 4a$

$b = \dfrac{Q - 4a}{4}$ or $b = \dfrac{Q}{4} - a$

The height is 7 m.

83. (a) $V = \dfrac{1}{3}\pi r^2 h$

$3V = \pi r^2 h$

$h = \dfrac{3V}{\pi r^2}$

(b) $h = \dfrac{3V}{\pi r^2}$

$h = \dfrac{3(47.8)}{\pi (3)^2}$

$h = \dfrac{143.4}{9\pi}$

$h \approx 5.1$ in.

85. Let x = angle and $90 - x$ = complementary angle.

(one angle) = (10 more than other angle)

$x = 10 + 90 - x$

$x = 100 - x$

$2x = 100$

$x = 50$

The measure of the angles are $40°$ and $50°$.

87. $2x + 25 = 4x - 15$

$25 = 2x - 15$

$40 = 2x$

$x = 20$

$(2x + 25)° = [2(20) + 25]°$

$= (40 + 25)°$

$= 65°$

The measure of the angle measure is $65°$.

Section 2.7

89. Let x km/hr = speed in stormy weather.
Then, the speed in good weather is $x + 15$.

	Distance	Rate	Time
stormy weather	$14x$ km	x km/hr	14 hr
good weather	$10.5(x + 15)$ km	$x + 15$ km/hr	10.5 hr

$14x = 10.5(x + 15)$

$14x = 10.5x + 157.5$

$3.5x = 157.5$

$x = 45$

$x + 15 = 45 + 15 = 60$

The truck travels 45 km/hr in stormy weather and 60 km/hr in good weather.

91. Let x hours = time that it will take for the cars to be 327.6 mi apart.

$$\begin{pmatrix} \text{Distance} \\ \text{traveled} \\ \text{by the} \\ \text{car heading} \\ \text{east} \end{pmatrix} + \begin{pmatrix} \text{Distance} \\ \text{traveled} \\ \text{by the} \\ \text{car heading} \\ \text{west} \end{pmatrix} = \begin{pmatrix} \text{total} \\ \text{distance} \\ \text{between} \\ \text{them} \end{pmatrix}$$

$55x + 62x = 327.6$

$117x = 327.6$

$x = 2.8$

$x = 2.8$ hr $= 2$ hr and $0.8(60) = 2$ hr 48 min

Two cars will be 327.6 mi apart after 2.8 hr or 2 hr and 48 min.

93. Let x = number of lb of ground beef. Then, the number of lb of a mixture is $x + 8$.

$$\begin{pmatrix} \text{Number} \\ \text{of lb of} \\ \text{fat in} \\ \text{ground} \\ \text{beef} \end{pmatrix} + \begin{pmatrix} \text{Number} \\ \text{of lb of} \\ \text{fat in} \\ \text{ground} \\ \text{sirloin} \end{pmatrix} = \begin{pmatrix} \text{Number} \\ \text{of lb of} \\ \text{fat in} \\ \text{the} \\ \text{mixture} \end{pmatrix}$$

$0.24x + 0.06(8) = 0.096(x + 8)$

$0.24x + 0.48 = 0.096x + 0.768$

$0.144x = 0.288$

$x = 2$

2 lb of ground beef with 24% fat

Section 2.8

95. $(-2, \infty)$

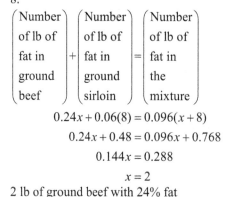

97. $(-1, 4]$

57

99. $c + 6 < 23$

$c < 17$

$\{c \mid c < 17\}, (-\infty, 17)$

101. $-2x - 7 \geq 5$

$-2x \geq 12$

$\dfrac{-2x}{-2} \geq \dfrac{12}{-2}$

$x \leq -6$

$\{x \mid x \leq -6\}, (-\infty, -6]$

103. $-\dfrac{3}{7}a \leq -21$

$\left(-\dfrac{7}{3}\right)\left(-\dfrac{3}{7}a\right) \leq \left(-\dfrac{7}{3}\right)(-21)$

$a \geq 49$

$\{a \mid a \geq 49\}, [49, \infty)$

105. $4k + 23 < 7k - 31$

$-3k < -54$

$k > 18$

$\{k \mid k > 18\}, (18, \infty)$

107. $-5x - 2(4x - 3) + 6 > 17 - 4(x - 1)$

$-5x - 8x + 6 + 6 > 17 - 4x + 4$

$-13x + 12 > -4x + 21$

$-13x > -4x + 9$

$-9x > 9$

$x < -1$

$\{x \mid x < -1\}, (-\infty, -1)$

109. $-2 \leq z + 4 \leq 9$

$-2 - 4 \leq z + 4 - 4 \leq 9 - 4$

$-6 \leq z \leq 5$

$\{z \mid -6 \leq z \leq 5\}, [-6, 5]$

111. Let x = number of wings.

$(\text{food} + \text{wings}) \leq \$11.25$

$2.50 + 2.50 + 1.75 + 0.25x \leq 11.25$

$6.75 + 0.25x \leq 11.25$

$0.25x \leq 4.50$

$x \leq 18$

Collette can have at most 18 wings.

Chapter 2 Test

1. (a) No; $4(-3) + 1 = 10$

$-12 + 1 = 10$

$-11 \neq 10$

(b) Yes; $6(-3 - 1) = -3 - 21$

$6(-4) = -24$

$-24 = -24$

(c) No; $5(-3) - 2 = 2(-3) + 1$

$-15 - 2 = -6 + 1$

$-17 \neq -5$

(d) Yes; $\dfrac{1}{3}(-3) + 1 = 0$

$-1 + 1 = 0$

$0 = 0$

3. $-3x + 5 = -2$

$-3x = -7$

$x = \dfrac{-7}{-3} = \dfrac{7}{3}$

5. $t + 3 = -13$

$t + 3 - 3 = -13 - 3$

$t = -16$

7. $8 = p - 4$

$8 + 4 = p - 4 + 4$

$p = 12$

9. $\dfrac{t}{8} = -\dfrac{2}{9}$

$(8)\dfrac{t}{8} = (8)\left(-\dfrac{2}{9}\right)$

$t = -\dfrac{16}{9}$

11. $-5(x+2)+8x = -2+3x-8$

$-5x-10+8x = -10+3x$

$3x-10 = 3x-10$

$0 = 0$

Identity; all real numbers

13. $0.5c-1.9 = 2.8+0.6c$

$-0.1c = 4.7$

$c = \dfrac{4.7}{-0.1} = -47$

15. $C = 2\pi r$

$\dfrac{C}{2\pi} = \dfrac{2\pi r}{2\pi}$

$r = \dfrac{C}{2\pi}$

17. (a) $(-\infty, 0)$

(b) $[-2, 5)$

19. $2(3-x) \geq 14$

$6-2x \geq 14$

$-2x \geq 8$

$x \leq -4$

$\{x \mid x \leq -4\}, (-\infty, -4]$

21. $-13 \leq 3p+2 \leq 5$

$-15 \leq 3p \leq 3$

$\dfrac{-15}{3} \leq \dfrac{3p}{3} \leq \dfrac{3}{3}$

$-5 \leq p \leq 1$

$\{p \mid -5 \leq p \leq 1\}, [-5, 1]$

23. Let x = original amount borrowed.

$A = P+Prt$

$8000 = x + x(6\%)(10)$

$8000 = x + 0.60x$

$8000 = 1.60x$

$x = 5000$

Clarita originally borrowed $5000.

25. Let x mph = speed of one family. Then, the speed of the other family is $x - 5$ mph.

$\begin{pmatrix} \text{distance} \\ \text{traveled} \\ \text{by one} \\ \text{family} \end{pmatrix} + \begin{pmatrix} \text{distance} \\ \text{traveled} \\ \text{by the} \\ \text{other} \\ \text{family} \end{pmatrix} = \begin{pmatrix} \text{distance} \\ \text{between} \\ \text{them} \end{pmatrix}$

$\begin{pmatrix} \text{speed} \\ \text{of one} \\ \text{family} \end{pmatrix}(2) + \begin{pmatrix} \text{speed} \\ \text{of the} \\ \text{other} \\ \text{family} \end{pmatrix}(2) = 210 \text{ mi.}$

$2x + 2(x-5) = 210$

$2x + 2x - 10 = 210$

$4x = 220$

$x = 55$

One family travels at 55 mph and the other family travels at 50 mph.

27. Let x = the first integer. Then, the next four consecutive integers are $x + 1$, $x + 2$, $x + 3$, $x + 4$. The perimeter is the sum of the sides.

(sum of the sides) = 315 in.

$$x + x + 1 + x + 2 + x + 3 + x + 4 = 315$$

$$5x + 10 = 315$$

$$5x = 305$$

$$x = 61$$

The sides are 61 in., 62 in., 63 in., 64 in., and 65 in.

29. Let x = price of a basketball ticket. Then, the price of a hockey ticket = $x + 4.32$.

$$\left(\begin{array}{c}\text{price of two} \\ \text{basketball} \\ \text{tickets}\end{array}\right) + \left(\begin{array}{c}\text{price of two} \\ \text{hockey} \\ \text{tickets}\end{array}\right) = \$153.92$$

$$2x + 2(x + 4.32) = 153.92$$

$$4x + 8.64 = 153.92$$

$$4x = 145.28$$

$$x = 36.32$$

Each basketball ticket was $36.32 and each hockey ticket was $40.64.

31. Let w = width of field. Then, $2w - 40$ is the length.

$$P = 2l + 2w$$

$$370 = 2(2w - 40) + 2w$$

$$370 = 4w - 80 + 2w$$

$$370 = 6w - 80$$

$$450 = 6w$$

$$w = 75$$

The field is 110 m long and 75 m wide.

Cumulative Review Exercises
Chapters 1–2

1. $\left| -\dfrac{1}{5} + \dfrac{7}{10} \right| = \left| -\dfrac{2}{10} + \dfrac{7}{10} \right| = \left| \dfrac{5}{10} \right| = \dfrac{1}{2}$

3. $-\dfrac{2}{3} + \left(\dfrac{1}{2} \right)^2 = -\dfrac{2}{3} + \dfrac{1}{4} = -\dfrac{8}{12} + \dfrac{3}{12} = -\dfrac{5}{12}$

5. $\sqrt{5 - (-20) - 3^2} = \sqrt{5 + 20 - 9} = \sqrt{16} = 4$

7. $-14 + 12 = -2$

9. $-4[2x - 3(x + 4)] + 5(x - 7)$

$$= -4[2x - 3x - 12] + 5x - 35$$

$$= 4x + 48 + 5x - 35$$

$$= 9x + 13$$

11. $-2.5x - 5.2 = 12.8$

$$-2.5x = 18$$

$$x = -7.2$$

13. $\dfrac{x + 3}{5} - \dfrac{x + 2}{2} = 2$

$$10\left(\dfrac{x + 3}{5} - \dfrac{x + 2}{2} \right) = 10(2)$$

$$2(x + 3) - 5(x + 2) = 20$$

$$2x + 6 - 5x - 10 = 20$$

$$-3x - 4 = 20$$

$$-3x = 24$$

$$x = -8$$

15. $-0.6w = 48$

$$w = -80$$

17. Let x = cost before tax.

$$\left(\begin{array}{c}\text{Cost of} \\ \text{suit}\end{array}\right) + \left(\begin{array}{c}7\% \text{ tax} \\ \text{on cost}\end{array}\right) = \$374.50$$

$$x + 0.07x = 374.50$$

$$1.07x = 374.50$$

$$x = 350$$

The cost before tax is $350.

19. $-3x - 3(x + 1) < 9$

$$-3x - 3x - 3 < 9$$

$$-6x - 3 < 9$$

$$-6x < 12$$

$$x > -2$$

$$\{x \mid x > -2\},\ (-2, \infty)$$

Chapter 3

Chapter 3 Review Your Skills

Across

5. 2.251 has a digit in the thousandths place. Therefore, 1000 would clear the decimal.

Down

1. $2a - 3 = 45$
$2a = 48$
$a = 24$

3. $-10^2 \cdot (-2)^3 \cdot 11$
$= -100 \cdot (-8) \cdot 11$
$= 800 \cdot 11$
$= 8800$

5. The lowest common multiple of 2, 4 & 3 is 12. Therefore, 12 would clear the fractions.

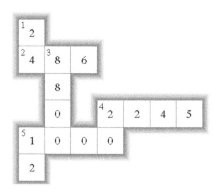

Section 3.1 Practice Exercises

1. **(a)** x; y-axis
   (b) ordered
   (c) origin; (0, 0)
   (d) quadrants
   (e) negative
   (f) III

3. **(a)** Month 10
   (b) 30
   (c) Between months 3 and 5; also between months 10 and 12

(d) Between months 8 and 9

(e) Month 3

(f) 80 patients

5. **(a)** On day 1 the price per share of the stock was $89.25.
   (b) $93.00 - $90.25 = $1.75
   (c) $90.25 - $93.00 = -$2.75

7.

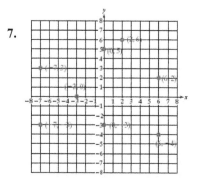

9.

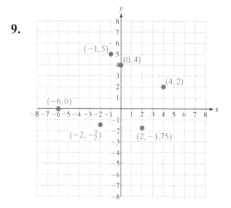

11. Quadrant IV

13. Quadrant II

15. Quadrant III

17. Quadrant I

19. (0, -5) lies on the y-axis.

21. $\left(\dfrac{7}{8}, 0\right)$ is located on the x-axis.

23. $A(-4, 2)$, $B\left(\dfrac{1}{2}, 4\right)$, $C(3, -4)$, $D(-3, -4)$,

$E(0, -3)$, $F(5, 0)$

25. (a) $A(400, 200)$, $B(200, -150)$,
$C(-300, -200)$, $D(-300, 250)$, $E(0, 450)$

(b) $250 \text{ m} - (-200 \text{ m}) = 450 \text{ m}$

27. (a) $(250, 225)$, $(175, 193)$, $(315, 330)$, $(220, 209)$, $(450, 570)$, $(400, 480)$, $(190, 185)$; the first ordered pair represents 250 people in attendance produces \$225 in popcorn sales.

(b)

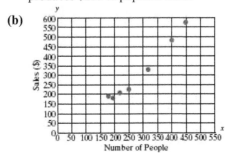

29. (a) $(1, -10.2)$, $(2, -9.0)$, $(3, -2.5)$, $(4, 5.7)$, $(5, 13.0)$, $(6, 18.3)$, $(7, 20.9)$, $(8, 19.6)$, $(9, 14.8)$, $(10, 8.7)$, $(11, 2.0)$, $(12, -6.9)$

(b)

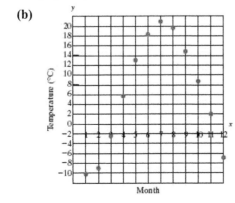

31. (a)

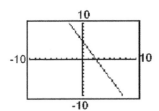

Percent of Males/Females with 4 or More Years of College, United States

Year ($x = 0$ corresponds to 1960)

(b) Increasing

(c) Increasing

Calculator Exercises

1. Set the viewing window at

```
WINDOW
 Xmin=-10
 Xmax=10
 Xscl=1
 Ymin=-10
 Ymax=10
 Yscl=1
 Xres=■
```

3. Set the viewing window at

```
WINDOW
 Xmin=-3
 Xmax=10
 Xscl=1
 Ymin=-6
 Ymax=3
 Yscl=■
 Xres=1
```

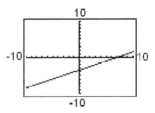

62

5. Set the viewing window at

```
WINDOW
 Xmin=-10
 Xmax=10
 Xscl=1
 Ymin=-10
 Ymax=10
 Yscl=1
 Xres=1
```

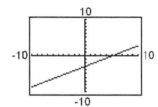

7. Set the viewing window at

```
WINDOW
 Xmin=-10
 Xmax=10
 Xscl=1
 Ymin=-10
 Ymax=10
 Yscl=1
 Xres=■
```

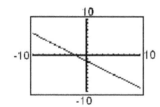

9.

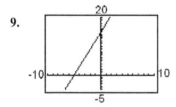

11.

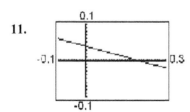

Section 3.2 Practice Exercises

1. (a) $Ax + By = C$

    (b) x-intercept

    (c) y-intercept

    (d) vertical

    (e) horizontal

3. $(-2, -2)$; quadrant III

5. $(-5, 0)$; x-axis

7. $(-3, 2)$; quadrant II

9. Yes; $8 - 2 = 6$

11. Yes; $4 = -\dfrac{1}{3}(-3) + 3$

13. No; $4\left(\dfrac{1}{4}\right) + 5\left(-\dfrac{2}{5}\right) \neq 20$

15. No; $6 \neq -2$

17. Yes; $-5 = -5$

19.

x	y
1	-3
-2	0
-3	1
-4	2

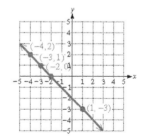

63

21.

x	y
-2	3
-1	0
-4	9

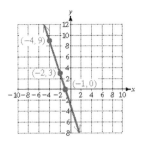

23.

x	y
0	4
2	0
3	-2

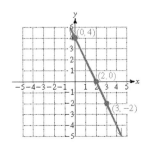

25.

x	y
0	-2
5	-5
10	-8

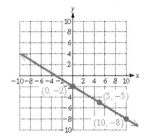

27.

x	y
0	-2
-3	-2
5	-2

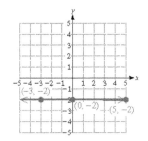

29.

x	y
$\frac{3}{2}$	-1
$\frac{3}{2}$	2
$\frac{3}{2}$	-3

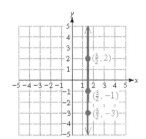

31.

x	y
0	4.6
1	3.4
2	2.2

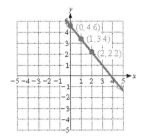

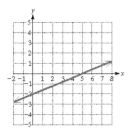

37. $y = -2x$

x	y
-2	4
0	0
2	-4

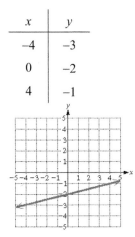

33. $x - y = 4$

x	y
2	-2
0	-4
-2	-6

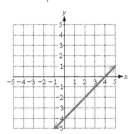

35. $2x - 5y = 10$

x	y
-5	-4
0	-2
5	0

39. $y = \frac{1}{4}x - 2$

x	y
-4	-3
0	-2
4	-1

65

41. $-x + y = 0$

x	y
-10	-10
-5	-5
0	0

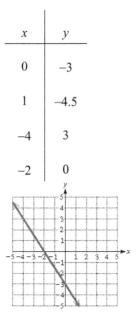

43. $-30x - 20y = 60$

x	y
0	-3
1	-4.5
-4	3
-2	0

45. Since the y-intercept is the point with x-coordinate $= 0$, it is on the y-axis.

47. x-intercept: $(-1, 0)$
y-intercept: $(0, -3)$

49. x-intercept: $(-4, 0)$
y-intercept: $(0, 1)$

51. $4x - 3y = -9$
x-intercept: $4x - 3(0) = -9$
$$4x = -9$$
$$x = -\frac{9}{4} \qquad \left(-\frac{9}{4}, 0\right)$$
y-intercept: $4(0) - 3y = -9$
$$-3y = -9$$
$$y = 3 \qquad (0, 3)$$

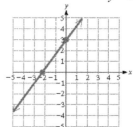

53. $y = -\frac{3}{4}x + 2$

x-intercept: $0 = -\frac{3}{4}x + 2$
$$-2 = -\frac{3}{4}x$$
$$\frac{8}{3} = x \qquad \left(\frac{8}{3}, 0\right)$$
y-intercept: $y = -\frac{3}{4}(0) + 2$
$$y = 2 \qquad (0, 2)$$

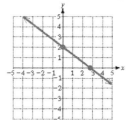

55. $2x + 8 = y$
x-intercept: $2x + 8 = 0$
$$2x = -8$$
$$x = -4 \qquad (-4, 0)$$
y-intercept: $2(0) + 8 = y$
$$8 = y \qquad (0, 8)$$

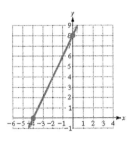

57. $2x - 2y = 0$

x-intercept: $2x - 2(0) = 0$

$$2x = 0$$

$$x = 0 \qquad (0, 0)$$

y-intercept: $2(0) - 2y = 0$

$$-2y = 0$$

$$y = 0 \qquad (0, 0)$$

Because the x-intercept and y-intercept are the same point, we need one additional point to graph the line. We arbitrarily let $x = 1$.

$$2(1) - 2y = 0$$

$$2 - 2y = 0$$

$$2 = 2y$$

$$1 = y \qquad (1, 1)$$

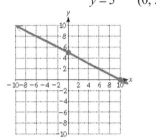

59. $20x = -40y + 200$

x-intercept: $20x = -40(0) + 200$

$$20x = 200$$

$$x = 10 \qquad (10, 0)$$

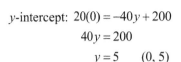

y-intercept: $20(0) = -40y + 200$

$$40y = 200$$

$$y = 5 \qquad (0, 5)$$

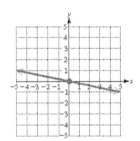

61. $x = -5y$

x-intercept: $x = -5(0)$

$$x = 0 \qquad (0, 0)$$

y-intercept: $\begin{aligned} 0 &= -5y \\ 0 &= y \qquad (0, 0) \end{aligned}$

Because the x-intercept and y-intercept are the same point, we need one additional point to graph the line. We arbitrarily let $x = 5$.

$$5 = -5y$$

$$-1 = y \qquad (5, -1)$$

63. True

65. True

67. $y = -1$

(a) Horizontal

(b)

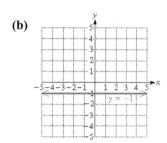

(c) no *x*-intercept; *y*-intercept: (0, −1)

69. $5x = 20$

$x = 4$

(a) Vertical

(b)

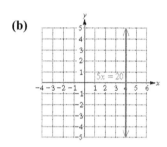

(c) *x*-intercept: (4, 0); no *y*-intercept

71. $y - 8 = -13$

$y = -5$

(a) Horizontal

(b)

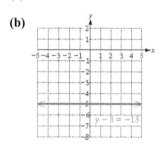

(c) no *x*-intercept; *y*-intercept: (0, −5)

73. $5x = 0$

$x = 0$

(a) Vertical

(b)

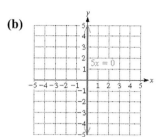

(c) All points on the *y*-axis are *y*-intercepts; *x*-intercept: (0, 0)

75. A horizontal line may not have an *x*-intercept. A vertical line may not have a *y*-intercept.

77. a,b,d

79. (a) $y = -1531(1) + 11,599 = 10,068$

   (b) $7006 = -1531x + 11,599$

$-4593 = -1531x$

$\dfrac{-4593}{-1531} = x$

$x = 3$

   (c) (1, 10,068) One year after purchase the value of the car is $10,068. (3, 7006) Three years after purchase the value of the car is $7,006.

Section 3.3 Practice Exercises

1. (a) slope; $\dfrac{y_2 - y_1}{x_2 - x_1}$

   (b) parallel

   (c) right

   (d) −1

   (e) undefined; horizontal

3. x-intercept   y-intercept

$x - 3(0) = 6$ $0 - 3y = 6$

$x = 6$ $y = -2$

(6, 0

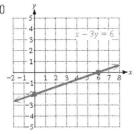

5. x-intercept   y-intercept

none $2y - 3 = 0$

$$y = \frac{3}{2}$$

$$\left(0, \frac{3}{2}\right)$$

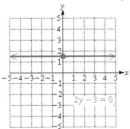

7. $m = \dfrac{8}{24} = \dfrac{1}{3}$

9. $m = \dfrac{3}{5.5} = \dfrac{3(2)}{5.5(2)} = \dfrac{6}{11}$

11. Undefined

13. Positive

15. Slope is negative because the line falls from left to right

17. Slope is zero because the line is horizontal

19. Slope is undefined because the line is vertical

21. Slope is positive because the line rises from left to right

23. Slope is negative because the line falls from left to right

25. (0, 1) and (2, 2) are on the line.

$$m = \frac{2-1}{2-0} = \frac{1}{2}$$

27. (0, 0) and (1, −3) are on the line.

$$m = \frac{-3-0}{1-0} = \frac{-3}{1} = -3$$

29. (0, −3) and (1, −3) are on the line.

$$m = \frac{-3-(-3)}{1-0} = \frac{-3+3}{1} = \frac{0}{1} = 0$$

31. (−2, 0) and (−2, 1) are on the line.

$$m = \frac{1-0}{-2-(-2)} = \frac{1}{-2+2} = \frac{1}{0}$$

The slope is undefined.

33. $m = \dfrac{3-4}{-1-2} = \dfrac{-1}{-3} = \dfrac{1}{3}$

35. $m = \dfrac{0-3}{-1-(-2)} = \dfrac{-3}{1} = -3$

37. $m = \dfrac{2-5}{(-4)-1} = \dfrac{-3}{-5} = \dfrac{3}{5}$

39. $m = \dfrac{3-3}{-2-5} = \dfrac{0}{-7} = 0$

41. $m = \dfrac{5-(-7)}{2-2} = \dfrac{12}{0}$ is undefined

43. $m = \dfrac{-\frac{4}{5}-\frac{3}{5}}{\frac{1}{4}-\frac{1}{2}} = \dfrac{-\frac{7}{5}}{-\frac{1}{4}} = \dfrac{28}{5}$

45. $m = \dfrac{-\frac{3}{4}-\frac{3}{1}}{-\frac{1}{2}-\left(-\frac{4}{3}\right)} = \dfrac{-\frac{15}{4}}{\frac{5}{6}} = -\dfrac{9}{2}$

47. $m = \dfrac{6-(-1)}{-5-3} = \dfrac{7}{-8} = -\dfrac{7}{8}$

49. $m = \dfrac{-1.3-0.5}{0.8-1.4} = \dfrac{-1.8}{-0.6} = 3$

51. $m = \dfrac{1.1-(-3.4)}{-3.2-6.8} = \dfrac{4.5}{-10} = -0.45$ or $-\dfrac{9}{20}$

Chapter 3

53. $m = \dfrac{2.6 - 3.5}{2000 - 1994} = \dfrac{-0.9}{6} = -0.15$ or $-\dfrac{3}{20}$

55. (a) $m = -2$

    (b) $m = -\dfrac{1}{-2} = \dfrac{1}{2}$

57. (a) $m = 0$

    (b) $m = -\dfrac{1}{0}$ is undefined

59. (a) $m = \dfrac{4}{5}$

    (b) $m = -\dfrac{1}{\frac{4}{5}} = -\dfrac{5}{4}$

61. $m_1 m_2 = -2\left(\dfrac{1}{2}\right) = -1$

The lines are perpendicular.

63. $m_1 = m_2 = 1$
The two lines are parallel.

65. $m_1 = \dfrac{2}{7} \neq m_2 = -\dfrac{2}{7}$

$m_1 m_2 = \dfrac{2}{7}\left(-\dfrac{2}{7}\right) = -\dfrac{4}{49} \neq -1$

The lines are neither parallel nor perpendicular.

67. $m_1 = \dfrac{-2 - 4}{-1 - 2} = \dfrac{-6}{-3} = 2$

$m_2 = \dfrac{5 - 7}{0 - 1} = \dfrac{-2}{-1} = 2$

Since $m_1 = m_2 = 2$, the two lines are parallel.

69. $m_1 = \dfrac{4 - 9}{0 - 1} = \dfrac{-5}{-1} = 5$

$m_2 = \dfrac{1 - 2}{10 - 5} = \dfrac{-1}{5}$

$m_1 m_2 = 5\left(-\dfrac{1}{5}\right) = -1$

The two lines are perpendicular.

71. $m_1 = \dfrac{3 - 4}{0 - 4} = \dfrac{1}{4}$

$m_2 = \dfrac{-1 - 7}{-1 - 1} = \dfrac{-8}{-2} = 4$

$m_1 \neq m_2$

$m_1 m_2 = \dfrac{1}{4}(4) = 1 \neq -1$

The lines are neither parallel nor perpendicular.

73. We use the ordered pairs (0, 32,000) and (15, 29,600) to find the slope.

$m = \dfrac{29,600 - 32,000}{15 - 0} = \dfrac{-2,400}{15} = -160$;
The average rate of change is –$160 per year.

The average rate of change is –$160 a year.

75. (a) We use the ordered pairs (1985, 539) and (2010, 1714) to find the slope.

$m = \dfrac{1714 - 539}{2010 - 1985} = \dfrac{1175}{25} = 47$

    (b) The slope $m = 47$ means that the number of male prisoners increased at a rate of 47 thousand per year during this time period.

77. (a) $d = 0.2t = 0.2(5 \text{ sec}) = 1$ mile

    (b) $d = 0.2t = 0.2(10 \text{ sec}) = 2$ miles

    (c) $d = 0.2t = 0.2(15 \text{ sec}) = 3$ miles

    (d) $m = \dfrac{(2 \text{ miles}) - (1 \text{ mile})}{(10 \text{ sec}) - (5 \text{ sec})} = \dfrac{1}{5} = 0.2$

The slope $m = 0.2$ means that the distance between a lightning strike and an observer increases by 0.2 miles for every additional second between seeing lightning and hearing thunder.

The slope means that the amount Michael owes decreases by $210 per month.

79. $m = \dfrac{3 \text{ units up}}{4 \text{ units right}} = \dfrac{3}{4}$

81. $m = \dfrac{0 \text{ units up}}{5 \text{ units right}} = \dfrac{0}{5} = 0$

83.

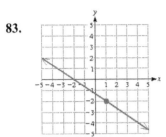

For example: $(4, -4)$ and $(-2, 0)$

From $(1, -2)$ going 2 units down and 3 units to the right yields the point $(4, -4)$. Or, going 2 units up and 3 units to the left yields the point $(-2, 0)$.

85.

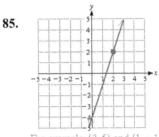

For example: $(3, 5)$ and $(1, -1)$

From $(2, 2)$ going 3 units up and 1 unit to the right yields the point $(3, 5)$. Or, going 3 units down and 1 unit to the left yields the point $(1, -1)$.

87.

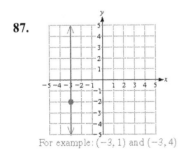

For example: $(-3, 1)$ and $(-3, 4)$

From $(-3, -2)$ going 3 units up yields $(-3, 1)$. Or, going 6 units up yields the point $(-3, 4)$.

For Exercises 89-94, answers will vary.

89.

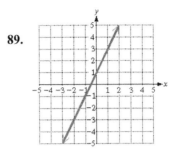

91.

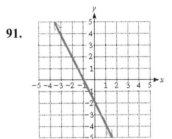

93.

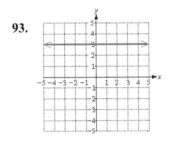

95. Label the points.

$(x_1, y_1) = (a + b, 4m - n)$

$(x_2, y_2) = (a - b, m + 2n)$

Now substitute into the slope formula.

$m = \dfrac{(m + 2n) - (4m - n)}{(a - b) - (a + b)}$

$= \dfrac{m + 2n - 4m + n}{a - b - a - b}$

$= \dfrac{m - 4m + 2n + n}{a - a - b - b}$

$= \dfrac{-3m + 3n}{-2b}$ or $\dfrac{3m - 3n}{2b}$

71

97. To find the x-intercept, substitute $y = 0$.

$$ax + b(0) = c$$
$$ax = c$$
$$x = \frac{c}{a}$$
$$\left(\frac{c}{a}, 0\right)$$

99. From $(2, -1)$ going 2 units up and 5 units to the right yields the point $(7, 1)$. Or, going 2 units down and 5 units to the left yields the point $(-3, -3)$.

Calculator Exercises

1. $l_1: y = -x + 1 \qquad m_1 = -1$

$l_2: y = x + 3 \qquad m_2 = 1$

$m_1 m_2 = -1(1) = -1$

The lines are perpendicular.

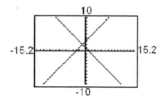

3. $l_1: y = 2x - 4 \qquad m_1 = 2$

$l_2: y = -\dfrac{3}{2}x + 2 \quad m_2 = -\dfrac{3}{2}$

$m_1 m_2 = 2\left(-\dfrac{3}{2}\right) = -3 \neq -1$

The lines are neither parallel nor perpendicular.

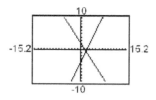

5. The lines may appear to coincide on a graph; however, they are not the same line because their y-intercepts are different.

Section 3.4 Practice Exercises

1. (a) $y = mx + b$,

(b) standard

3. x-intercept: $y = 0$ y-intercept: $x = 0$
$\quad x - 5(0) = 10 \qquad\qquad 0 - 5y = 10$
$\qquad\quad x = 10 \qquad\qquad\qquad y = -2$
$\quad (10, 0) \qquad\qquad\qquad (0, -2)$

5. x-intercept: $y = 0$ y-intercept: $x = 0$
$\quad$ none $\qquad\qquad\qquad\quad 3y = -9$
$\qquad\qquad\qquad\qquad\qquad\quad y = -3$
$\qquad\qquad\qquad\qquad\quad (0, -3)$

7. x-intercept: $y = 0$ y-intercept: $x = 0$
$\quad -4x = 6(0) \qquad\qquad -4(0) = 6y$
$\qquad\quad x = 0 \qquad\qquad\qquad\quad y = 0$
$\quad (0, 0) \qquad\qquad\qquad\quad (0, 0)$

9. x-intercept: $y = 0$ y-intercept: $x = 0$
$\quad 5x = 20 \qquad\qquad\qquad$ none
$\quad\ x = 4$

$\quad (4, 0)$

11. $y = -2x + 3$
$m = -2$; y-intercept $= (0, 3)$

13. $y = x - 2$
$m = 1$; y-intercept $= (0, -2)$

15. $y = -x$
$m = -1$; y-intercept $= (0, 0)$

17. $y = \dfrac{3}{4}x - 1$

$m = \dfrac{3}{4}$; y-intercept $= (0, -1)$

19. $2x - 5y = 4$

$-5y = -2x + 4$

$y = \dfrac{-2}{-5}x + \dfrac{4}{-5}$

$y = \dfrac{2}{5}x - \dfrac{4}{5}$

$m = \dfrac{2}{5};\ y\text{-intercept} = \left(0, -\dfrac{4}{5}\right)$

21. $3x - y = 5$

$y = 3x - 5$

$m = 3;\ y\text{-intercept} = (0, -5)$

23. $x + y = 6$

$y = -x + 6$

$m = -1;\ y\text{-intercept} = (0, 6)$

25. $x + 6 = 8$

$x = 2$

It is a vertical line at $x = 2$; slope is undefined and no y-intercept.

27. $-8y = 2$

$y = -\dfrac{1}{4}$

It is a horizontal line; $m = 0$, y-intercept = $\left(0, -\dfrac{1}{4}\right)$

29. $3y - 2x = 0$

$3y = 2x$

$y = \dfrac{2}{3}x$

$m = \dfrac{2}{3};\ y\text{-intercept} = (0, 0)$

31. $15x + 5y = -20$

$5y = -15x - 20$

$y = -3x - 4$

$m = -3;\ y\text{-intercept} = (0, -4)$

33. $4y - 11x = -3$

$4y = 11x - 3$

$y = \dfrac{11}{4}x - \dfrac{3}{4}$

$m = \dfrac{11}{4};\ y\text{-intercept} = \left(0, -\dfrac{3}{4}\right)$

35.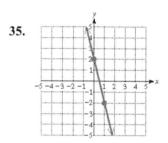

37.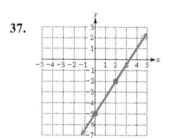

39. The slope is positive and the y-intercept is positive, so the equation matches graph b.

41. The slope is negative and the y-intercept is positive, so the equation matches graph e.

43. The slope is positive and the y-intercept is $(0, 0)$, so the equation matches graph c.

45. $2x + y = 9$

$y = -2x + 9$

73

47. $x - 2y = 6$

$-2y = -x + 6$

$y = \dfrac{1}{2}x - 3$

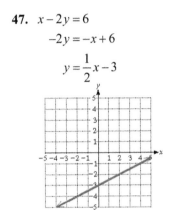

49. $2x = -4y + 6$

$-4y = 2x - 6$

$y = -\dfrac{1}{2}x + \dfrac{3}{2}$

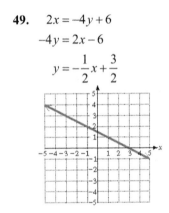

51. $x + y = 0$

$y = -x$

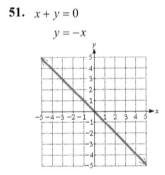

53. $5y = 4x$

$y = \dfrac{4}{5}x$

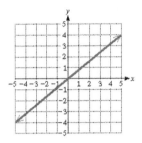

55. $3y + 2 = 0$

$y = -\dfrac{2}{3}$

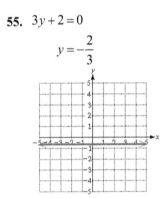

57. $l_1: m_1 = -2$

$l_2: m_2 = \dfrac{1}{2}$

$m_1 m_2 = -2\left(\dfrac{1}{2}\right) = -1$

The lines are perpendicular.

59. $l_1: m_1 = \dfrac{4}{5}$

$l_2: m_2 = \dfrac{5}{4}$

$m_1 m_2 = \dfrac{4}{5}\left(\dfrac{5}{4}\right) = 1 \neq -1$

$m_1 \neq m_2$

The lines are neither parallel nor perpendicular.

61. $l_1: m = -9$

$l_2: m = -9$

The lines are parallel.

63. Vertical and horizontal lines; they are perpendicular.

65. Both are vertical lines and are parallel.

67. l_1: $2x + 5y = 6$

$$3y = -2x + 6$$

$$y = -\frac{2}{3}x + 2$$

$$m_1 = -\frac{2}{3}$$

l_2: $3x - 2y = 12$

$$-2y = -3x + 12$$

$$y = \frac{3}{2}x - 6$$

$$m_2 = \frac{3}{2}$$

$$m_1 m_2 = -\frac{2}{3}\left(\frac{3}{2}\right) = -1$$

The lines are perpendicular.

69. l_1: $4x + 2y = 6$

$$2y = -4x + 6$$

$$y = -2x + 3$$

$$m_1 = -2$$

l_2: $4x + 8y = 16$

$$8y = -4x + 16$$

$$y = -\frac{1}{2}x + 2$$

$$m_2 = -\frac{1}{2}$$

$$m_1 m_2 = -2\left(-\frac{1}{2}\right) = 1 \neq -1$$

$$m_1 \neq m_2$$

The lines are neither parallel nor perpendicular.

71. l_1: $m = \frac{1}{5}$

l_2: $2x - 10y = 20$

$$-10y = -2x + 20$$

$$y = \frac{1}{5}x - 2$$

$$m_2 = \frac{1}{5}$$

$$m_1 = m_2$$

The lines are parallel.

73. $y = mx + b$

$$y = -\frac{1}{3}x + 2$$

75. $y = mx + b$
$$y = 10x - 19$$

77. $y = mx + b$

$$-2 = 6(1) + b$$

$$-2 = 6 + b$$

$$-8 = b$$

$$y = 6x - 8$$

79. $y = mx + b$

$$-5 = \frac{1}{2}(-4) + b$$

$$b = -3$$

$$y = \frac{1}{2}x - 3$$

81. $y = mx + b$
$$y = 0x + (-11)$$
$$y = -11$$

83. $y = 5x + 0$

$$y = 5x$$

85. $y = mx + b$

$$m = \frac{3 - (-1)}{0 - 2} = \frac{4}{-2} = -2$$

$$3 = -2(0) + b$$

$$b = 3$$

$$y = -2x + 3$$

87. $y = mx + b$

$$m = \frac{3-1}{-3-3} = \frac{2}{-6} = -\frac{1}{3}$$

$$3 = -\frac{1}{3}(-3) + b$$

$$b = 2$$

$$y = -\frac{1}{3}x + 2$$

89. $y = mx + b$

$$m = \frac{-9-(3)}{-2-1} = \frac{-12}{-3} = 4$$

$$3 = 4(1) + b$$

$$b = -1$$

$$y = 4x - 1$$

91. (a) $m = 1203$; The slope represents the rate of the increase in the number of cases of Lyme disease per year.

(b) (0, 10,006); in 1993 there were 10,006 cases of Lyme disease

(c) $y = 1203(17) + 10,006 = 30,457$; Approximately. 30,457 cases of Lyme disease

(d) $36,472 = 1203x + 10,006$
$x = 22$; the year 2015

93. $y = -\frac{A}{B}x + \frac{C}{B}$; the slope is $-\frac{A}{B}$

95. $m = -\frac{6}{7}$

97. $m = \frac{11}{8}$

Problem Recognition Exercises

a. $y = 5x$: $m = 5$, passes through the origin

b. $2x + 3y = 12$:

$$m = -\frac{2}{3}; y\text{-intercept} = (0, 4);$$

x-intercept = (6, 0)

c. $y = \frac{1}{2}x - 5$: $m = \frac{1}{2}$; y-intercept = (0, −5);

x-intercept = (10, 0)

d. $3x - 6y = 10$: $m = \frac{1}{2}$;

$$y\text{-intercept} = \left(0, -\frac{5}{3}\right); x\text{-intercept} = \left(\frac{10}{3}, 0\right)$$

e. $2y = -8$: $m = 0$; y-intercept = (0, −4);

no x-intercept

f. $y = -2x + 4$: $m = -2$; y-intercept = (0, 4);

x-intercept = (2, 0)

g. $3x = 1$; slope is undefined; no y-intercept;

$$x\text{-intercept} = \left(\frac{1}{3}, 0\right)$$

h. $x + 2y = 6$: $m = -\frac{1}{2}$; y-intercept = (0,3);

x-intercept = (6, 0)

1. Line whose slope is positive: a, c, d

3. Line that passes through the origin: a

5. Line whose y-intercept is (0, 4): b, f

7. Line whose slope is $\frac{1}{2}$: c, d

9. Line whose slope is 0: e

11. Line that is parallel to the line with the equation $y = -\frac{2}{3}x + 4$: b

13. Line that is vertical: g

15. Line whose x-intercept is (10, 0): c

17. Line that is parallel to the x-axis: e

19. Line with a negative slope and positive y-intercept: b, f, h

Section 3.5 Practice Exercises

1. (a) $Ax + By = C$

(b) horizontal

(c) vertical

(d) slope; y-intercept

(e) $y - y_1 = m(x - x_1)$,

3. $2x - 3y = -3$

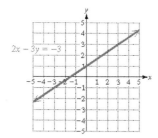

5. $3 - y = 9$

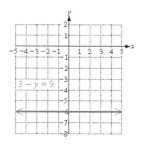

7. $m = \dfrac{-3 - 6}{1 - 2} = \dfrac{-9}{-1} = 9$

9. $m = \dfrac{5 - 5}{-2 - 5} = \dfrac{0}{-7} = 0$

11. $y - y_1 = m(x - x_1)$
$y - 1 = 3(x - (-2))$
$y - 1 = 3(x + 2)$
$y - 1 = 3x + 6$
$\quad y = 3x + 7$ or $3x - y = -7$

13. $y - y_1 = m(x - x_1)$
$y - (-2) = -4[x - (-3)]$
$y + 2 = -4(x + 3)$
$y + 2 = -4x - 12$
$\quad y = -4x - 14$ or $4x + y = -14$

15. $y - y_1 = m(x - x_1)$
$y - 0 = -\dfrac{1}{2}[x - (-1)]$
$y = -\dfrac{1}{2}(x + 1)$
$y = -\dfrac{1}{2}x - \dfrac{1}{2}$

$Or, \dfrac{1}{2}x + y = -\dfrac{1}{2}$
$x + 2y = -1$

$Or, \dfrac{3}{4}x + y = \dfrac{3}{2}$
$3x + 4y = 6$

17. $m = \dfrac{-6 - 0}{-2 - 1} = \dfrac{-6}{-3} = 2$
$y - y_1 = m(x - x_1)$
$y - (-6) = 2(x - (-2))$
$y + 6 = 2(x + 2)$
$y + 6 = 2x + 4$
$\quad y = 2x - 2$ or $2x - y = 2$

19. $m = \dfrac{-4 - (-3)}{0 - (-1)} = \dfrac{-1}{1} = -1$
$y - y_1 = m(x - x_1)$
$y - (-4) = -1(x - 0)$
$y + 4 = -x$
$\quad y = -x - 4$ or $x + y = -4$

$Or, \dfrac{5}{8}x + y = -\dfrac{19}{8}$
$5x + 8y = -19$

21. $m = \dfrac{-3.3 - (-5.3)}{2.2 - 12.2} = \dfrac{2}{-10} = -0.2$

$$y - y_1 = m(x - x_1)$$
$$y - (-3.3) = -0.2(x - 2.2)$$
$$y + 3.3 = -0.2x + 0.44$$
$$y = -0.2x - 2.86$$

$Or, 0.2x + y = -2.86$

$20x + 100y = -286$

23. Two points $(-2, 5)$ and $(-1, 3)$

$$m = \dfrac{5 - 3}{-2 - (-1)} = \dfrac{2}{-1} = -2$$
$$y - y_1 = m(x - x_1)$$
$$y - 3 = -2[x - (-1)]$$
$$y - 3 = -2(x + 1)$$
$$y - 3 = -2x - 2$$
$$y = -2x + 1$$

25. Two points $(-2, 0)$ and $(0, 4)$

$$m = \dfrac{0 - 4}{-2 - 0} = \dfrac{-4}{-2} = 2$$
$$y - y_1 = m(x - x_1)$$
$$y - 4 = 2(x - 0)$$
$$y - 4 = 2x$$
$$y = 2x + 4$$

27. Two points $(-4, -3)$ and $(4, 1)$

$$m = \dfrac{-3 - 1}{-4 - 4} = \dfrac{-4}{-8} = \dfrac{1}{2}$$
$$y - y_1 = m(x - x_1)$$
$$y - 1 = \dfrac{1}{2}(x - 4)$$
$$y - 1 = \dfrac{1}{2}x - 2$$
$$y = \dfrac{1}{2}x - 1$$

29. $l: y = 4x + 3$: slope $= 4$

Since the two lines are parallel, $m = 4$.

$$y - y_1 = m(x - x_1)$$
$$y - 1 = 4[x - (-3)]$$
$$y - 1 = 4(x + 3)$$
$$y - 1 = 4x + 12$$
$$y = 4x + 13 \text{ or } 4x - y = -13$$

31. $3x + 2y = 8$

$$2y = -3x + 8$$
$$y = -\dfrac{3}{2}x + 4$$

Slope $= -\dfrac{3}{2}$

Since the two lines are parallel, $m = -\dfrac{3}{2}$

$$y - y_1 = m(x - x_1)$$
$$y - 0 = -\dfrac{3}{2}(x - 4)$$
$$y = -\dfrac{3}{2}x + 6 \text{ or } 3x + 2y = 12$$

33. The slope of the given line is $\dfrac{1}{2}$. The slope of a line perpendicular to the given line is the negative reciprocal of $\dfrac{1}{2}$ or -2.

$$y - y_1 = m(x - x_1)$$
$$y - 2 = -2(x - (-5))$$
$$y - 2 = -2(x + 5)$$
$$y - 2 = -2x - 10$$
$$y = -2x - 8 \text{ or } 2x + y = -8$$

35. Write the given line in slope-intercept form to determine the slope of the line.
$$-5x + y = 4$$
$$y = 5x + 4$$

The slope of the given line is 5. The slope of a line perpendicular to the given line is the negative reciprocal of 5 or $-\frac{1}{5}$.

$$y - y_1 = m(x - x_1)$$
$$y - (-6) = -\frac{1}{5}(x - 0)$$
$$y + 6 = -\frac{1}{5}x$$
$$y = -\frac{1}{5}x - 6$$

$$Or, \frac{1}{5}x + y = -6$$
$$x + 5y = -30$$

37. iv Vertical line

39. vi Slope formula

41. iii Horizontal line

43. A line parallel to a horizontal line is a horizontal line and its equation has the form y equal to a constant.
$$y = 1$$

45. A line perpendicular to a horizontal line is a vertical line and its equation has the form x equal to a constant.
$$x = 2$$

47. A line perpendicular to $x = 0$, a vertical line, is a horizontal line and its equation has the form y equal to k, a constant.
$$y = 2$$

49. $y - y_1 = m(x - x_1)$
$$y - 6 = \frac{1}{4}(x - (-8))$$
$$y - 6 = \frac{1}{4}x + 2$$
$$y = \frac{1}{4}x + 8 \text{ or } x - 4y = -32$$

51. Write the given line in slope-intercept form to determine the slope of the line.

$$3x - y = 6$$
$$-y = -3x + 6$$
$$y = 3x + 6$$
The slope of the given line is 3. The slope of a line parallel to the given line is 3.
$$y - y_1 = m(x - x_1)$$
$$y - 4 = 3(x - 4)$$
$$y - 4 = 3x - 12$$
$$y = 3x - 8 \text{ or } 3x - y = 8$$

53. .
$$y - y_1 = m(x - x_1)$$
$$y - (-2.2) = 4.5(x - 5.2)$$
$$y + 2.2 = 4.5x - 23.4$$
$$y = 4.5x - 25.6 \text{ or } 45x - 10y = 256$$

55. For undefined slope, we give the x-coordinate: $x = -6$.

57. For a slope of 0, we give the y-coordinate: $y = -2$.

59. $y = mx + b$
$$m = \frac{3 - 0}{-4 - (-4)} = \frac{3}{0}, \text{undefined}$$
$$x = -4$$

61. Using the points $(0, -4)$ and $(1, -2)$:
$$y = mx + b$$
$$m = \frac{-2 - (-4)}{1 - 0} = \frac{2}{1} = 2$$
$$y = 2x - 4$$

63. Using the points $(0, 1)$ and $(2, 0)$:
$$y = mx + b$$
$$m = \frac{0 - 1}{2 - 0} = \frac{-1}{2} = -\frac{1}{2}$$
$$y = -\frac{1}{2}x + 1$$

Calculator Exercises

1.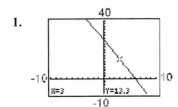

3.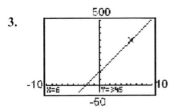

Section 3.6 Practice Exercises

1. $5x + 2y = -6$

$$2y = -5x - 6$$
$$y = \frac{-5}{2}x - \frac{6}{2}$$
$$y = -\frac{5}{2}x - 3$$
$$m = -\frac{5}{2}$$

3. $5x + 6y = 30$

x-intercept: $5x + 6(0) = 30$
$$5x = 30$$
$$x = 6 \quad (6, 0)$$

y-intercept: $5(0) + 6y = 30$
$$6y = 30$$
$$y = 5 \quad (0, 5)$$

5. $y = -2x - 4$

x-intercept: $0 = -2x - 4$
$$2x = -4$$
$$x = -2 \quad (-2, 0)$$

y-intercept: $(0, -4)$

7. $y = -9$
x-intercept: none
y-intercept: $(0, -9)$

9. **(a)** 1980 corresponds to $x = 1980 - 1970$
$= 10$

$y = 0.14(10) + 1.60 = 3.00$

Minimum wage in the year 1980 was $3.00 per hour.

(b) 2015 corresponds to $x = 45$
$y = 0.14x + 1.60$
$y = 0.14(45) + 1.60$
$y = 7.90$
$7.90 per hour

(c) The y-intercept is $(0, 1.6)$. This indicates that the minimum hourly wage was $1.60 per hour in the year 1970.

(d) The slope is 0.14. This indicates that the minimum wage has risen approximately $0.14 per year during this period.

11. **(a)** $m = \dfrac{46 - 42}{84 - 70} = \dfrac{4}{14} = \dfrac{2}{7}$

(b) $m = \dfrac{48 - 40}{84 - 70} = \dfrac{8}{14} = \dfrac{4}{7}$

(c) $m = \dfrac{2}{7}$ means that Grindel's weight increased at a rate of 2 oz in 7 days.

$m = \dfrac{4}{7}$ means that Frisco's weight increased at a rate of 4 oz in 7 days.

(d) Frisco gained weight more rapidly.

13. $y = 0.095x + 11.95 \qquad x \geq 0$

(a) $y = 0.095(1000) + 11.95$
$y = \$106.95$

(b) $y = 0.095(2000) + 11.95$
$y = \$201.95$

(c) y-intercept: $y = 0.095(0) + 11.95$
$y = 11.95 \quad (0, 11.95)$
For 0 kilowatt-hours used, the cost is $11.95.

(d) $m = 0.095$
The cost increases by $0.095 for each kilowatt-hour used.

15. (a) $m = \dfrac{902 - 976}{150 - 75} = \dfrac{-74}{75} = -1.0$

(b) $y = mx + b$
$976 = -1(75) + b$
$b = 1051$
$y = -x + 1051$

(c) $y = -x + 1051$
$y = -(130) + 1051$
$y = 921$
The minimum pressure was 921 mb.

17. (a) Use the points $(0, 57)$ and $(4, 143)$ to determine the slope.

$m = \dfrac{143 - 57}{4 - 0} = \dfrac{86}{4} = 21.5$

(b) The slope means that the consumption of wind energy in the United States increased by 21.5 trillions of Btu per year.

(c) $y - y_1 = m(x - x_1)$

$y - 57 = 21.5(x - 0)$
$y - 57 = 21.5x$
$y = 21.5x + 57$

(d) Year 2010 corresponds to $x = 10$.

$y = 21.5(10) + 57 = 272$

The consumption of wind energy will be 272 trillion Btu in the year 2010.

19. (a) $y = 0.20x + 39.99$

(b) $y = 0.20(40) + 39.99$
$y = 8 + 39.99$
$y = 47.99$
It will cost $47.99 For Andre to send or receive 40 text messages.

21. (a) $y = 90x + 105$

(b) $y = 90(12) + 105$
$y = 1080 + 105$
$y = 1185$
It will cost $1185 to rent the unit for 1 year.

23. (a) $y = 0.8x + 100$

(b) $y = 0.8(200) + 100$
$y = 160 + 100$
$y = 260$
It will cost $260 to produce 200 loaves of bread in one day.

Group Activity

Answers will vary throughout this exercise.

Chapter 3 Review Exercises

Section 3.1

1.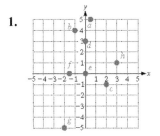

3. III

5. IV

7. IV

9. x-axis

11. (a) On day 1, the price per share of a stock was $26.25.

(b) Day 2

(c) $28.50 - $26.25 = $2.25

Section 3.2

81

13. $5(0) - 3(4) = 12$
$$-12 \neq 12$$
No

15. $1 = \frac{1}{3}(9) - 2$
$$1 = 3 - 2 = 1$$
Yes

17.

x	y
2	1
3	4
1	-2

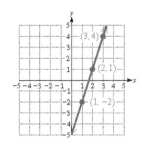

19.

x	y
0	-1
3	1
-6	-5

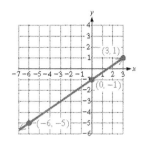

21.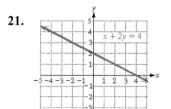

23.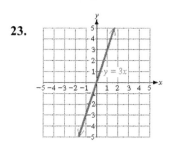

25. Vertical

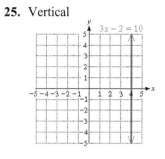

27. Horizontal

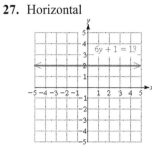

29. x-intercept $\qquad$ y-intercept
$-4x + 8(0) = 12 \qquad -4(0) + 8y = 12$
$-4x = 12 \qquad\qquad 8y = 12$
$x = -3 \qquad\qquad y = \frac{12}{8} = \frac{3}{2}$

$(-3, 0) \qquad\qquad \left(0, \frac{3}{2}\right)$

31.
x-intercept	y-intercept
$0 = 8x$	$y = 8(0)$
$x = 0$	$y = 0$
$(0, 0)$	$(0, 0)$

33.
x-intercept	y-intercept
none	$6y = -24$
	$y = -4$
	$(0, -4)$

35.
x-intercept	y-intercept
$2x + 5 = 0$	none
$x = -\dfrac{5}{2}$	
$\left(-\dfrac{5}{2}, 0\right)$	

Section 3.3

37. $m = \dfrac{12 \text{ ft}}{5 \text{ ft}} = \dfrac{12}{5}$

39. $m = \dfrac{-1 - (-9)}{-5 - 7} = \dfrac{8}{-12} = -\dfrac{2}{3}$

41. $m = \dfrac{-7 - 0}{3 - 3} = \dfrac{-7}{0}$ is undefined

43. (a) $m = -5$

(b) $m = -\dfrac{1}{-5} = \dfrac{1}{5}$

45. $m_1 = \dfrac{5 - 7}{0 - 3} = \dfrac{-2}{-3} = \dfrac{2}{3}$

$m_2 = \dfrac{-3 - 3}{-3 - 6} = \dfrac{-6}{-9} = \dfrac{2}{3}$

Since $m_1 = m_2$, the two lines are parallel.

47. $m_1 = \dfrac{0 - \frac{5}{6}}{2 - 0} = \dfrac{-\frac{5}{6}}{2} = -\dfrac{5}{12}$

$m_2 = \dfrac{0 - \frac{6}{5}}{-\frac{1}{2} - 0} = \dfrac{-\frac{6}{5}}{-\frac{1}{2}} = \dfrac{12}{5}$

Since $m_1 m_2 = -1$, the two lines are perpendicular.

49. (a) $m = \dfrac{37005 - 35955}{31 - 1} = \dfrac{1050}{30} = 35$

(b) The number of kilowatt-hours increased at a rate of 35 kilowatt-hours per day.

Section 3.4

51. $5x - 2y = 10$

$-2y = -5x + 10$

$y = \dfrac{5}{2}x - 5$

$m = \dfrac{5}{2}$ y-intercept $(0, -5)$

53. $x - 3y = 0$

$3y = x$

$y = \dfrac{1}{3}x$

$m = \dfrac{1}{3}$ y-intercept $(0, 0)$

55. $2y = -5$

$y = -\dfrac{5}{2}$

$m = 0$ y-intercept $\left(0, -\dfrac{5}{2}\right)$

57. $m_1 = \dfrac{3}{5}$

$m_2 = \dfrac{5}{3}$

$m_1 \neq m_2$

$m_1 m_2 = 1 \neq -1$

The lines are neither parallel nor perpendicular.

59. $m_1 = -\dfrac{3}{2}$

$m_2 = -\dfrac{3}{2}$

Since $m_1 = m_2$, the lines are parallel.

61. m_1 is undefined.

$m_2 = 0$

Since one line is vertical and the other one is horizontal, they are perpendicular.

63. $y = -\dfrac{4}{3}x - 1$ or $4x + 3y = -3$

65. $y = mx + b$

$2 = -\dfrac{4}{3}(-6) + b$

$b = -6$

$y = -\dfrac{4}{3}x - 6$ or $4x + 3y = -18$

Section 3.5

67. Answers may vary. For example: $y = 3x + 2$

69. $m = \dfrac{y_1 - y_2}{x_1 - x_2} = \dfrac{y_2 - y_1}{x_2 - x_1}$

71. Answers may vary. For example: $x = 6$

73. $y - 8 = -6(x - (-1))$

$y - 8 = -6(x + 1)$

$y - 8 = -6x - 6$

$y = -6x + 2$ or $6x + y = 2$

75. First, use the points to determine the slope.

$m = \dfrac{-4 - (-2)}{0 - 8} = \dfrac{-2}{-8} = \dfrac{1}{4}$

Use this value, one of the points, and the point-slope formula to determine the equation of the line.

$y - (-4) = \dfrac{1}{4}(x - 0)$

$y + 4 = \dfrac{1}{4}x$

$y = \dfrac{1}{4}x - 4$

Or, $\dfrac{1}{4}x - y = 4$

$x - 4y = 16$

77. The slope of the given line is $-\dfrac{5}{6}$. The slope of a perpendicular line is the negative reciprocal, $\dfrac{6}{5}$. Use this value, the point, and the point-slope formula to determine the equation.

$y - 12 = \dfrac{6}{5}(x - 5)$

$y - 12 = \dfrac{6}{5}x - 6$

$y = \dfrac{6}{5}x + 6$

Or, $\dfrac{6}{5}x - y = -6$

$6x - 5y = -30$

Section 3.6

79. (a) $y = 2.4x + 31$

$y = 2.4(7) + 31$

$y = 47.8$ inches

(b) The slope is 2.4 and indicates that the average height for girls increases at a rate of 2.4 in. per year.

81. (a) $y = 20x + 55$

(b) $y = 20(9) + 55$

$y = 180 + 55$

$y = 235$

The total cost of renting the system for nine months is $235.

Chapter 3 Test

1. (a) II

(b) IV

(c) III

3. 0

   (c) Approximately 35 in.

5. (a) $2(0) - 6 = 6$

          $-6 \neq 6$

    No

   (b) $2(4) - 2 = 6$

            $6 = 6$

    Yes

   (c) $2(3) - 0 = 6$

            $6 = 6$

    Yes

   (d) $2\left(\dfrac{9}{2}\right) - 3 = 6$

            $6 = 6$

    Yes

7. Begin at the y-intercept $(0, 2)$. Go up 3 units and across 1 unit to yield $(1, 5)$. Connect the points.

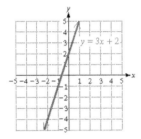

9. $3x + 2y = 8$

    $2y = -3x + 8$

    $y = -\dfrac{3}{2}x + 4$

Begin at the y-intercept $(0, 4)$. Go down 3 units and across 2 units to yield $(2, 1)$. Connect the points.

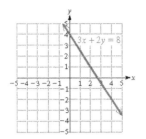

11. $y = -3$; horizontal

13.

x-intercept	y-intercept
$-4x + 3(0) = 6$	$-4(0) + 3y = 6$
$-4x = 6$	$3y = 6$
$x = -\dfrac{3}{2}$	$y = 2$
$\left(-\dfrac{3}{2},\, 0\right)$	$(0, 2)$

15.

x-intercept	y-intercept
$x = 4$	none
$(4, 0)$	

17. $\dfrac{4 \text{ ft}}{10 \text{ ft}} = \dfrac{2}{5}$

19. (a) $x + 4y = -16$

        $y = -\dfrac{1}{4}x - 4$

        $m = -\dfrac{1}{4}$

   (b) $m = 4$

21. (a) $m = \dfrac{89.95 - 41.95}{100 - 20} = \dfrac{48}{80} = 0.6$

   (b) The cost of renting the truck increases at \$0.60 per mile.

23. $2y = 3x - 3$ $\qquad$ $4x = -6y + 1$

$y = \dfrac{3}{2}x - \dfrac{3}{2}$ $\qquad$ $6y = -4x + 1$

$\qquad\qquad\qquad$ $y = -\dfrac{2}{3}x + \dfrac{1}{6}$

$m_1 = \dfrac{3}{2}$ $\quad m_2 = -\dfrac{2}{3};$

Since $m_1 m_2 = -1$, the lines are perpendicular.

25. First, use the points to determine the slope of the line.

$m = \dfrac{8-1}{2-4} = \dfrac{7}{-2} = -\dfrac{7}{2}$

Use this value, one of the points, and the point-slope formula to determine the equation.

$y - 8 = -\dfrac{7}{2}(x - 2)$

$y - 8 = -\dfrac{7}{2}x + 7$

$y = -\dfrac{7}{2}x + 15$

Or, $\dfrac{7}{2}x + y = 15$

$7x + 2y = 30$

27. First, write the given equation in slope-intercept form to find the slope of the given line.

$x + 3y = 9$

$3y = -x + 9$

$y = -\dfrac{1}{3}x + 3$

The slope of the given line is $-\dfrac{1}{3}$. The slope of a line perpendicular to the given line is the negative reciprocal of $-\dfrac{1}{3}$ or 3.

Use this value, the point, and the point-slope formula to determine the equation of the perpendicular line.

$y - (-1) = 3(x - (-3))$

$y + 1 = 3(x + 3)$

$y + 1 = 3x + 9$

$y = 3x + 8$ or $3x - y = -8$

29. $y = mx + b$

$2 = -1(-5) + b$

$b = -3$

$y = -x - 3$

31. **(a)** $m = \dfrac{814 - 614}{20 - 10} = \dfrac{200}{10} = 20$

The slope indicates that there is an increase of 20 thousand medical doctors per year.

(b) $y - 614 = 20(x - 10)$

$y - 614 = 20x - 200$

$y = 20x + 414$

(c) 2015 corresponds to $x = 2015 - 1980 = 35$

$y = 20(35) + 414 = 1114$

The number of medical doctors in the United States for the year 2015 will be approximately 1114 thousand or, equivalently, 1,114,000.

Cumulative Review Exercises
Chapters 1-3

1. **(a)** Rational

(b) Rational

(c) Irrational

(d) Rational

3. $32 \div 2 \cdot 4 + 5 = 16 \cdot 4 + 5 = 69$

5. $16 - 5 - (-7) = 16 - 5 + 7 = 18$

7. $(-2.1)(-6); \ (-2.1)(-6) = 12.6$

9. $6x - 10 = 14$

$6x = 24$

$x = 4$

11. $\dfrac{2}{3}y - \dfrac{1}{6} = y + \dfrac{4}{3}$

$6\left(\dfrac{2}{3}y - \dfrac{1}{6}\right) = 6\left(y + \dfrac{4}{3}\right)$

$4y - 1 = 6y + 8$

$-2y = 9$

$y = -\dfrac{9}{2}$

13. Let x = the area of Maine.

712 less than $29x = 267{,}277$

$29x - 712 = 267{,}277$

$29x = 267{,}989$

$x = 9241$

The area of Maine is 9241 mi^2.

15.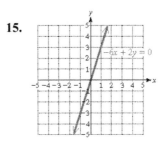

17. $3x + 2y = -12$

$2y = -3x - 12$

$y = -\dfrac{3}{2}x - 6$

$m = -\dfrac{3}{2}$ y-intercept $(0, -6)$

19. $y - y_1 = m(x - x_1)$

$y - (-5) = -3(x - 2)$

$y + 5 = -3x + 6$

$y = -3x + 1$ or $3x + y = 1$

Chapter 4

Chapter 4 Review Your Skills

$y = x$
$y = 1$
$(1, 1)$

$7 = x$
$(7, 7)$

$x = 4$
$(4, 9)$

$(4, 0)$

$x + y = 8$
$x + 7 = 8$
$x = 1$
$(1, 7)$

$7 + y = 8$
$y = 1$
$(7, 1)$

$y = 4$
$(0, 4)$

$(9, 4)$

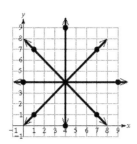

The lines intersect at (4, 4).

Calculator Exercises

1.

The solution is (2, 1).

3. Equation 1
   $x + y = 4$
   $\quad y = -x + 4$

   Equation 2
   $-2x + y = -5$
   $\quad y = 2x - 5$

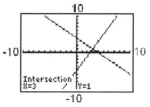

The solution is (3, 1).

5. To graph the lines, write each in slope-intercept form.

   Equation 1
   $-x + 3y = -6$
   $\quad 3y = x - 6$
   $\quad y = \frac{1}{3}x - 2$

   Equation 2
   $6y = 2x + 6$
   $\quad y = \frac{1}{3}x + 1$

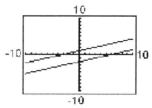

Since the two lines have the same slopes but different y-intercept, they are parallel. The system has no solution. The system is inconsistent.

Section 4.1 Practice Exercises

1. (a) system
   (b) solution
   (c) intersect
   (d) consistent
   (e) inconsistent
   (f) dependent
   (g) independent

3. $3x - y = 7$
   $x - 2y = 4$ point: $(2, -1)$
   Substitute the given point into both equations.
   $3(2) - (-1) \overset{?}{=} 7 \checkmark$
   $2 - 2(-1) \overset{?}{=} 4 \checkmark$

Because the given point is a solution to each equation, it is a solution to the system of equations.

Yes

5. $4y = -3x + 12$

$y = \dfrac{2}{3}x - 4$ point: $(0, 4)$

Substitute the given point into both equations.

$4(4) \overset{?}{=} -3(0) + 12$

$4 \overset{?}{=} \dfrac{2}{3}(0) - 4$

Because the given point is not a solution of either equation it is not a solution of the system of equations.   No

7. $3x - 6y = 9$

$x - 2y = 3$ point: $\left(4, \dfrac{1}{2}\right)$

Substitute the given point into both equations.

$3(4) - 6\left(\dfrac{1}{2}\right) \overset{?}{=} 9 \checkmark$

$4 - 2\left(\dfrac{1}{2}\right) \overset{?}{=} 3 \checkmark$

Because the given point is a solution to each equation, it is a solution to the system of equations.

Yes

9. $\dfrac{1}{3}x = \dfrac{2}{5}y - \dfrac{4}{5}$

$\dfrac{3}{4}x + \dfrac{1}{2}y = 2$ point: $(0, 2)$

Substitute the given point into both equations.

$\dfrac{1}{3}(0) \overset{?}{=} \dfrac{2}{5}(2) - \dfrac{4}{5}$ $\checkmark$

$\dfrac{3}{4}(0) + \dfrac{1}{2}(2) \overset{?}{=} 2$

Because the given point is not a solution to the second equation, it is not a solution to the system of equations.
No

11. **b** Since parallel lines do not intersect, parallel lines represent a system with no solution.

13. **d** The lines intersect at the origin or (0, 0).

15. **(a)** $y = 2x - 3$
$y = 2x + 5$

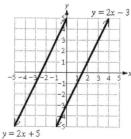

(b) $y = 2x + 1$
$y = 4x - 5$

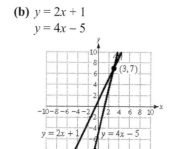

(c) $y = 3x - 5$
$y = 3x - 5$

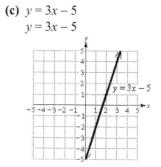

17. **c** Coinciding lines have the same slope and the same y-intercept.

19. **a** An inconsistent system graphs as parallel lines.

21. a   The graph of the system is parallel lines.

23. b   Distinct intersecting lines always have different slopes.

25. c   Coinciding lines with the same slopes and same y-intercept have infinitely many solutions.

27. $y = -x + 4$
$y = x - 2$

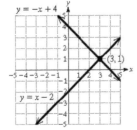

(3, 1); consistent; independent system

29. $2x + y = 0$

$3x + y = 1$

Rewrite in slope-intercept form.
$y = -2x$
$y = -3x + 1$

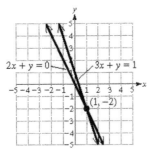

(1, −2); consistent; independent system

31. $2x + y = 6$

$x = 1$

Rewrite the first equation in slope-intercept form. The second equation is a vertical line through the point (1, 0).
$y = -2x + 6$

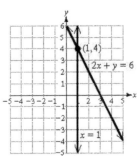

(1, 4); consistent; independent system

33. $-6x - 3y = 0$

$4x + 2y = 4$

Rewrite both equations in slope-intercept form.
$y = -2x$
$y = -2x + 2$

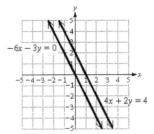

No solution; inconsistent; independent

35. $-2x + y = 3$

$6x - 3y = -9$

Rewrite both equations in slope-intercept form.
$y = 2x + 3$
$y = 2x + 3$

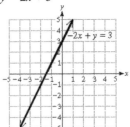

Infinitely many solutions
$\{(x, y) | y = 2x + 3\}$;
consistent; dependent equations

37. $y = 6$

$2x + 3y = 12$

Rewrite the second equation in slope-intercept form. The first equation is a horizontal line through the point (0, 6).

$y = -\dfrac{2}{3}x + 4$

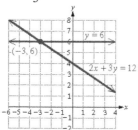

(−3, 6); consistent; independent system

39. $x = 4 + y$

$3y = -3x$

Rewrite both equations in slope-intercept form.

$y = x - 4$

$y = -x$

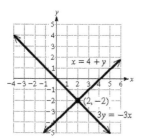

(2, −2); consistent; independent system

41. $-x + y = 3$

$4y = 4x + 6$

Rewrite both equations in slope-intercept form.

$y = x + 3$

$y = x + \dfrac{3}{2}$

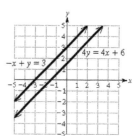

No solution; inconsistent; independent system

43. $x = 4$

$2y = 4$

Solve the second equation for y.

$x = 4$

$y = 2$

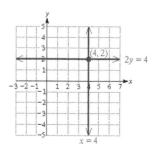

(4, 2); consistent; independent system

45. $2x + y = 4$

$4x - 2y = 0$

Rewrite both equations in slope-intercept form.

$y = -2x + 4$

$y = 2x$

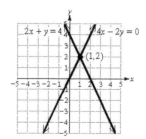

(1, 2); consistent; independent system

47.
$$y = 0.5x + 2$$
$$-x + 2y = 4$$
Rewrite the second equation in slope-intercept form.
$$y = 0.5x + 2$$
$$y = 0.5x + 2$$

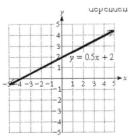

Infinitely many solutions
$\{(x, y) | y = 0.5x + 2\}$;
consistent; dependent equations

49. $x - 3y = 0$
$$y = -x - 4$$
Rewrite both equations in slope-intercept form.
$$y = \frac{1}{3}x$$
$$y = -x - 4$$

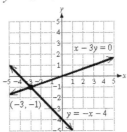

$(-3, -1)$; consistent; independent system

51. The point of intersection gives the answer (2500, 50). The same cost occurs when $2500 of merchandise is purchased.

53. The point of intersection is below the x-axis and cannot have a positive y-coordinate.

55. Answers may vary. For example:
$$4x + y = 9$$
$$-2x - y = -5$$

These equations were found by simply putting together some combination of x and y and substituting $x = 2$ and $y = 1$ into them to determine the constant value.

57. Answers may vary. For example:
$$2x + 2y = 1$$
This equation was found by multiplying each side of the given equation by a different value.

Section 4.2 Practice Exercises

1. $2x - y = 4$
$$-2y = -4x + 8$$
Rewrite both equations in slope-intercept form.
$$y = 2x - 4$$
$$y = 2x - 4$$
Since the slopes and the y-intercepts are equal, they are coinciding lines.

3. $2x + 3y = 6$
$$x - y = 5$$
Rewrite both equations in slope-intercept form.
$$y = -\frac{2}{3}x + 2$$
$$y = x - 5$$

Since the slopes are unequal, the lines intersect.

5. $2x = \frac{1}{2}y + 2$
$$4x - y = 13$$

Rewrite both equations in slope-intercept form.
$$y = 4x - 4$$
$$y = 4x - 13$$

Since the slopes are equal but the y-intercepts are different, the lines are parallel.

7. $3x + 2y = -3$

$y = 2x - 12$

The second equation is solved for y. Substitute this value for y into the first equation.

$$3x + 2y = -3$$
$$3x + 2(2x - 12) = -3$$
$$3x + 4x - 24 = -3$$
$$7x - 24 = -3$$
$$7x = 21$$
$$x = 3$$

Substitute $x = 3$ into the second equation to solve for y.

$y = 2x - 12$
$y = 2(3) - 12$
$y = -6$
Solution: $(3, -6)$

9. $x = -4y + 16$

$3x + 5y = 20$

The first equation is solved for x. Substitute this value for x into the second equation.

$$3x + 5y = 20$$
$$3(-4y + 16) + 5y = 20$$
$$-12y + 48 + 5y = 20$$
$$-7y + 48 = 20$$
$$-7y = -28$$
$$y = 4$$

Substitute $y = 4$ into the first equation to solve for x.

$x = -4y + 16$
$x = -4(4) + 16$
$x = 0$

Solution: $(0, 4)$

11. (a) y in the second equation is easiest to isolate because its coefficient is 1.

(b) $4x - 2y = -6$

$3x + y = 8$

Solving the second equation for y gives $y = -3x + 8$. Substitute this value for y into the first equation.

$$4x - 2y = -6$$
$$4x - 2(-3x + 8) = -6$$
$$4x + 6x - 16 = -6$$
$$10x - 16 = -6$$
$$10x = 10$$
$$x = 1$$
$$3x + y = 8$$
$$3(1) + y = 8$$
$$3 + y = 8$$
$$y = 5$$
Solution: $(1, 5)$

13. $x = 3y - 1$

$2x - 4y = 2$

The first equation is solved for x. Substitute this value for x into the second equation.

$$2x - 4y = 2$$
$$2(3y - 1) - 4y = 2$$
$$6y - 2 - 4y = 2$$
$$2y - 2 = 2$$
$$2y = 4$$
$$y = 2$$
$$x = 3y - 1$$
$$x = 3(2) - 1$$
$$x = 5$$
Solution: $(5, 2)$

15. $-2x + 5y = 5$

$x - 4y = -10$

Solving the second equation for x gives $x = 4y - 10$. Substitute this value for x into the first equation.

93

$$-2x + 5y = 5$$
$$-2(4y - 10) + 5y = 5$$
$$-8y + 20 + 5y = 5$$
$$-3y + 20 = 5$$
$$-3y = -15$$
$$y = 5$$
$$x - 4y = -1$$
$$x - 4(5) = -10$$
$$x = 10$$
Solution: (10, 5)

17. $4x - y = -1$

$2x + 4y = 13$

Solving the first equation for y gives $y = 4x + 1$. Substitute this value for y into the second equation.

$$2x + 4y = 13$$
$$2x + 4(4x + 1) = 13$$
$$2x + 16x + 4 = 13$$
$$18x + 4 = 13$$
$$18x = 9$$
$$x = \frac{1}{2}$$
$$4x - y = -1$$
$$4\left(\frac{1}{2}\right) - y = -1$$
$$2 - y = -1$$
$$-y = -3$$
$$y = 3$$
Solution: $\left(\frac{1}{2}, 3\right)$

19. $4x - 3y = 11$

$x = 5$

The second equation is solved for x. Substitute this value for x into the first equation.

$$4(5) - 3y = 11$$
$$20 - 3y = 11$$
$$-3y = -9$$
$$y = 3$$
Solution: (5, 3)

21. $2y = x - 1$

$5x - 3y = 5$

Solving the first equation for x gives $x = 2y + 1$. Substitute this value for x into the second equation.

$$5x - 3y = 5$$
$$5(2y + 1) - 3y = 5$$
$$10y + 5 - 3y = 5$$
$$7y + 5 = 5$$
$$7y = 0$$
$$y = 0$$
$$4x = 8y + 4$$
$$4x = 8(0) + 4$$
$$4x = 4$$
$$x = 1$$
Solution: (1, 0)

23. $x - 3y = -11$

$6x - y = 2$

Solving the first equation for x gives $x = 3y - 11$. Substitute this value for x into the second equation.

$$6x - y = 2$$
$$6(3y - 11) - y = 2$$
$$18y - 66 - y = 2$$
$$17y - 66 = 2$$
$$17y = 68$$
$$y = 4$$
$$x - 3y = -11$$
$$x - 3(4) = -11$$
$$x - 12 = -11$$
$$x = 1$$
Solution: (1, 4)

25. $3x + 2y = -1$

$\dfrac{3}{2}x + y = 4$

Solving the second equation for y gives $y = -\dfrac{3}{2}x + 4$. Substitute this value for y into the first equation.

$$3x + 2y = -1$$
$$3x + 2\left(-\frac{3}{2}x + 4\right) = -1$$
$$3x - 3x + 8 = -1$$
$$8 \neq -1$$
No solution; inconsistent system

27. $10x - 30y = -10$
$\quad 2x - 6y = -2$

None of the variables in either equation have a coefficient of 1. You may choose either equation to solve for either variable. Solving the first equation for x gives $x = 3y - 1$. Substitute this value for x into the second equation.
$$2x - 6y = -2$$
$$2(3y - 1) - 6y = -2$$
$$6y - 2 - 6y = -2$$
$$-2 = -2 \quad \text{Identity}$$
Infinitely may solutions
Solution: $\{(x, y) | 2x - 6y = -2\}$;
dependent equations

29. $2x + y = 3$
$\quad\quad y = -7$

A value for y has been determined. To find a value for x, substitute $y = -7$ into the first equation.
$$2x + y = 3$$
$$2x - 7 = 3$$
$$2x = 10$$
$$x = 5$$
Solution: $(5, -7)$

31. $x + 2y = -2$
$\quad 4x = -2y - 17$

Solving the first equation for x gives $x = -2y - 2$. Substitute the value for x into the second equation.

$$4x = -2y - 17$$
$$4(-2y - 2) = -2y - 17$$
$$-8y - 8 = -2y - 17$$
$$-6y - 8 = -17$$
$$-6y = -9$$
$$y = \frac{3}{2}$$
$$x + 2y = -2$$
$$x + 2\left(\frac{3}{2}\right) = -2$$
$$x + 3 = -2$$
$$x = -5$$
Solution: $\left(-5, \frac{3}{2}\right)$

33. $y = -\frac{1}{2}x - 4$
$\quad y = 4x - 13$

The first equation is solved for y. Substitute this value into the second equation.
$$y = 4x - 13$$
$$-\frac{1}{2}x - 4 = 4x - 13$$
$$-x - 8 = 8x - 26$$
$$-9x - 8 = -26$$
$$-9x = -18$$
$$x = 2$$
$$y = -\frac{1}{2}x - 4$$
$$y = -\frac{1}{2}(2) - 4$$
$$y = -5$$
Solution: $(2, -5)$

35. $\quad\quad y = 6$
$\quad y - 4 = -2x - 6$

A value for y has been determined. To find a value for x, substitute $y = 6$ into the second equation.

$$y - 4 = -2x - 6$$
$$6 - 4 = -2x - 6$$
$$2 = -2x - 6$$
$$2x = -8$$
$$x = -4$$

Solution: $(-4, 6)$

37. $y = \dfrac{3}{2}x + 2$

$$3x + 4y = -1$$

The first equation is solved for y. Substitute this value into the second equation.

$$3x + 4y = -1$$
$$3x + 4\left(\dfrac{3}{2}x + 2\right) = -1$$
$$3x + 6x + 8 = -1$$
$$9x + 8 = -1$$
$$9x = -9$$
$$x = -1$$
$$y = \dfrac{3}{2}(-1) + 2$$
$$y = -\dfrac{3}{2} + \dfrac{4}{2}$$
$$y = \dfrac{1}{2}$$

Solution: $\left(-1, \dfrac{1}{2}\right)$

39. $y = 0.25x + 1$

$$-x + 4y = 4$$

The first equation is solved for y. Substitute this value into the second equation.

$$-x + 4y = 4$$
$$-x + 4(0.25x + 1) = 4$$
$$-x + x + 4 = 4$$
$$4 = 4 \quad \text{Identity}$$

Infinitely many solutions
Solution: $\{(x, y) | y = 0.25x + 1\}$; dependent equations.

41. $0.2x + 0.1y = 1.1$
$$1.5x - 0.2y = 5.4$$

We can work without decimals by multiplying every term by 10. This gives:
$$2x + y = 11$$
$$15x - 2y = 54$$

None of the variables in either equation has a coefficient of 1. You may choose either equation to solve for either variable. Solving the first equation for y gives $y = -2x + 11$. Substitute this value for y into the second equation.

$$15x - 2y = 54$$
$$15x - 2(-2x + 11) = 54$$
$$15x + 4x - 22 = 54$$
$$19x - 22 = 54$$
$$19x = 76$$
$$x = 4$$
$$0.2(4) + 0.1y = 1.1$$
$$0.8 + 0.1y = 1.1$$
$$0.1y = 0.3$$
$$y = 3$$

Solution: $(4, 3)$

43. $x + 2y = 4$
$$4y = -2x - 8$$

Solving the first equation for x gives $x = -2y + 4$. Substitute the value for x into the second equation.
$$4y = -2x - 8$$
$$4y = -2(-2y + 4) - 8$$
$$4y = 4y - 8 - 8$$
$$4y = 4y - 16$$
$$0 \neq -16$$

Contradiction; no solution; inconsistent system

45. $2x = 3 - y$
$$x + y = 4$$

Solving the second equation for y gives $y = -x + 4$. Substitute the value for y into the first equation.

$2x = 3 - y$

$2x = 3 - (-x + 4)$

$2x = 3 + x - 4$

$2x = x - 1$

$x = -1$

$x + y = 4$

$-1 + y = 4$

$y = 5$

Solution: $(-1, 5)$

47. $\dfrac{x}{3} + \dfrac{y}{2} = -4$

$x - 3y = 6$

Solving the second equation for x gives $x = 3y + 6$. Substitute the value for x into the first equation.

$$\frac{x}{3} + \frac{y}{2} = -4$$

$$\frac{3y + 6}{3} + \frac{y}{2} = -4$$

$$2(3y + 6) + 3y = -24$$

$$6y + 12 + 3y = -24$$

$$9y + 12 = -24$$

$$9y = -36$$

$$y = -4$$

$$x - 3y = 6$$

$$x - 3(-4) = 6$$

$$x = -6$$

Solution: $(-6, -4)$

49. Let x represent one number. Let y represent the other number. The statement "two numbers have a sum of 106" translates to the equation $x + y = 106$. The statement "one number is 10 less than the other" translates to the equation $x = y - 10$.

$x + y = 106$

$x = y - 10$

The second equation is solved for x. Substitute this value into the first equation.

$x + y = 106$

$y - 10 + y = 106$

$2y - 10 = 106$

$2y = 116$

$y = 58$

$x = y - 10$

$x = 58 - 10$

$x = 48$

The numbers are 48 and 58.

51. Let x represent one positive number. Let y represent the other positive number. The statement "the difference between two positive numbers is 26" translates to the equation $x - y = 26$. The statement "the larger number is three times the smaller" translates to the equation $x = 3y$.

$x - y = 26$

$x = 3y$

The second equation is solved for x. Substitute this value into the first equation.

$3y - y = 26$

$2y = 26$

$y = 13$

$x = 3(13)$

$x = 39$

The numbers are 13 and 39.

53. Let x represent the measure of one angle. Let y represent the measure of the second angle. The statement "two angles are supplementary" translates to the equation $x + y = 180$. The statement "one angle is 15° more than 10 times the other angle" translates to the equation $x = 10y + 15$.

$x + y = 180$

$x = 10y + 15$

The second equation is solved for x. Substitute this value into the first equation.

$x + y = 180$

$10y + 15 + y = 180$

$11y + 15 = 180$

$11y = 165$

$y = 15$

$x = 10y + 15$

$x = 10(15) + 15$
$x = 165$
The measures of the angles are 165° and 15°.

55. Let x represent the measure of the first angle. Let y represent the measure of the second angle. The statement "two angles are complementary" translates to the equation
$x + y = 90$. The statement "one angle is 10° more than 3 times the other angle" translates to the equation $x = 3y + 10$.

$x + y = 90$
$x = 3y + 10$

The second equation is solved for x. Substitute this value into the first equation.

$x + y = 90$
$3y + 10 + y = 90$
$4y = 80$
$y = 20$

$x = 3y + 10$
$x = 3(20) + 10$
$x = 70$

The measures of the angles are 70° and 20°.

57. Let x represent the measure of one of the acute angles. Let y represent the measure of the other acute angle. The sum of the measures of the angles of a triangle is 180°. In a right triangle one of the angles is a right angle measuring 90°. Thus the sum of the measures of the other two angles is 90°. From this information comes the first equation: $x + y = 90$. The statement "one of the acute angles is 6° less than the other acute angle" translates to the equation
$x = y - 6$.

$x + y = 90$
$x = y - 6$

The second equation is solved for x. Substitute this value into the first equation.

$x + y = 90$
$y - 6 + y = 90$
$2y - 6 = 90$
$2y = 96$
$y = 48$

$x = y - 6$
$x = 48 - 6$
$x = 42$

The measures of the angles are 42° and 48°.

59. $y = 2x + 3$
$-4x + 2y = 6$

Answers may vary. It is stated that the equations are dependent. Therefore, x can equal any real number. Pick a value for x, substitute that value into one of the equations and determine a corresponding value for y.

If $x = 0$, then $y = 2(0) + 3 = 3$.   (0, 3)
If $x = 1$, then $y = 2(1) + 3 = 5$.   (1, 5)
If $x = -1$, then $y = 2(-1) + 3 = 1$.   (−1, 1)

Section 4.3 Practice Exercises

1. $-\dfrac{3}{4}x + 2y = -10$

$x - \dfrac{1}{2}y = 7$ Ordered pair: $(8, -2)$

$-\dfrac{3}{4}(8) + 2(-2) \overset{?}{=} -10$ ✓

$8 - \dfrac{1}{2}(-2) \overset{?}{=} 7$ Not a solution

No; since the ordered pair is not a solution to the second equation, it is not a solution of the system.

3. $x = y + 1$

$-x + 2y = 0$ Ordered pair: $(3, 2)$

$3 \overset{?}{=} 2 + 1$ ✓

$-3 + 2(2) \overset{?}{=} 0$ Not a solution

No; since the ordered pair is not a solution to the second equation, it is not a solution of the system.

5.

$$x = 2y - 11$$
$$-x + 5y = 23 \quad \text{Ordered pair: } (-3, 4)$$

$$-3 \overset{?}{=} 2(4) - 11 \checkmark$$

$$-(-3) + 5(4) \overset{?}{=} 23 \checkmark \quad \text{A solution}$$

Yes; since the ordered pair is a solution to both equations, it is a solution of the system.

7. (a) True. This will create opposite coefficients on the x-variable.

(b) False. Remember, the goal is to create opposite coefficients on the y-variable. Multiply the first equation by 8 and the second equation by 5.

9. (a) x-variable is easier, because one coefficient is a multiple of the other

(b) $-2x + 5y = -15$
$$6x - 7y = 21$$

Both equations are already written in standard form. There are no fractions or decimals. Neither set of coefficients is opposites. Multiplying the first equation by 3 will create opposite coefficients on x.

$$-6x + 15y = -45$$
$$\underline{6x - 7y = 21}$$
$$8y = -24$$
$$y = -3$$
$$-2x + 5(-3) = -15$$
$$-2x - 15 = -15$$
$$-2x = 0$$
$$x = 0$$
Solution: $(0, -3)$

11. $2x - 3y = 11$
$$-4x + 3y = -19$$

Both equations are already written in standard form. Coefficients on y are opposites. Add the two equations and then solve the resulting equation.

$$2x - 3y = 11$$
$$\underline{-4x + 3y = -19}$$
$$-2x = -8$$
$$x = 4$$

$$2(4) - 3y = 11$$
$$8 - 3y = 11$$
$$-3y = 3$$
$$y = -1$$
Solution: $(4, -1)$

13. $-2u + 6v = 10$
$$-2u + v = -5$$

Both equations are already written in standard form. Neither set of coefficients are opposites. Multiplying the first equation by -1 will result in opposite coefficients on u.

$$2u - 6v = -10$$
$$\underline{-2u + v = -5}$$
$$-5v = -15$$
$$v = 3$$

$$-2u + 6(3) = 10$$
$$-2u + 18 = 10$$
$$-2u = -8$$
$$u = 4$$
Solution: $(4, 3)$

15. $5m - 2n = 4$
$$3m + n = 9$$

Both equations are already written in standard form. Neither set of coefficients is opposites. Multiplying the second equation by 2 will result in opposite coefficients on n.

$$5m - 2n = 4$$
$$\underline{6m + 2n = 18}$$
$$11m = 22$$
$$m = 2$$

$$5(2) - 2n = 4$$
$$10 - 2n = 4$$
$$-2n = -6$$
$$n = 3$$
Solution: (2, 3)

17. $7a + 2b = -1$

$3a - 4b = 19$

Both equations are already written in standard form. Neither set of coefficients is opposites. Multiplying the first equation by 2 will result in opposite coefficients on b.

$$14a + 4b = -2$$
$$\underline{3a - 4b = 19}$$
$$17a = 17$$
$$a = 1$$
$$7(1) + 2b = -1$$
$$7 + 2b = -1$$
$$2b = -8$$
$$b = -4$$
Solution: (1, −4)

19. $2s + 3t = -1$

$5s = 2t + 7$

Writing the second equation in standard form yields $5s - 2t = 7$.

$$2s + 3t = -1$$
$$5s - 2t = 7$$

Neither set of coefficients is opposites. Multiplying the first equation by 2 and the second equation by 3 will result in opposite coefficients on t.

$$4s + 6t = -2$$
$$\underline{15s - 6t = 21}$$
$$19s = 19$$
$$s = 1$$
$$2s + 3t = -1$$
$$2(1) + 3t = -1$$
$$2 + 3t = -1$$
$$3t = -3$$
$$t = -1$$
Solution: (1, −1)

21. $4k - 2r = -4$

$3k - 5r = 18$

Both equations are already written in standard form. Neither set of coefficients is opposites. Multiplying the first equation by 5 and the second equation by −2 will result in opposite coefficients on r.

$$20k - 10r = -20$$
$$\underline{-6k + 10r = -36}$$
$$14k = -56$$
$$k = -4$$

$$4k - 2r = -4$$
$$4(-4) - 2r = -4$$
$$-16 - 2r = -4$$
$$-2r = 12$$
$$r = -6$$
Solution: (−4, −6)

23. $6x + 6y = 8$

$9x - 18y = -3$

Both equations are already written in standard form. Neither set of coefficients is opposites. Multiplying the first equation by 3 will result in opposite coefficients on y.

$$18x + 18y = 24$$
$$\underline{9x - 18y = -3}$$
$$27x = 21$$
$$x = \frac{21}{27}$$
$$x = \frac{7}{9}$$

$$6x + 6y = 8$$

$$6\left(\frac{7}{9}\right) + 6y = 8$$

$$\frac{14}{3} + 6y = 8$$

$$6y = \frac{10}{3}$$

$$y = \frac{10}{18}$$

$$y = \frac{5}{9}$$

Solution: $\left(\frac{7}{9}, \frac{5}{9}\right)$

25. Since the statement is a contradiction, the system will have no solutions. The lines are parallel.

27. Since the statement is an identity, the system will have infinitely many solutions. The lines coincide.

29. The system will have one solution. The lines intersect at a point whose x-coordinate is 0.

31. $-2x + y = -5$

$8x - 4y = 12$

Both equations are already written in standard form. Neither set of coefficients is opposites. Multiplying the first equation by 4 will result in opposite coefficients on y.

$$-8x + 4y = -20$$
$$\underline{8x - 4y = 12}$$
$$0 = -8$$

No solution; inconsistent system

33. $x + 2y = 2$

$-3x - 6y = -6$

Both equations are already written in standard form. Neither set of coefficients is opposites. Multiplying the first equation by 3 will result in opposite coefficients on x.

$$3x + 6y = 6$$
$$\underline{-3x - 6y = -6}$$
$$0 = 0 \qquad \text{Identity}$$

Infinitely many solutions

Solution: $\{(x, y) | x + 2y = 2\}$; dependent equation

35. $3a + 2b = 11$

$7a - 3b = -5$

Both equations are already written in standard form. Neither set of coefficients is opposites. However, the signs of the coefficients on b are opposites. Choose to eliminate b by multiplying the first equation by 3 and the second equation by 2 to obtain opposite coefficients on b.

$$9a + 6b = 33$$
$$\underline{14a - 6b = -10}$$
$$23a = 23$$

$$a = 1$$

$$3(1) + 2b = 11$$
$$3 + 2b = 11$$
$$2b = 8$$
$$b = 4$$

Solution: $(1, 4)$

37. $3x - 5y = 7$

$5x - 2y = -1$

Both equations are already written in standard form. Neither set of coefficients is opposites. Neither set of coefficients has opposite signs. Choose either variable to eliminate. x will be the chosen variable in this solution. To obtain opposite coefficients on x multiply the first equation by 5 and the second equation by -3.

$$15x - 25y = 35$$
$$\underline{-15x + 6y = 3}$$
$$-19y = 38$$
$$y = -2$$

$$3x - 5(-2) = 7$$
$$3x + 10 = 7$$
$$3x = -3$$
$$x = -1$$
Solution: $(-1, -2)$

39. $2x + 2 = -3y + 9$

$3x - 10 = -4y$

Write both equations in standard form.
$$2x + 3y = 7$$

$$3x + 4y = 10$$

Neither set of coefficients are opposites of each other. Neither variable has coefficients with opposite signs. Choose either variable to eliminate. x will be the chosen variable in this solution. To obtain opposite coefficients on x multiply the first equation by -3 and the second equation by 2.
$$-6x - 9y = -21$$
$$6x + 8y = 20$$
$$\overline{-y = -1}$$
$$y = 1$$

$$3x - 10 = -4(1)$$
$$3x - 10 = -4$$
$$3x = 6$$
$$x = 2$$
Solution: $(2, 1)$

41. $4x - 5y = 0$

$8(x - 1) = 10y$

Write the second equation in standard form.
$$4x - 5y = 0$$
$$8x - 10y = 8$$

The coefficient on x in the second equation is a multiple of the coefficient on x in the first equation. Multiply the first equation by -2 to obtain opposite coefficients on x.
$$-8x + 10y = 0$$
$$8x - 10y = 8$$
$$\overline{0 = 8}$$

No solution; inconsistent system

43. $5x - 2y = 4$

$y = -3x + 9$

Since the second equation is solved for y, use the substitution method by substituting this value for y into the first equation and solving this new equation for x.
$$5x - 2(-3x + 9) = 4$$
$$5x + 6x - 18 = 4$$
$$11x - 18 = 4$$
$$11x = 22$$
$$x = 2$$

$$y = -3(2) + 9$$
$$y = 3$$
Solution: $(2, 3)$

45. $0.1x + 0.1y = 0.6$

$0.1x - 0.1y = 0.1$

Both equations are written in standard form. The coefficients on y are opposites. Use the addition method. Add the equations and solve the resulting equation.
$$0.1x + 0.1y = 0.6$$
$$0.1x - 0.1y = 0.1$$
$$\overline{0.2x = 0.7}$$
$$x = 3.5$$

$$0.1x + 0.1y = 0.6$$
$$0.1(3.5) + 0.1y = 0.6$$
$$0.35 + 0.1y = 0.6$$
$$0.1y = 0.25$$
$$y = 2.5$$
Solution: $(3.5, 2.5)$

47. $3x = 5y - 9$

$2y = 3x + 3$

Since neither equation is solved for x or y, consider using the addition method. Write both equations in standard form.
$$3x - 5y = -9$$
$$-3x + 2y = 3$$

The coefficients on x are opposites. Add the two equations and then solve the resulting equation.

$$3x - 5y = -9$$

$$-3x + 2y = 3$$

$$\overline{ -3y = -6}$$

$$y = 2$$

$$3x = 5(2) - 9$$

$$3x = 1$$

$$x = \frac{1}{3}$$

Solution: $\left(\dfrac{1}{3}, 2\right)$

49. $\dfrac{1}{10}y = -\dfrac{1}{2}x - \dfrac{1}{2}$

$$\dfrac{3}{2}x - \dfrac{3}{4} = -\dfrac{3}{4}y$$

We can work without the fractions by multiplying the first equation by 10 and the second equation by 4. This gives:

$$y = -5x - 5$$

$$6x - 3 = -3y$$

Since the first equation is solved for y, use the substitution method. Substitute the value of y into the second equation and solve the resulting equation for x.

$$6x - 3 = -3(-5x - 5)$$

$$6x - 3 = 15x + 15$$

$$-9x = 18$$

$$x = -2$$

$$y = -5x - 5$$

$$y = -5(-2) - 5$$

$$y = 5$$

Solution: $(-2, 5)$

51. $x = -\dfrac{1}{2}$

$$6x - 5y = -8$$

Since the first equation is solved for x, use the substitution method by substituting this value for x into the second equation and solving this new equation for y.

$$6\left(-\dfrac{1}{2}\right) - 5y = -8$$

$$-3 - 5y = -8$$

$$-5y = -5$$

$$y = 1$$

Solution: $\left(-\dfrac{1}{2}, 1\right)$

53. $0.02x + 0.04y = 0.12$

$$0.03x - 0.05y = -0.15$$

Both equations are written in standard form. Neither set of coefficients are opposites of each other. Multiplying the first equation by 5 and the second equation by 4 will result in opposite coefficients on y.

$$0.1x + 0.2y = 0.6$$

$$\underline{0.12x - 0.2y = -0.6}$$

$$0.22x = 0$$

$$x = 0$$

$$0.02x + 0.04y = 0.12$$

$$0.02(0) + 0.04y = 0.12$$

$$y = 3$$

Solution: $(0, 3)$

55. $8x - 16y = 24$

$$2x - 4y = 0$$

Since both equations are in standard form, use the addition method. The coefficient on x in the first equation is a multiple of the coefficient on x in the second equation. Multiply the second equation by -4 to obtain opposite coefficients on x.

$$8x - 16y = 24$$

$$\underline{-8x + 16y = 0}$$

$$0 = 24$$

No solution

57. $\dfrac{m}{2} + \dfrac{n}{5} = \dfrac{13}{10}$

$$3m - 3n = m - 10$$

Since neither equation is solved for m and n, consider using the addition method.

Write both equations in standard form.

$5m + 2n = 13$

$2m - 3n = -10$

Neither set of coefficients are opposites of each other. The n-variable has coefficients with opposite signs. To obtain opposite coefficients on n multiply the first equation by 3 and the second equation by 2.

$15m + 6n = 39$

$\underline{4m - 6n = -20}$

$19m = 19$

$m = 1$

$3(1 - n) = 1 - 10$

$3 - 3n = -9$

$-3n = -12$

$n = 4$

Solution: $(1, 4)$

59. $2m - 6n = m + 4$

$3m + 8 = 5m - n$

Since neither equation is solved for m or n, consider using the addition method. Write both equations in standard form.

$m - 6n = 4$

$-2m + n = -8$

The signs of the m-variable are opposite. Multiply the first equation by 2 to obtain opposite coefficients on m.

$2m - 12n = 8$

$\underline{-2m + n = -8}$

$-11n = 0$

$n = 0$

$3m + 8 = 5m - 0$

$-2m + 8 = 0$

$-2m = -8$

$m = 4$

Solution: $(4, 0)$

61. $9a - 2b = 8$

$18a + 6 = 4b + 22$

Since neither equation is solved for a or b,

consider using the addition method. Write both equations in standard form.

$9a - 2b = 8$

$18a - 4b = 16$

Neither variable has opposite signs nor opposite coefficients. The coefficient on a in the second equation is a multiple of the coefficient in the first equation. Multiply the first equation by -2 to obtain opposite coefficients on a.

$-18a + 4b = -16$

$\underline{18a - 4b = 16}$

$0 = 0$ \qquad Identity

Infinitely many solutions

Solution: $\{(a, b) | 9a - 2b = 8\}$; dependent equations

63. $6x - 5y = 7$

$4x - 6y = 7$

Since neither equation is solved for x or y, consider using the addition method. Neither variable has opposite signs nor opposite coefficients. The least common multiple of the coefficients of x is 12. Multiply the first equation by 2 and the second equations by -3 to obtain opposite coefficients on x. Add the two equations and solve the resulting equation.

$12x - 10y = 14$

$-12x + 18y = -21$

$8y = -7$

$y = -\dfrac{7}{8}$

$6x - 5\left(-\dfrac{7}{8}\right) = 7$

$6x + \dfrac{35}{8} = 7$

$6x = \dfrac{21}{8}$

$x = \dfrac{7}{16}$

Solution: $\left(\dfrac{7}{16}, -\dfrac{7}{8}\right)$

65. Let x represent the first positive number. Let y represent the second positive number. The statement "difference of two positive numbers is 2" translates to the equation $x - y = 2$. The statement "sum of the numbers is 36" translates to the equation $x + y = 36$.

$$x + y = 36$$
$$x - y = 2$$

The coefficients on y are opposites. Add the two equations and solve the resulting equation.

$$\begin{aligned} x + y &= 36 \\ \underline{x - y} &= \underline{2} \\ 2x &= 38 \\ x &= 19 \end{aligned}$$

$$19 + y = 36$$
$$y = 17$$

The two positive numbers are 19 and 17.

67. Let x represent the smaller number. Let y represent the larger number. The statement "six times the smaller of two numbers minus the larger number is –9" translates to the equation $6x - y = -9$. The statement "ten times the smaller number plus five times the larger number is 5" translates to the equation $10x + 5y = 5$.

$$6x - y = -9$$
$$10x + 5y = 5$$

Since the signs on y are opposites, multiply the first equation by 5 to obtain opposite coefficients on y.

$$\begin{aligned} 30x - 5y &= -45 \\ \underline{10x + 5y} &= \underline{5} \\ 40x &= -40 \\ x &= -1 \end{aligned}$$

$$6(-1) - y = -9$$
$$-y = -3$$
$$y = 3$$

The two numbers are –1 and 3.

69. "The difference between an angle and twice another angle is 42°" yields the equation:

$$x - 2y = 42$$

Complementary angles have the equation: $x + y = 180$. By multiplying the second equation by 2, the y coefficients will be opposites:

$$\begin{aligned} x - 2y &= 42 \\ 2x + 2y &= 360 \\ 3x &= 402 \\ x &= 134 \end{aligned}$$

$$(134) - 2y = 42$$
$$-2y = -92$$
$$y = 46$$

Therefore the angles are 134° and 46°

71.
$$3x + y = 6$$
$$-2x + 2y = 4$$

(a) Graphing Method: Write the equations in slope-intercept form.

$$y = -3x + 6$$
$$y = x + 2$$

The two lines have unequal slopes and they intersect.

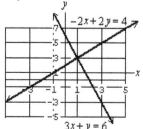

Solution: $(1, 3)$

(b)
$$3x + y = 6$$
$$-2x + 2y = 4$$

Substitution method: Solving the first equation for y gives $y = -3x + 6$.

Substitute this value of y into the second equation and solve the resulting equation for x.

$$-2x + 2y = 4$$
$$-2x + 2(-3x + 6) = 4$$
$$-2x - 6x + 12 = 4$$
$$-8x = -8$$
$$x = 1$$

Substitute this value of x into the first equation and solve for y.
$$3x + y = 6$$
$$3(1) + y = 6$$
$$3 + y = 6$$
$$y = 3$$
Solution: $(1, 3)$

(c) $\quad 3x + y = 6$
$$-2x + 2y = 4$$

Multiply the first equation by -2 to obtain opposite coefficients on y. Solve for x.
$$-6x - 2y = -12$$
$$\underline{-2x + 2y = 4}$$
$$-8x = -8$$
$$x = 1$$

Substitute this value of x into the first equation and solve for y.
$$3x + y = 6$$
$$3(1) + y = 6$$
$$3 + y = 6$$
$$y = 3$$
Solution: $(1, 3)$

73. One line within the system would have to "bend" for the system to have exactly two points of intersection. This is not possible.

75. $4x + Ay = -32$
$\quad\;\; Bx + 6y = 18$

If $(-3, 4)$ is a solution to the system then substitute $x = -3$ and $y = 4$ into the equations and solve for A and B.

First equation:

$$4(-3) + A(4) = -32$$
$$-12 + 4A = -32$$
$$4A = -20$$
$$A = -5$$

Second equation:
$$B(-3) + 6(4) = 18$$
$$-3B + 24 = 18$$
$$-3B = -6$$
$$B = 2$$

Problem Recognition Exercises

1. Since the equations represent the same line, there are infinitely many solutions.

3. Since the equations represent intersecting lines, there is one solution.

5. Since the equations represent two parallel lines, there is no solution.

7. $2x - 5y = -11$
$\quad 7x + 5y = -16$
Add the equations to eliminate the y-variables.
$$2x - 5y = -11$$
$$7x + 5y = -16$$
$$9x = -27$$
$$x = -3$$

$$2(-3) - 5y = -11$$
$$-6 - 5y = -11$$
$$-5y = -5$$
$$y = 1$$
The solution is $(-3, 1)$

9. $\quad x = -3y + 4$
$\quad 5x + 4y = -2$
Because the variable x is already isolated in the first equation, we can substitute into

the second equation.

$$5(-3y+4)+4y=-2$$
$$-15y+20+4y=-2$$
$$-11y+20=-2$$
$$-11y=-22$$
$$y=2$$

$$x=-3(2)+4$$
$$x=-6+4$$
$$x=-2$$

The solution is $(-2, 2)$

11.
$$x=-2y+5$$
$$2x-4y=10$$

Substitute the value of x into the second equation and solve the resulting equation.

$$2x-4y=10$$
$$2(-2y+5)-4y=10$$
$$-4y+10-4y=10$$
$$-8y+10=10$$
$$-8y=0$$
$$y=0$$
$$x=-2y+5$$
$$x=-2(0)+5$$
$$x=5$$

Solution: $(5, 0)$

13. $3x-2y=22$
$$5x+2y=10$$

Add the two equations and solve the resulting equation.
$$3x-2y=22$$
$$5x+2y=10$$
$$\overline{}$$
$$8x=32$$
$$x=4$$

$$3(4)-2y=22$$
$$12-2y=22$$
$$-2y=10$$
$$y=-5$$

Solution: $(4, -5)$

15. $\dfrac{1}{3}x+\dfrac{1}{2}y=\dfrac{2}{3}$

$$-\dfrac{2}{3}x+y=-\dfrac{4}{3}$$

Rewrite the equations in standard form. Clear the fractions in the first equation by multiplying by 6, in the second equation by multiplying by 3.
$$2x+3y=4$$
$$-2x+3y=-4$$
$$\overline{}$$
$$6y=0$$
$$y=0$$

$$-\dfrac{2}{3}x+y=-\dfrac{4}{3}$$
$$-\dfrac{2}{3}x+0=-\dfrac{4}{3}$$
$$-\dfrac{2}{3}x=-\dfrac{4}{3}$$
$$x=2$$

Solution: $(2, 0)$

17. $2c+7d=-1$
$$c=2$$

A value for c has been determined. To find a value for d, substitute $c=2$ into the first equation.
$$2c+7d=-1$$
$$2(2)+7d=-1$$
$$4+7d=-1$$
$$7d=-5$$
$$d=-\dfrac{5}{7}$$

Solution: $\left(2, -\dfrac{5}{7}\right)$

19. $y=0.4x-0.3$

$$-4x+10y=20$$

Substitute the value for y into the second equation and solve the resulting equation.
$$-4x+10(0.4x-0.3)=20$$
$$-4x+4x-3=20$$
$$-3\neq 20$$

No solution; inconsistent system

21. $3a + 7b = -3$

$-11a + 3b = 11$

Either method is appropriate. Since the equations are in standard form, the addition method will be used. To obtain opposite coefficients on a, multiply the first equation by 11 and the second equation by 3.

$33a + 77b = -33$

$\underline{-33a + 9b = 33}$

$86b = 0$

$b = 0$

$3a + 7b = -3$

$3a + 7(0) = -3$

$3a = -3$

$a = -1$

Solution: $(-1, 0)$

23. $y = 2x - 14$

$4x - 2y = 28$

Rewrite the equations in standard form,

$-2x + y = -14$

$4x - 2y = 28$

To obtain opposite coefficients on x, multiply the first equation by 2.

$-4x + 2y = -28$

$\underline{4x - 2y = 28}$

$ 0 = 0 \qquad \text{Identity}$

Infinitely many solutions

Solution: $\{(x, y) | y = 2x - 14\}$; dependent equations

25. $x + y = 3200$

$0.06x + 0.04y = 172$

Rewrite the equations in standard form. Clear the decimals in the second equation by multiplying it by 100.

$x + y = 3200$

$6x + 4y = 17,200$

To obtain opposite coefficients on x,

multiply the first equation by –6.

$-6x - 6y = -19,200$

$\underline{6x + 4y = 17,200}$

$ -2y = -2000$

$ y = 1000$

$x + y = 3200$

$x + 1000 = 3200$

$x = 2200$

Solution: $(2200, 1000)$

27. $3x + y - 7 = x - 4$

$3x - 4y + 4 = -6y + 5$

Rewrite the equations in standard form.

$2x + y = 3$

$3x + 2y = 1$

To obtain opposite coefficients on y, multiply the first equation by –2.

$-4x - 2y = -6$

$\underline{3x + 2y = 1}$

$-x = -5$

$x = 5$

$3x + y - 7 = x - 4$

$3(5) + y - 7 = 5 - 4$

$y + 8 = 1$

$y = -7$

Solution: $(5, -7)$

29. $3x - 6y = -1$

$9x + 4y = 8$

Either method is appropriate. Since the equations are in standard form, the addition method will be used. To obtain opposite coefficients on x, multiply the first equation by –1.

$-9x + 18y = 3$

$\underline{9x + 4y = 8}$

$22y = 11$

$y = \dfrac{1}{2}$

$$3x - 6y = -1$$

$$3x - 6\left(\frac{1}{2}\right) = -1$$

$$3x - 3 = -1$$

$$3x = 2$$

$$x = \frac{2}{3}$$

Solution: $\left(\frac{2}{3}, \frac{1}{2}\right)$

Section 4.4 Practice Exercises

1. $-2x + y = 6$
   $2x + y = 2$

   (a) Graphing Method: The x- and y-intercepts of $-2x + y = 6$ are $(-3, 0)$ and $(0, 6)$. The x- and y-intercepts of $2x + y = 2$ are $(1, 0)$ and $(0, 2)$. Use these points to graph the equations.

   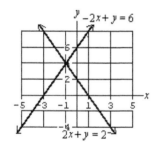

   Point of intersection: $(-1, 4)$

   (b) $-2x + y = 6$
   $2x + y = 2$
   Substitution method: Solving the first equation for y gives $y = 2x + 6$.

   Substitute this value of y into the second equation and solve the resulting equation for x.
   $$2x + y = 2$$
   $$2x + 2x + 6 = 2$$
   $$4x + 6 = 2$$
   $$4x = -4$$
   $$x = -1$$

$$-2(-1) + y = 6$$
$$2 + y = 6$$
$$y = 4$$
Solution: $(-1, 4)$

(c) $-2x + y = 6$
$2x + y = 2$
Addition method: The equations of the system are written in standard form. The coefficients on x are opposites. Add the two equations together and solve the resulting equation.

$$-2x + y = 6$$
$$\underline{2x + y = 2}$$
$$2y = 8$$
$$y = 4$$

$$-2x + y = 6$$
$$-2x + 4 = 6$$
$$-2x = 2$$
$$x = -1$$
Solution: $(-1, 4)$

3. $y = -2x + 6$
   $4x - 2y = 8$

   (a) Graphing method: The x- and y-intercepts of $y = -2x + 6$ are $(3, 0)$ and $(0, 6)$. the x- and y-intercepts of $4x - 2y = 8$ are $(2, 0)$ and $(0, -4)$. Use these points to graph the equations.

   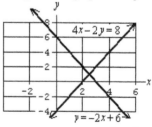

   Point of intersection: $\left(\frac{5}{2}, 1\right)$

   (b) Substitution method: Since the first equation is solved for y, substitute this value into the second equation and

109

solve the resulting equation for x.

$$4x - 2y = 8$$

$$4x - 2(-2x + 6) = 8$$

$$4x + 4x - 12 = 8$$

$$8x - 12 = 8$$

$$8x = 20$$

$$x = \frac{20}{8} = \frac{5}{2}$$

$$y = -2x + 6$$

$$y = -2\left(\frac{5}{2}\right) + 6$$

$$y = 1$$

Solution: $\left(\frac{5}{2}, 1\right)$

(c) Addition method: Rewrite the equations in standard form.

$$2x + y = 6$$

$$4x - 2y = 8$$

Since the y-variables have opposite signs, multiply the first equation by 2 to obtain opposite coefficients on y.

$$4x + 2y = 12$$

$$\underline{4x - 2y = 8}$$

$$8x = 20$$

$$x = \frac{20}{8} = \frac{5}{2}$$

$$4x - 2y = 8$$

$$4\left(\frac{5}{2}\right) - 2y = 8$$

$$10 - 2y = 8$$

$$-2y = -2$$

$$y = 1$$

Solution: $\left(\frac{5}{2}, 1\right)$

5. Let x represent one number. Let y represent the other number. The statement "one number is eight more than twice another" translates to the equation $x = 2y + 8$. The statement "their sum is 20"

translates to the equation $x + y = 20$.

$$x = 2y + 8$$

$$x + y = 20$$

Since one equation is solved for x, use the substitution method. Substitute this value for x into the second equation and solve the resulting equation for y.

$$x + y = 20$$

$$2y + 8 + y = 20$$

$$3y + 8 = 20$$

$$3y = 12$$

$$y = 4$$

$$x = 2y + 8$$

$$x = 2(4) + 8$$

$$x = 8 + 8$$

$$x = 16$$

The numbers are 4 and 16.

7. Let x represent the measure of one of the angles. Let y represent the measure of the other angle. The statement "two angles are complementary" translates to the equation $x + y = 90$. The statement "one angle is 10° less than 9 times the other" translates to the equation $x = 9y - 10$.

$$x + y = 90$$

$$x = 9y - 10$$

Since one equation is solved for x, use the substitution method. Substitute this value for x into the first equation and solve the resulting equation for y.

$$x + y = 90$$

$$9y - 10 + y = 90$$

$$10y - 10 = 90$$

$$10y = 100$$

$$y = 10$$

$$x = 9y - 10$$

$$x = 9(10) - 10$$

$$x = 80$$

The measures of the angles are 80° and 10°.

9. Let v represent the cost of one video game. Let d represent the cost of one DVD.

$$\left(\begin{array}{c}\text{Cost of} \\ \text{two video games}\end{array}\right) + \left(\begin{array}{c}\text{Cost of} \\ \text{three DVDs}\end{array}\right) = \left(\begin{array}{c}\text{Total} \\ \text{cost}\end{array}\right)$$

$$\left(\begin{array}{c}\text{Cost of} \\ \text{one video game}\end{array}\right) + \left(\begin{array}{c}\text{Cost of} \\ \text{two DVDs}\end{array}\right) = \left(\begin{array}{c}\text{Total} \\ \text{cost}\end{array}\right)$$

$2v + 3d = 34.10$

$v + 2d = 19.80$

Since both equations are in standard form, use the addition method. To obtain opposite coefficients on v, multiply the second equation by -2.

$2v + 3d = 34.10$

$-2v - 4d = -39.60$

$-d = -5.5$

$d = 5.50$

$v + 2(5.50) = 19.80$

$v + 11 = 19.80$

$v = 8.80$

Video games cost $8.80 each and DVDs cost $5.50 each.

11. Let t represent the cost of one share of technology stock. Let m represent the cost of one share of mutual fund.

$$\left(\begin{array}{c}\text{Cost of} \\ \text{100 shares} \\ \text{technology} \\ \text{stock}\end{array}\right) + \left(\begin{array}{c}\text{Cost of} \\ \text{200 shares} \\ \text{mutual} \\ \text{fund}\end{array}\right) = \left(\begin{array}{c}\text{Total} \\ \text{cost}\end{array}\right)$$

$$\left(\begin{array}{c}\text{Cost of} \\ \text{300 shares} \\ \text{technology} \\ \text{stock}\end{array}\right) + \left(\begin{array}{c}\text{Cost of} \\ \text{50 shares} \\ \text{mutual} \\ \text{fund}\end{array}\right) = \left(\begin{array}{c}\text{Total} \\ \text{cost}\end{array}\right)$$

$100t + 200m = 3800$

$300t + 50m = 5350$

Since the equations are written in standard form, use the addition method. To obtain opposite coefficients on t, multiply the first equation by -3.

$-300t - 600m = -11400$

$300t + 50m = 5350$

$-550m = -6050$

$m = 11$

$100t + 200m = 3800$

$100t + 200(11) = 3800$

$100t + 2200 = 3800$

$100t = 1600$

$t = 16$

Technology stock costs $16 per share and the mutual fund costs $11 per share.

13. Let x represent the number of 45¢ stamps. Let y represent the number of 65¢ stamps.

$$\left(\begin{array}{c}\text{Cost of} \\ \text{x} \\ \text{45¢ stamps}\end{array}\right) + \left(\begin{array}{c}\text{Cost of} \\ \text{y} \\ \text{65¢ stamps}\end{array}\right) = \left(\begin{array}{c}\text{Total} \\ \text{cost}\end{array}\right)$$

$$\left(\begin{array}{c}\text{Number of} \\ \text{45¢} \\ \text{stamps}\end{array}\right) + \left(\begin{array}{c}\text{Number of} \\ \text{65¢} \\ \text{stamps}\end{array}\right) = \left(\begin{array}{c}\text{Total number} \\ \text{of stamps}\end{array}\right)$$

$0.45x + 0.65y = 24.90$

$x + y = 50$

Since the equations are in standard form and the coefficient on x in the second equation is one, use the addition method by multiplying the second equation by -0.45.

$0.45x + 0.65y = 24.90$

$-0.45x - 0.46y = -23.0$

$0.19y = 1.90$

$y = 10$

$x + y = 50$

$x + 10 = 50$

$x = 40$

Mylee bought forty 45¢ stamps and ten 65¢ stamps.

15. Let x represent the amount Shanelle invested in the 10% account. Let y represent the amount Shanelle invested in the 7% account.

	10% Account	7% Account	Total
Principal invested	x	y	$10,000
Interest earned	0.10x	0.07y	$805

$$x + y = 10,000$$
$$0.1x + 0.07y = 805$$

Since the equations are in standard form and the coefficient on x in the first equation is one, use the addition method by multiplying the second equation by -0.1.

$$-0.1x - 0.1y = -1000$$
$$0.1x + 0.07y = 805$$
$$\overline{-0.03y = -195}$$
$$y = 6500$$
$$x + 6500 = 10,000$$
$$x = 3500$$

Shanelle invested $3500 in the 10% account and $6500 in the 7% account.

17. Let x represent the amount borrowed at 9%. Let y represent the amount borrowed at 6%.

	9% account	6% account	Total
Amount borrowed	x	y	$12,000
Interest charged	0.09x	0.06y	$810

$$x + y = 12,000$$
$$0.09x + 0.06y = 810$$

Since the equations are in standard form and the coefficient on x in the first equation is 1, use the addition method by multiplying the first equation by -0.09.

$$-0.09x + (-0.09y) = -1080$$
$$0.09x + 0.06y = 810$$
$$\overline{-0.03y = -270}$$
$$y = 9000$$
$$x + 9000 = 12,000$$
$$x = 3000$$

$9000 is borrowed at 6%, and $3000 is borrowed at 9%.

19. Let x represent the amount you invest in the bond fund that returns 8%. Let y represent the amount you invest in the stock fund that will return 12%.

	8% bond fund	12% stock fund	Total
Principle invested	x	y	$30000
Interest earned	0.08x	0.12y	$3120

$$x + y = 30000$$
$$0.08x + 0.12y = 3120$$

Since the equations are in standard form and the coefficient on x in the first equation is 1, use the addition method by multiplying the first equation by -0.08.

$$-0.08x - 0.08y = -2400$$
$$0.08x + 0.12y = 3120$$
$$\overline{0.04y = 720}$$
$$y = 18000$$
$$x + 18000 = 30000$$
$$x = 12000$$

$12000 must be invested in the bond fund that returns 8% and $18000 invested in the stock fund that returns 12%.

21. Let x represent the amount of 50% disinfectant solution. Let y represent the amount of 40% disinfectant solution.

	50% Mixture	40% Mixture	46% Mixture
Amount of solution	x	y	25
Amount of disinfectant	$0.50x$	$0.40y$	$0.46(25)$

$$x + y = 25$$
$$0.5x + 0.4y = 11.5$$

Since the equations are in standard form and the coefficient on x is one, use the addition method by multiplying the first equation by -0.5.

$$
\begin{array}{r}
-0.5x - 0.5y = -12.5 \\
0.5x + 0.4y = 11.5 \\
\hline
-0.1y = -1
\end{array}
$$
$$y = 10$$

$$x + 10 = 25$$
$$x = 15$$

15 gallons of 50% mixture should be mixed with 10 gallons of 40% mixture.

23. Let x represent the amount of 45% disinfectant solution. Let y represent the amount of 30% disinfectant solution.

	45% Mixture	30% Mixture	39% Mixture
Amount of solution	x	y	20
Amount of disinfectant	$0.45(x)$	$0.30(y)$	$0.39(20)$

$$x + y = 20$$
$$0.45x + 0.3y = 7.8$$

Since the equations are in standard from and the coefficient on x is one, use the addition method by multiplying the first equation by -0.45.

$$
\begin{array}{r}
-0.45x - 0.45y = -9.0 \\
0.45x + 0.3y = 7.8 \\
\hline
-0.15y = -1.2
\end{array}
$$
$$y = 8$$

$$x + 8 = 20$$
$$x = 12$$

12 gallons of the 45% disinfectant solution should be mixed with 8 gallons of the 30% disinfectant mixture.

25. Let x represent the amount of 13% salt solution. Let y represent the amount of 18% salt solution.

	13% Mixture	18% Mixture	16% Mixture
Amount of solution	x	y	50
Amount of salt	$0.13x$	$0.18y$	$50(0.16)$

$$x + y = 50$$
$$0.13x + 0.18y = 8$$

Since the equations are in standard form and the coefficient on x is one, use the addition method by multiplying the first equation by -0.13.

$$
\begin{array}{r}
-0.13x - 0.13y = -6.5 \\
0.13x + 0.18y = 8 \\
\hline
0.05y = 1.5
\end{array}
$$
$$y = 30$$

$$x + 30 = 50$$
$$x = 20$$

20 mL of the 13% salt solution should be mixed with 30 mL of the 18% salt solution.

27. Let x represent the amount of pure antifreeze. Let y represent the amount of 40% antifreeze solution.

	Pure	40% Mixture	50% Mixture
Amount of solution	x	y	6
Amount of antifreeze	x	$0.4y$	$6(0.5)$

$$x + y = 6$$
$$x + 0.4y = 3$$

Since the equations are in standard form and the coefficient on x is one, use the addition method by multiplying the second equation by -1.

$$x + y = 6$$
$$-x - 0.4y = -3$$
$$\overline{0.6y = 3}$$
$$y = 5$$
$$x + 5 = 6$$
$$x = 1$$

1 L of the pure antifreeze should be mixed with 5 L of the 40% antifreeze solution to get 6 L of 50% antifreeze solution.

29. Let x represent the speed of the boat in still water. Let y represent the speed of the current.

	Distance	Rate	Time
Downstream	16	$x+y$	2
Upstream	16	$x-y$	4

$$2x + 2y = 16$$
$$4x - 4y = 16$$

Since the equations are in standard form and the coefficients on y have opposite signs, use the addition method by

multiplying the first equation by 2.

$$4x + 4y = 32$$
$$4x - 4y = 16$$
$$\overline{8x = 48}$$
$$x = 6$$

$$2(6) + 2y = 16$$
$$2y = 4$$
$$y = 2$$

The speed of the boat in still water is 6 mph and the speed of the current is 2 mph.

31. Let x represent the speed of the plane in still air. Let y represent the speed of the wind.

	Distance	Rate	Time
Flying with the wind	960	$x+y$	3
Flying against the wind	840	$x-y$	3

$$3x + 3y = 960$$
$$3x - 3y = 840$$

Since the equations are in standard form and the coefficients on y are opposites, use the addition method.

$$3x + 3y = 960$$
$$3x - 3y = 840$$
$$\overline{6x = 1800}$$
$$x = 300$$

$$3(300) + 3y = 960$$
$$900 + 3y = 960$$
$$3y = 60$$
$$y = 20$$

The speed of the plane in still air is 300 mph, and the wind is 20 mph.

33. Let p represent the speed of the plane in still air. Let w represent the speed of the wind.

	Distance	Rate	Time
With the wind	3600	$p + w$	6
Against the wind	3600	$p - w$	8

$3600 = 6(p + w)$
$3600 = 8(p - w)$

$3600 = 6p + 6w$

$3600 = 8p - 8w$

Multiply the first equation by 4. Multiply the second equation by -3.

$14400 = 24p + 24w$

$-10800 = -24p + 24w$
_____
$3600 = 48w$

$75 = w$

$3600 = 6p + 6(75)$

$3600 = 6p + 450$

$3150 = 6p$

$525 = p$

The speed of the plane in still air is 525 mph. The speed of the wind is 75 mph.

35. Let x represent the number of dimes. Let y represent the number of nickels.

$\begin{pmatrix} \text{number} \\ \text{of nickels} \end{pmatrix} = \begin{pmatrix} \text{number} \\ \text{of dimes} \end{pmatrix} + 5$

$\begin{pmatrix} \text{value of} \\ \text{dimes} \end{pmatrix} + \begin{pmatrix} \text{value of} \\ \text{nickels} \end{pmatrix} = \begin{pmatrix} \text{total} \\ \text{value} \end{pmatrix}$

$y = x + 5$

$0.10x + 0.05y = 2.80$

Since the first equation is solved for y, use the substitution method by substituting this value into the second equation.

$0.10x + 0.05(x + 5) = 2.80$

$0.10x + 0.05x + 0.25 = 2.80$

$0.15x + 0.25 = 2.80$

$0.15x = 2.55$

$x = 17$

$y = x + 5$
$y = 17 + 5$
$y = 22$

There are 17 dimes and 22 nickels.

37. Let x represent the amount of water. Let y represent the amount of 30% vinegar solution.

	Water	30% Mixture	25% Mixture
Amount of solution	x	y	12
Amount of vinegar	0	$0.3y$	$12(0.25)$

$x + y = 12$

$0.3y = 3$

We can solve the second equation to find y.

$0.3y = 3$

$y = 10$

$x + 10 = 12$

$x = 2$

2 qt of water should be mixed with 10 qt of 30% solution to get 12 qt of 25% vinegar solution.

39. (a) Let x represent the number of free throws. Let y represent the number of field goals. The statement "made 2432 baskets...some were free throws and some were field goals" translates to the equation $x + y = 2432$. The statement "the number of field goals was 762 more than the number of free throws" translates to the equation $y = x + 762$.

$x + y = 2432$

$y = x + 762$

Since the second equation is solved

for y, use the substitution method substituting this value into the first equation.

$$x + x + 762 = 2432$$
$$2x + 762 = 2432$$
$$2x = 1670$$
$$x = 835$$

$$y = x + 762$$
$$y = 835 + 762$$
$$y = 1597$$

He made 835 free throws and 1597 field goals.

(b) $835(1) + 1597(2) = 835 + 3194 = 4029$

He scored 4029 points.

(c) $\dfrac{4029}{80} = 50.3625$

He averaged approximately 50 points per game.

41. Let x represent the speed of the plane in still air. Let y represent the speed of the wind.

	Distance	Rate	Time
Flying with a tailwind	350	$x + y$	$1\frac{3}{4}$
Flying with a headwind	210	$x - y$	$1\frac{3}{4}$

$$\frac{7}{4}x + \frac{7}{4}y = 350$$
$$\frac{7}{4}x - \frac{7}{4}y = 210$$

Clear fractions by multiplying both equations by 4.
$$7x + 7y = 1400$$
$$7x - 7y = 840$$

Since the equations are in standard form and the coefficients on y are opposites, use the addition method.

$$7x + 7y = 1400$$
$$\underline{7x - 7y = 840}$$
$$14x = 2240$$
$$x = 160$$

$$\frac{7}{4}x + \frac{7}{4}y = 350$$
$$\frac{7}{4}(160) + y = 350$$

$$1120 + 7y = 1400$$
$$7y = 280$$
$$y = 40$$

The speed of the plane in still air is 160 mph and the wind is 40 mph.

43. Let x represent the amount invested in the first account. Let y represent the amount invested in the second account.

	First Account	Second Account	Total
Principal	x	y	60,000
Interest	0.055x	0.065y	3750

$$x + y = 60,000$$
$$0.055x + 0.065y = 3750$$

Since the equations are written in standard form, use the addition method by multiplying the first equation by -0.055.
$$-0.055x - 0.055y = -3300$$
$$\underline{0.055x + 0.065y = 3750}$$
$$0.010y = 450$$
$$y = 45,000$$

$$x + y = 60,000$$
$$x + 45,000 = 60,000$$
$$x = 15,000$$

$15,000 is invested in the 5.5% account and $45,000 is invested in the 6.5% account.

45. Let x represent the number of pounds of candy needed. Let y represent the number of pounds of nuts needed.

	Candy	Nuts	Total
Number of	x	y	20
Value of	$1.80x$	$1.20y$	$1.56(20)$

$$x + y = 20$$
$$1.80x + 1.20y = 31.20$$

Since the equations are written in standard form, use the addition method by multiplying the first equation by -1.20.

$$-1.20x - 1.20y = -24.00$$
$$\underline{1.80x + 1.20y = 31.20}$$
$$0.60x = 7.20$$
$$x = 12$$

$$x + y = 20$$
$$12 + y = 20$$
$$y = 8$$

12 pounds of candy should be mixed with 8 pounds of nuts.

47. Let d represent the number of points scored by the Dallas Cowboys. Let b represent the number of points scored by the Buffalo Bills. The statement "the Dallas Cowboys scored four more than twice the number of points scored by the Buffalo Bills" translates to the equation $d = 2b + 4$. The statement "total number of points scored by both teams was 43" translates to the equation $d + b = 43$.

$$d = 2b + 4$$
$$d + b = 43$$

Since one equation is solved for d, use the substitution method. Substitute this value for d into the first equation and solve the resulting equation for b.

$$d + b = 43$$
$$2b + 4 + b = 43$$
$$3b + 4 = 43$$
$$3b = 39$$
$$b = 13$$

$$d = 2b + 4$$
$$d = 2(13) + 4$$
$$d = 30$$

Dallas scored 30 points, and Buffalo scored 13 points.

49. Let x represent the number of women college students. Let y represent the number of men college students. The statement "500 college students" translates to the equation $x + y = 500$. The statement "340 said that the campus lacked adequate lighting" together with "$\frac{4}{5}$ of the women and $\frac{1}{2}$ of the men said that they thought the campus lacked adequate lighting" translates to the equation $\frac{4}{5}x + \frac{1}{2}y = 340$.

$$x + y = 500$$
$$\frac{4}{5}x + \frac{1}{2}y = 340$$

Clear fractions from the second equation by multiplying it by 10.

$$x + y = 500$$
$$8x + 5y = 3400$$

Since the equations are written in standard form, use the addition method by multiplying the first equation by -5.

$$-5x - 5y = -2500$$
$$\underline{8x + 5y = 3400}$$
$$3x = 900$$
$$x = 300$$

117

$$x + y = 500$$
$$300 + y = 500$$
$$y = 200$$

There were 300 women and 200 men in the survey.

Section 4.5 Practice Exercises

1. (a) linear

   (b) intersect or overlap

3. Graph the line $y = \dfrac{3}{5}x + 2$ using the point-slope form.

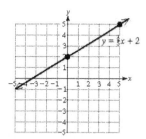

5. When the inequality symbol is $\leq$ or $\geq$, a solid line is used in the graph of a linear inequality in two variables.

7. All of the points in the shaded region are solutions to the inequality.

9. (a) $-2x - y \leq 2$

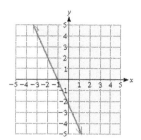

11. $3x + 2y \overset{?}{>} 1$

$3(3) + 2(-1) \overset{?}{>} 1$

$7 > 1 \checkmark$

$(3, -1)$ is a solution to the inequality. True.

13. $y \overset{?}{<} -2x + 4$

$0 \overset{?}{<} -2(2) + 4$

$0 = 0$

$(0, 2)$ is not a solution to the inequality. False.

15. $x + 10y \overset{?}{<} 1$

$(-3) + 10(0) \overset{?}{<} 1$

$-3 < 1 \checkmark$

$(-3, 0)$ is a solution to the inequality. True.

17. $y \geq -x + 5$

Using the slope-intercept method of graphing, graph the related equation $y = -x + 5$ with a solid line. Using $(0, 0)$ as a test point, a false statement is obtained. Shade the region not containing $(0, 0)$.

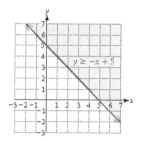

Answers may vary. For example: $(0, 5)$, $(2, 7)$, $(-1, 8)$

19. $y < 4x$

Using the slope-intercept method of graphing, graph the related equation $y = 4x$ with a dashed line. Using $(1, 1)$ as a test point, a true statement is obtained. Shade the region containing $(1, 1)$.

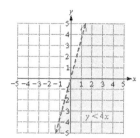

Answers may vary. For example: (1, –1), (3, 0), (–2, –9)

21. $3x + 7y \leq 14$

Using the intercepts method of graphing, graph the related equation $3x + 7y = 14$ with a solid line. Using (0, 0) as a test point, a true statement is obtained. Shade the region containing (0, 0).

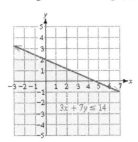

Answers may vary. For example: (0, 0), (0, 2), (–1, –3)

23. $x - y > 6$

Using the intercepts method of graphing, graph the related equation $x - y = 6$ with a dashed line. Using (0, 0) as a test point, a false statement is obtained. Shade the region not containing (0, 0).

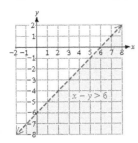

25. $x \geq -1$

Graph the related equation $x = -1$ with a solid vertical line. Using (0, 0) as a test point, a true statement is obtained. Shade the region containing (0, 0).

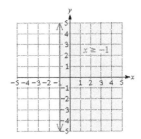

27. $y < 3$

Graph the related equation $y = 3$ with a dashed horizontal line. Using (0, 0) as a test point, a true statement is obtained. Shade the region containing (0, 0).

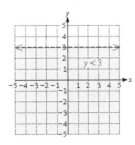

29. $y \leq -\dfrac{3}{4}x + 2$

Using the slope-intercept method of graphing, graph the related equation $y = -\dfrac{3}{4}x + 2$ with a solid line. Using (0, 0) as a test point, a true statement is obtained. Shade the region containing (0, 0).

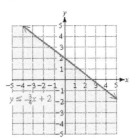

31. $y - 2x > 0$

Using the slope-intercept method of graphing, graph the related equation $y = 2x$ with a dashed line. Using (1, 1) as a test point, a false statement is obtained. Shade the region not containing (1, 1).

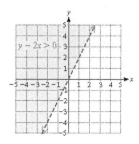

33. $x \le 0$

Graph the related equation $x = 0$ or the y-axis with a solid vertical line. Using $(1, 1)$ as a test point, a false statement is obtained. Shade the region not containing $(1, 1)$.

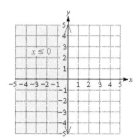

35. $y \ge 0$

Graph the related equation $y = 0$ or the x-axis with a solid horizontal line. Using $(1, 1)$ as a test point, a true statement is obtained. Shade the region containing $(1, 1)$.

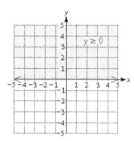

37. $-x \le \frac{1}{2}y - 2$

$-x + 2 \le \frac{1}{2}y$

$-2x + 4 \le y$

$y \ge -2x + 4$

Using the slope-intercept method of graphing, graph the related equation $y = -2x + 4$ with a solid line. Using $(0, 0)$ as a test point, a false statement is

obtained. Shade the region not containing $(0, 0)$.

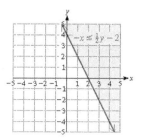

39. $2x > 3y$

$\frac{2}{3}x > y$

$y < \frac{2}{3}x$

Using the slope-intercept method of graphing, graph the related equation $y = \frac{2}{3}x$ with a dashed line. Using $(1, 0)$ as a test point, a true statement is obtained. Shade the region containing $(1, 0)$.

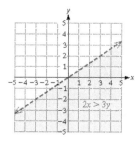

41. (a) $x + y > 4$

The graph of the inequality is a set of ordered pairs above the dashed line $x + y = 4$, for example, $(6, 3)$, $(-2, 8)$, $(0, 5)$. Answers will vary.

(b) $x + y = 4$

The graph of the equation is a set of ordered pairs on the line $x + y = 4$, for example $(0, 4)$, $(4, 0)$, $(2, 2)$. Answers will vary.

(c) $x + y < 4$

The graph of the inequality is a set of ordered pairs below the dashed line $x + y = 4$, for example, $(0, 0)$, $(-2, 1)$, $(3, 0)$. Answers will vary.

43. $2x + y < 3$

Using the intercepts method of graphing, graph the related equation $2x + y = 3$ with a dashed line. Substituting the test point $(0, 0)$ into the inequality results in a true statement. Shade the region below the dashed line.

$y \geq x + 3$

Using the slope-intercept method of graphing, graph the related equation $y = x + 3$ with a solid line. Substituting the test point $(0, 0)$ into the inequality results in a false statement. Shade the region above the solid line.

Putting the two regions on the same graph, the intersection is the solution to the system of inequalities.

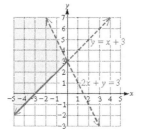

45. $x + y \geq -3$

Using the intercepts method of graphing, graph the related equation $x + y = -3$ with a solid line. Substituting the test point $(0, 0)$ into the inequality results in a true statement. Shade the region above the solid line.

$x - 2y \geq 6$

Using the intercepts method of graphing, graph the related equation $x - 2y = 6$ with a solid line. Substituting the test point $(0, 0)$ into the inequality results in a false statement. Shade the region below the solid line.

Putting the two regions on the same graph, the intersection is the solution to the system of inequalities

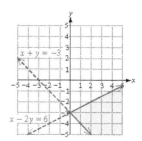

47. $2x + 3y < 6$

Using the intercepts method of graphing, graph the related equation $2x + 3y = 6$ with a dashed line. Substituting the test point $(0, 0)$ into the inequality results in a true statement. Shade the region below the dashed line.

$3x + y > -5$

Using the intercepts method of graphing, graph the related equation $3x + y = -5$ with a dashed line. Substituting the test point $(0, 0)$ into the inequality results in a true statement. Shade the region above the dashed line.

Putting the two regions on the same graph, the intersection is the solution to the system of inequalities.

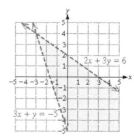

49. $y > 2x$

Using the slope-intercept method of graphing, graph the related equation $y = 2x$ with a dashed line. Substituting the test point $(1, 1)$ into the inequality results in a false statement. Shade the region above the dashed line.

$y > -4x$

Using the slope-intercept method of graphing, graph the related equation $y = -4x$ with a dashed line. Substituting the test point $(1, 1)$ into the inequality results in a true statement. Shade the region above the dashed line.

Putting the two regions on the same graph, the intersection is the solution to the system of inequalities.

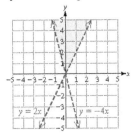

51. $y < \frac{1}{2}x - 1$

Using the slope-intercept method of graphing, graph the related equation $y = \frac{1}{2}x - 1$ with a dashed line. Substituting the test point (0, 0) into the inequality results in a false statement. Shade the region below the solid line.

$x + y \le -4$

Using the intercepts method of graphing, graph the related equation $x + y \le -4$ with a solid line. Substituting the test point (0, 0) into the inequality results in a false statement. Shade the region below the solid line.

Putting the two regions on the same graph, the intersection is the solution to the system of inequalities.

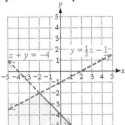

53. $y < 4$

Graph the related equation $y = 4$ with a dashed horizontal line. Substituting the test point (0, 0) into the inequality results in a true statement. Shade the region below the dashed line.

$4x + 3y \ge 12$

Using the intercepts method of graphing, graph the related equation $4x + 3y = 12$ with a solid line. Substituting the test point (0, 0) into the inequality results in a false statement. Shade the region above the solid line.

Putting the two regions on the same graph, the intersection is the solution to the system of inequalities.

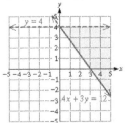

55. $x > -4$

Graph the related equation $x = -4$ with a dashed vertical line. Substituting the test point (0, 0) into the inequality results in a true statement. Shade the region to the right of the dashed vertical line.

$y \le 3$

Graph the related equation $y = 3$ with a solid horizontal line. Substituting the test point (0, 0) into the inequality results in a true statement. Shade the region below the solid line.

Putting the two regions on the same graph, the intersection is the solution to the system of inequalities.

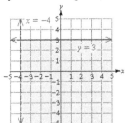

57. $2x \ge 5$

Graph the related equation $x = 2.5$ with a solid vertical line. Substituting the test point (0, 0) into the inequality results in a false statement. Shade the region to the right of the solid vertical line.

$6 > 3y$

Graph the related equation $2 = y$ with a dashed horizontal line. Substituting the test point $(0, 0)$ into the inequality results in a true statement. Shade the region below the dashed horizontal line.

Putting the two regions on the same graph, the intersection is the solution to the system of inequalities.

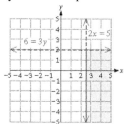

59. $x \geq -4$

Graph the related equation $x = -4$ with a solid vertical line. Substituting the test point $(0, 0)$ into the inequality results in a true statement. Shade the region to the right of the solid vertical line.

$x \leq 1$

Graph the related equation $x = 1$ with a solid vertical line. Substituting the test point $(0, 0)$ into the inequality results in a true statement. Shade the region to the left of the solid line.

Putting the two regions on the same graph, the intersection is the solution to the system of inequalities.

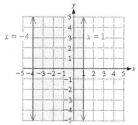

Group Activity

Answers will vary throughout this exercise.

Chapter 4 Review Exercises

Section 4.1

1. $x - 4y = -4$

$x + 2y = 8$ ordered pair: $(4, 2)$

$(4) - 4(2) \overset{?}{=} -4$ ✓

$(4) + 2(2) \overset{?}{=} 8$ ✓

The ordered pair $(4, 2)$ is a solution.

3. $3x + y = 9$

$y = 3$ ordered pair: $(1, 3)$

$3(1) + 3 \overset{?}{=} 9$ No

$3 \overset{?}{=} 3$ ✓

The ordered pair $(1, 3)$ is not a solution.

5. $y = -\dfrac{1}{2}x + 4$

$y = x - 1$

Since the slopes are different, the lines intersect.

7. $y = -\dfrac{4}{7}x + 3$

$y = -\dfrac{4}{7}x - 5$

Since the lines have the same slope but different y-intercepts, the lines are parallel.

9. $y = 9x - 2$

$9x - y = 2$

Rewrite the second equation in slope-intercept form.

$y = 9x - 2$

$y = 9x - 2$

Since the lines have the same slope and same y-intercept, the lines coincide.

123

11. $y = -\dfrac{2}{3}x - 2$

$-x + 3y = -6$

Rewrite the second equation in slope-intercept form.

$y = -\dfrac{2}{3}x - 2$

$y = \dfrac{1}{3}x - 2$

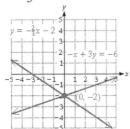

$(0, -2)$ consistent and independent system

13. $4x = -2y + 10$

$2x + y = 5$

Rewrite the equations in slope-intercept form.

$y = -2x + 5$

$y = -2x + 5$

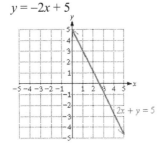

Infinitely many solutions
$\{(x, y) | y = -2x + 5\}$
consistent and dependent equations

Infinitely many solutions
$\{(x, y) | -x + 5y = -5\}$
consistent and dependent equations

15. $6x - 3y = 9$

$y = -1$

Rewrite the first equation in slope-intercept form.

$y = 2x - 3$

$y = -1$

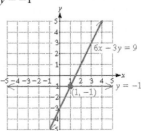

$(1, -1)$ consistent and independent system

17. $x - 7y = 14$

$-2x + 14y = 14$

Rewrite the equations in slope-intercept form.

$y = \dfrac{1}{7}x - 2$

$y = \dfrac{1}{7}x + 1$

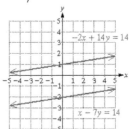

No solution; inconsistent and independent system

19. $y_c = 20 + 0.25x$

$y_m = 30 + 0.20x$

As both equations are solved for y we can set them equal and solve for x.

$20 + 0.25x = 30 + 0.20x$

$0.05x = 10$

$x = 200$

$y_c = 20 + (0.25(200)) = 70$

$y_m = 30 + (0.20(200)) = 70$

Section 4.2

21. $2x + 3y = -5$
    $x = y - 5$

Substitute the value for x into the first equation and solve for y.
$2(y - 5) + 3y = -5$
$2y - 10 + 3y = -5$
$5y - 10 = -5$
$5y = 5$
$y = 1$

$x = y - 5$
$x = 1 - 5$
$x = -4$
Solution: $(-4, 1)$

23. $4x + 2y = 4$
    $y = -2x + 2$

Substitute the value for y into the first equation and solve for x.
$4x + 2(-2x + 2) = 4$
$4x - 4x + 4 = 4$
$4 = 4$ Identity

Infinitely many solutions
$\{(x, y) \mid y = -2x + 2\}$; dependent equations

25. **(a)** y in the second equation is easiest to isolate for because its coefficient is 1.

(b) $4x - 3y = 9$
$2x + y = 12$

Solving the second equation for y gives
$y = -2x + 12$. Substitute this value into the first equation and solve for x.
$4x - 3(-2x + 12) = 9$
$4x + 6x - 36 = 9$
$10x - 36 = 9$
$10x = 45$
$x = \dfrac{45}{10} = \dfrac{9}{2}$

$2x + y = 12$
$2\left(\dfrac{9}{2}\right) + y = 12$
$9 + y = 12$
$y = 3$
Solution: $\left(\dfrac{9}{2}, 3\right)$

27. $x + 5y = 20$
    $3x + 2y = 8$

Solving the first equation for x gives $x = -5y + 20$. Substitute this value into the second equation and solve for y.
$3(-5y + 20) + 2y = 8$
$-15y + 60 + 2y = 8$
$-13y + 60 = 8$
$-13y = -52$
$y = 4$

$x + 5y = 20$
$x + 5(4) = 20$
$x + 20 = 20$
$x = 0$
Solution: $(0, 4)$

29. $-3x + y = 15$
    $6x - 2y = 12$

Solving the first equation for y gives $y = 3x + 15$. Substitute this value into the second equation and solve for y.
$6x - 2(3x + 15) = 12$
$6x - 6x - 30 = 12$
$-30 = 12$
No solution; inconsistent and independent system

31. Let x represent the measure of one of the acute angles. Let y represent the measure of the second acute angle. The sum of the acute angles in a right triangle is 90°. This translates to the equation $x + y = 90$. The statement "one of the acute angles is 6° more than the other acute angle" translates to the equation $y = x + 6$.

$$x + y = 90$$
$$y = x + 6$$

Since the second equation is solved for y, use the substitution method by substituting the value for y into the first equation and solving for x.

$$x + (x + 6) = 90$$
$$x + x + 6 = 90$$
$$2x + 6 = 90$$
$$2x = 84$$
$$x = 42$$

$$y = x + 6$$
$$y = 42 + 6$$
$$y = 48$$

The measures of the angles are $42°$ and $48°$.

Section 4.3

33. 1. Write both equations in standard form.

2. Multiply one or both equations by a constant to create opposite coefficients for one of the variables.

3. Add the equations to eliminate the variable.

4. Solve for the remaining variable.

5. Substitute the known variable value into an original equation to solve for the other variable.

35. (a) Answers may vary. For example, y is easier to eliminate because the signs are opposite.

(b) $9x - 2y = 14$
$\quad 4x + 3y = 14$

Multiply the first equation by 3 and the second equation by 2 to obtain opposite coefficients on y.

$$27x - 6y = 42$$
$$\underline{8x + 6y = 28}$$
$$35x = 70$$
$$x = 2$$

$$9x - 2y = 14$$
$$9(2) - 2y = 14$$
$$18 - 2y = 14$$
$$-2y = -4$$
$$y = 2$$
Solution: $(2, 2)$

37. $\quad x + 3y = 0$
$\quad -3x - 10y = -2$

Multiply the first equation by 3 to obtain opposite coefficients on x.

$$3x + 9y = 0$$
$$\underline{-3x - 10y = -2}$$
$$-y = -2$$
$$y = 2$$

$$x + 3y = 0$$
$$x + 3(2) = 0$$
$$x + 6 = 0$$
$$x = -6$$
Solution: $(-6, 2)$

39. $12x = 5y + 5$
$\quad 5y = -1 - 4x$

Rewrite the equations in standard form.

$$12x - 5y = 5$$
$$\underline{4x + 5y = -1}$$
$$16x = 4$$
$$x = \frac{1}{4}$$

$$5y = -1 - 4x$$
$$5y = -1 - 4\left(\frac{1}{4}\right)$$
$$5y = -1 - 1$$
$$5y = -2$$
$$y = -\frac{2}{5}$$

Solution: $\left(\frac{1}{4}, -\frac{2}{5}\right)$

41. $-8x - 4y = 16$
$10x + 5y = 5$

Multiply the first equation by 5 and the second equation by 4 to obtain opposite coefficients on y.
$-40x - 20y = 80$
$40x + 20y = 20$
_____
$0 = 100$ Contradiction

No solution; inconsistent and independent system

43. $0.5x - 0.2y = 0.5$
$0.4x + 0.7y = 0.4$

Multiply both equations by 10 to clear the decimals.
$5x - 2y = 5$
$4x + 7y = 4$

Multiply the first equation by 7 and the second equation by 2 to obtain opposite coefficients on y.
$35x - 14y = 35$
$8x + 14y = 8$
_____
$43x = 43$
$x = 1$

$0.5x - 0.2y = 0.5$
$0.5(1) - 0.2y = 0.5$
$0.5 - 0.2y = 0.5$
$-0.2y = 0$
$y = 0$
Solution: $(1, 0)$

45. (a) Answers may vary. For example, use the addition method because the equations are written in standard form.

(b) $5x - 8y = -2$
$3x - y = -5$

Multiply the second equation by -8 to obtain opposite coefficients on y.

$5x - 8y = -2$
$-24x + 8y = 40$
_____
$-19x = 38$
$x = -2$
$3x - y = -5$
$3(-2) - y = -5$
$-6 - y = -5$
$-y = 1$
$y = -1$
Solution: $(-2, -1)$

Section 4.4

47. Let x represent the amount of money invested at 12%. Let y represent the amount of money invested at 4%.

	12% Account	4% Account	Total
Principal	x	y	20,000
Interest	$0.12x$	$0.04y$	30000

$x + y = 600,000$
$0.12x + 0.04y = 30,000$

Multiply the second equation by 100 to eliminate the decimals.
$x + y = 600,000$
$12x + 4y = 3,000,000$

Multiply the first equation by -12 to obtain opposite coefficients on x.
$-12x - 12y = -7,200,000$
$12x + 4y = \ \ 3,000,000$
_____
$-8y = -4,200,000$
$y = 525,000$
$x + y = 600,000$
$x + 525,500 = 600,000$
$x = 75,000$

He invested \$75,000 in the 12% account and \$525,000 in the 4% account.

49. Let b represent the speed of the boat in still water. Let c represent the speed of the current.

	Distance	Rate	Time
Downstream	80	$b+c$	4
Upstream	80	$b-c$	5

$4b + 4c = 80$

$5b - 5c = 80$

Multiply the first equation by 5 and the second equation by 4 to obtain opposite coefficients on c.

$20b + 20c = 400$

$\underline{20b - 20c = 320}$

$40b = 720$

$b = 18$

$4(b + c) = 80$

$4b + 4c = 80$

$4(18) + 4c = 80$

$72 + 4c = 80$

$4c = 8$

$c = 2$

The speed of the boat in still water is 18 mph and the speed of the current is 2 mph.

51. Let x represent the cost of a soft drink. Let y represent the cost of a hot dog. The statement "the total cost of a soft drink and a hot dog is \$8.00" translates to the equation $x + y = 8.00$. The statement "the price of the hot dog is \$1.00 more than the cost of the soft drink" translates to the equation

$y = x + 1$.

$x + y = 8.00$

$y = x + 1$

Since the second equation is solved for y, use the substitution method by substituting the value for y into the first equation and solving for x.

$x + x + 1 = 8.00$

$2x + 1 = 8.00$

$2x = 7.00$

$x = 3.50$

$y = x + 1$

$y = 3.50 + 1$

$y = 4.50$

A hot dog costs \$4.50 and a drink costs \$3.50.

53. $y < 3x - 1$

The related equation is in slope-intercept form with a slope of 3 and y-intercept of $(0, -1)$. Since the sign of inequality does not include an equal sign, the line will be dashed. Using $(-2, 0)$ as a test point, a false statement is obtained. Using $(2, 0)$ as a test point, a true statement is obtained. Shade the region containing $(2, 0)$.

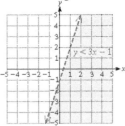

Answers may vary. Three possible solutions are: $(1, -1)$, $(0, -4)$, $(2, 0)$

55. $-2x - 3y \geq 8$

Rewriting the related equation in slope-intercept form gives $y = -\dfrac{2}{3}x - \dfrac{8}{3}$ with a slope of $-\dfrac{2}{3}$ and a y-intercept of $\left(0, -\dfrac{8}{3}\right)$.

Since the sign of inequality includes an equal sign, the line will be solid. Using $(0, 0)$ as a test point, a false statement is obtained. Using $(-6, 0)$ as a test point, a true statement is obtained. Shade the region containing $(-6, 0)$.

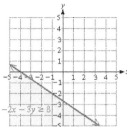

Three possible solutions are: $(-4, 0)$, $(-2, -2)$, and $(1, -4)$

57. $x - 5y \geq 0$

Rewriting the related equation in slope-intercept form gives $y = \frac{1}{5}x$ with a slope of $\frac{1}{5}$ and a y-intercept of $(0, 0)$. Since the sign of inequality includes an equal sign, the line will be solid. Using $(-2, 0)$ as a test point, a false statement is obtained. Using $(2, 0)$ as a test point, a true statement is obtained. Shade the region containing $(2, 0)$.

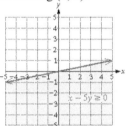

59. $x > 5$

The related equation, $x = 5$, is a vertical line through the point $(5, 0)$. Since the sign of inequality does not include an equal sign, the line will be dashed. Using $(0, 0)$ as a test point, a false statement is obtained. Using $(6, 0)$ as a test point, a true statement is obtained. Shade the region containing $(6, 0)$.

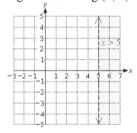

61. $y \geq 0$

The related equation, $y = 0$, is the x-axis. Since the sign of inequality does include an equal sign, the line will be solid. Using $(0, 1)$ as a test point, a true statement is obtained. Shade the region containing $(0, 1)$.

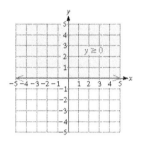

63. $2x - y \geq 8$

Using the intercepts method of graphing, graph the related equation $2x - y = 8$ with a solid line. Substituting the test point $(0, 0)$ into the inequality results in a false statement. Shade the region below the solid line.

$x + y \leq 3$

Using the intercepts method of graphing, graph the related equation $x + y = 3$ with a solid line. Substituting the test point $(0, 0)$ into the inequality results in a true statement. Shade the region below the solid line.

Putting the two regions on the same graph, the intersection is the solution to the system of inequalities.

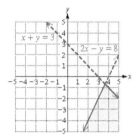

65. $y \leq 2x$

Using the slope-intercept method of graphing, graph the related equation $y \leq 2x$ with a solid line. Substituting the test point $(1, 0)$ into the inequality results in a true statement. Shade the region below the solid line.

$-2x - y > -3$

Using the intercepts method of graphing, graph the related equation $-2x - y > -3$ with a dashed line. Substituting the test point $(0, 0)$ into the inequality results in a true statement. Shade the region below the dashed line.

Putting the two regions on the same graph, the intersection is the solution to the system of inequalities.

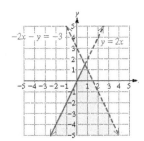

Chapter 4 Test

1. $5x + 2y = -6$

$-\dfrac{5}{2}x - y = -3$

Rewriting each equation in slope-intercept form gives

$y = -\dfrac{5}{2}x - 3$

$y = -\dfrac{5}{2}x + 3$

Since the slopes of the two lines are equal and their y-intercepts are unequal, the lines are parallel.

3. Graph each line using the slope-intercept method.
   $y = 2x - 4$

   $y = \dfrac{2}{3}x$

Since the slopes of the two lines are unequal, the lines intersect.

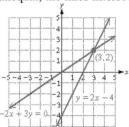

Solution: $(3, 2)$.

5. $x = 5y - 2$

$2x + y = -4$

Since the first equation is solved for x, substitute this value for x into the second equation and solve for y.

$$2x + y = -4$$
$$2(5y - 2) + y = -4$$
$$10y - 4 + y = -4$$
$$11y - 4 = -4$$
$$11y = 0$$
$$y = 0$$

$x = 5y - 2$
$x = 5(0) - 2$
$x = -2$
Solution: $(-2, 0)$

7. $\dfrac{1}{3}x + y = \dfrac{7}{3}$

$x = \dfrac{3}{2}y - 11$

Since the second equation is solved for x, use the substitution method by substituting this value for x into the first equation and solving for y.

$$\dfrac{1}{3}x + y = \dfrac{7}{3}$$
$$\dfrac{1}{3}\left(\dfrac{3}{2}y - 11\right) + y = \dfrac{7}{3}$$
$$\dfrac{1}{2}y - \dfrac{11}{3} + y = \dfrac{7}{3}$$

Clear the fractions by multiplying both sides of the equation by 6.

$$3y - 22 + 6y = 14$$
$$9y - 22 = 14$$
$$9y = 36$$
$$y = 4$$

$$x = \frac{3}{2}y - 11$$
$$x = \frac{3}{2}(4) - 11$$
$$x = 6 - 11$$
$$x = -5$$

Solution: $(-5, 4)$

9. $3x - 4y = 29$
$2x + 5y = -19$

Since the equations are in standard form and the signs on y are opposite, use the addition method. Multiply the second equation by 4 and the first equation by 5 to obtain opposite coefficients on y. Add these equations together and solve for x.

$$15x - 20y = 145$$
$$\underline{8x + 20y = -76}$$
$$23x = 69$$
$$x = 3$$
$$3x - 4y = 29$$
$$3(3) - 4y = 29$$
$$9 - 4y = 29$$
$$-4y = 20$$
$$y = -5$$

Solution: $(3, -5)$

11. $-0.25 - 0.05y = 0.2$
$10x + 2y = -8$

Clear the decimals from the first equation by multiplying the terms by 100.

$$-25x - 5y = 20$$
$$10x + 2y = -8$$

The equations are now in standard form. Use the addition method by multiplying the first equation by 2 and the second equation by 5 to obtain opposite coefficients on x. Add the equations and solve for y.

$$-50x - 10y = 40$$
$$\underline{50x + 10y = -40}$$
$$0 = 0 \qquad \text{Identity}$$

Infinitely many solutions

$\{(x, y)| \ 10x + 2y = -8\}$; dependent system

13. $5x - y \geq -6$

Using the intercept method of graphing, graph the related equation $5x - y = -6$ with a solid line. Using $(0, 0)$ as a test point, a true statement is obtained. Shade the region containing the point $(0, 0)$.

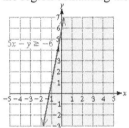

15. Let x represent the number of points scored by Sheryl Swoopes. Let y represent the number of points scored by Lauren Jackson. The statement "together they scored a total 1211 points" translates to the equation $x + y = 1211$. The statement "Swoopes ... scored 17 points more than ... Jackson" translates to the equation $x = y + 17$.

$$x + y = 1211$$
$$x = y + 17$$

Since the second equation is solved for x, substitute this value for x into the first equation and solve for y.

131

$$x + y = 1211$$
$$y + 17 + y = 1211$$
$$2y + 17 = 1211$$
$$2y = 1194$$
$$y = 597$$
$$x = y + 17$$
$$x = 597 + 17$$
$$x = 614$$

Swoopes scored 614 points and Jackson scored 597 points.

17. Let x represent the number of milliliters of 50% acid solution. Let y represent the number of milliliters of 20% acid solution.

	50% Solution	20% Solution	30% Solution
Amount of solution	x	y	36
Amount of acid	$0.5x$	$0.2y$	$0.3(36)$

$$x + y = 36$$
$$0.5x + 0.2y = 10.8$$

Since the equations are written in standard form, use the addition method. Multiply the first equation by -0.5 to obtain opposite coefficients on x.

$$-0.5x - 0.5y = -18$$
$$\underline{0.5x + 0.2y = 10.8}$$
$$-0.3y = -7.2$$
$$y = 24$$
$$x + y = 36$$
$$x + 24 = 36$$
$$x = 12$$

12 milliliters of the 50% acid solution should be mixed with 24 milliliters of the 20% acid solutions.

19. Let x represent the amount borrowed at 10%. Let y represent the amount borrowed at 8%.

	10% account	8% account	Total
Amount borrowed	x	y	$5000
Interest	$0.10x$	$0.08y$	$424

$$x + y = 5000$$
$$0.10x + 0.08y = 424$$

Multiply the second equation by -10 to obtain opposite coefficients on x.

$$x + y = 5000$$
$$\underline{-x - 0.8y = -4240}$$
$$0.2y = 760$$
$$y = 3800$$
$$x + y = 5000$$
$$x + 3800 = 5000$$
$$x = 1200$$

$1200 was borrowed at 10%, and $3800 was borrowed at 8%.

21. Let p represent the speed of the plane in still air. Let w represent the speed of the wind.

	Rate	Time	Distance
With the wind	$p + w$	2	1090
Against the wind	$p - w$	2	910

$$2(p + w) = 1090$$
$$2(p - w) = 910$$

Rewrite the equations in standard form.

$$2p + 2w = 1090$$
$$2p - 2w = 910$$

Since the coefficients on w are opposites, add the equations and solve for p.

$$2p + 2w = 1090$$
$$\underline{2p - 2w = 910}$$
$$4p = 2000$$
$$p = 500$$

$$2(p + w) = 1090$$
$$2p + 2w = 1090$$
$$2(500) + 2w = 1090$$
$$1000 + 2w = 1090$$
$$2w = 90$$
$$w = 45$$

The speed of the plane in still air is 500 mph and the speed of the wind is 45 mph.

23. Let x represent the number of mL of 10% acid. Let y represent the number of mL of 25% acid.

	10% Solution	25% Solution	16% Solution
Amount of solution	x	y	100
Amount of antifreeze	$0.10x$	$0.25y$	0.16 (100)

$$x + y = 100$$
$$0.10x + 0.25y = 16$$

Since the equations are written in standard form, use the addition method by multiplying the first equation by -0.10 to obtain opposite coefficients on x. Add the equations and solve for y.

$$-0.10x - 0.10y = -10$$
$$\underline{0.10x + 0.25y = 16}$$
$$0.15y = 6$$
$$y = 40$$

$$x + y = 100$$
$$x + 40 = 100$$
$$x = 60$$

60 mL of the 10% acid solution should be mixed with 40 mL of the 25% acid solution.

Cumulative Review Exercises, Chapters 1–4

1. $\dfrac{|2 - 5| + 10 \div 2 + 3}{\sqrt{10^2 - 8^2}} = \dfrac{|2 - 5| + 10 \div 2 + 3}{\sqrt{100 - 64}}$

$$= \frac{|-3| + 5 + 3}{\sqrt{36}}$$
$$= \frac{3 + 5 + 3}{6}$$
$$= \frac{11}{6}$$

3.

$$-4(a + 3) + 2 = -5(a + 1) + a$$
$$-4a - 12 + 2 = -5a - 5 + a$$
$$-4a - 10 = -4a - 5$$
$$-10 = -5 \qquad \text{Contradiction}$$

No solution

5. $z = \dfrac{x - m}{5}$

$$5z = x - m$$
$$x = 5z + m$$

7. Let x represent the measure of one of the remaining angles. Let y represent the measure of the other remaining angle. The sum of the angles in a triangle is 180. Thus the equation is $110 + x + y = 180$ or $x + y = 70$. The statement "one is 4° less than the other angle" translates to the equation $y = x - 4$.

$$x + y = 70$$
$$y = x - 4$$

Since the second equation is solved for y, use the substitution method by substituting this value into the first equation and solving for x.

$$x + y = 70$$
$$x + x - 4 = 70$$
$$2x - 4 = 70$$
$$2x = 74$$
$$x = 37$$

$y = x - 4$
$y = 37 - 4$
$y = 33$
The measures of the angles are 37°, 33°, and 110°.

9. 37% of 2,060,000 votes
$0.37(2,060,000) = 762,200$
Jesse Ventura received 762,200 votes.

11. Let x represent the measure of one angle. Let y represent the measure of the other angle. The statement "two angles are complementary" translates to the equation $x + y = 90$.
The statement "one angle measures 17° larger than the other angle" translates to the equation $x = y + 17$.
$x + y = 90$
$x = y + 17$
Since the second equation is solved for x, substitute this value for x in the first equation and solve for y.
$x + y = 90$
$y + 17 + y = 90$
$2y + 17 = 90$
$2y = 73$
$y = 36.5$
$x = y + 17$
$x = 36.5 + 17$
$x = 53.5$
The angles are 36.5° and 53.5°.

13. (a) The slope of a line parallel to the given line is the same as the given line: $-\dfrac{2}{3}$.

(b) The slope of a line perpendicular to the given line is the negative reciprocal of the given line: $\dfrac{3}{2}$

15. (a) & (b)

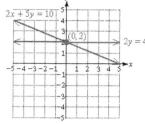

(c) Point of intersection: $(0, 2)$
$2(0) + 5(2) \overset{?}{=} 10$ ✓
$2(2) \overset{?}{=} 4$ ✓

17. (a) $2x + y = 3$ Use the intercepts method to graph. Substituting $x = 0$ into the equation gives $y = 3$. The y-intercept is $(0, 3)$. Substituting $y = 0$ into the equation gives $x = \dfrac{3}{2}$. The x-intercept is $\left(\dfrac{3}{2}, 0\right)$.

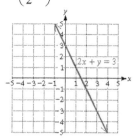

(b) Using the intercept method of graphing, graph the related equation $2x + y = 3$ with a dashed line. Using $(0, 0)$ as a test point, a true statement is obtained. Shade the region containing the point $(0, 0)$.

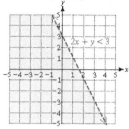

(c) Part (a) represents the solutions to an equation. Part (b) represents the solutions to a strict inequality.

134

19. Since $x°$ and $y°$ are complementary angles, we have x + y = 90. Since $x°$ and $(3y - 36)°$ are supplementary angles, we have

$$x° + (3\underline{y}° - 36°) = 180°$$
$$x + y = 90$$

$$x + (3y - 36) = 180$$

Solving the first equation for x gives $x = 90 - y$. Use the substitution method by substituting this value into the second equation to solve for y.

$$90 - y + 3y - 36 = 180$$
$$2y + 54 = 180$$
$$2y = 126$$
$$y = 63$$

$$x + y = 90$$
$$x + 63 = 90$$
$$x = 27$$

x is 27° and y is 63°.

Chapter 5

Chapter 5 Chapter Opener Puzzle

1. $2x - 9$

3. $10x + 3$

5. $2x + 24$

A polynomial that has two terms is called a <u>BINOMIAL.</u>

Section 5.1 Calculator Connection,

1. 1.06^5
       1.338225578

3. 5000(1.06)^5
       6691.127888

5. 3000(1+0.06)²
       3370.8

Section 5.1 Practice Exercises

1. **(a)** exponent

   (b) base; exponent

   (c) 1

   (d) $I = Prt$

   (e) compound interest

3. Base: x; exponent: 4

5. Base: 3; exponent: 5

7. Base: -1; exponent: 4

9. Base: 13; exponent: 1

11. Base: 10; exponent: 3

13. Base: t; exponent: 6

15. v

17. 1

19. $(-6b)(-6b) = (-6b)^2$

21. $-6 \cdot b \cdot b = -6b^2$

23. $(y+2)(y+2)(y+2)(y+2) = (y+2)^4$

25. $\dfrac{-2}{t \cdot t \cdot t} = \dfrac{-2}{t^3}$

27. No; $-5^2 = -25$ and $(-5)^2 = 25$

29. Yes; $-2^5 = -32$ and $(-2)^5 = -32$

31. Yes; $\left(\dfrac{1}{2}\right)^3 = \dfrac{1}{8}$ and $\dfrac{1}{2^3} = \dfrac{1}{8}$

33. Yes; $-(-2)^4 = -16$ and $-(2)^4 = -16$

35. $16^1 = 16$

37. $(-1)^{21} = (-1)(-1)^{20} = (-1)(1) = -1$

39. $\left(-\dfrac{1}{3}\right)^2 = \left(-\dfrac{1}{3}\right)\left(-\dfrac{1}{3}\right) = \dfrac{1}{9}$

41. $-\left(\dfrac{2}{5}\right)^2 = -\left(\dfrac{2}{5}\right)\left(\dfrac{2}{5}\right) = -\dfrac{4}{25}$

43. $3 \cdot 2^4 = 3 \cdot 16 = 48$

45. $-4(-1)^7 = -4(-1) = 4$

47. $6^2 - 3^3 = 36 - 27 = 9$

49. $2 \cdot 3^2 + 4 \cdot 2^3 = 2 \cdot 9 + 4 \cdot 8 = 18 + 32 = 50$

51. $-4b^2 = -4(5)^2 = -4(25) = -100$

53. $(-4b)^2 = (-4 \cdot 5)^2 = (-20)^2 = 400$

55. $(a+b)^2 = [-4+5]^2 = (1)^2 = 1$

57. $a^2 + 2ab + b^2 = (-4)^2 + 2(-4)(5) + 5^2$
$= 16 + (-40) + 25$
$= 1$

59. $-10ab^2 = -10(-4)(5^2)$
$= -10(-4)(25)$
$= 1000$

61. $-10a^2b = -10(-4)^2(5) = -10(16)(5) = -800$

63. (a) $x^4 \cdot x^3 = (x \cdot x \cdot x \cdot x)(x \cdot x \cdot x) = x^7$

   (b) $5^4 \cdot 5^3 = (5 \cdot 5 \cdot 5 \cdot 5)(5 \cdot 5 \cdot 5) = 5^7$

65. $z^5 z^3 = z^{5+3} = z^8$

67. $a \cdot a^8 = a^{1+8} = a^9$

69. $4^5 \cdot 4^9 = 4^{5+9} = 4^{14}$

71. $\left(\dfrac{2}{3}\right)^3 \left(\dfrac{2}{3}\right) = \left(\dfrac{2}{3}\right)^4$

73. $c^5 c^2 c^7 = c^{5+2+7} = c^{14}$

75. $x \cdot x^4 \cdot x^{10} \cdot x^3 = x^{1+4+10+3} = x^{18}$

77. (a) $\dfrac{p^8}{p^3} = \dfrac{p \cdot p \cdot p \cdot p \cdot p \cdot \cancel{p} \cdot \cancel{p} \cdot \cancel{p}}{\cancel{p} \cdot \cancel{p} \cdot \cancel{p}} = p^5$

   (b) $\dfrac{8^8}{8^3} = \dfrac{8 \cdot 8 \cdot 8 \cdot 8 \cdot 8 \cdot \cancel{8} \cdot \cancel{8} \cdot \cancel{8}}{\cancel{8} \cdot \cancel{8} \cdot \cancel{8}} = 8^5$

79. $\dfrac{x^8}{x^6} = x^{8-6} = x^2$

81. $\dfrac{a^{10}}{a} = a^{10-1} = a^9$

83. $\dfrac{7^{13}}{7^6} = 7^{13-6} = 7^7$

85. $\dfrac{5^8}{5} = 5^{8-1} = 5^7$

87. $\dfrac{y^{13}}{y^{12}} = y^{13-12} = y$

89. $\dfrac{h^3 h^8}{h^7} = \dfrac{h^{3+8}}{h^7} = h^{11-7} = h^4$

91. $\dfrac{7^2 \cdot 7^6}{7} = \dfrac{7^{2+6}}{7} = 7^{8-1} = 7^7$

93. $\dfrac{10^{20}}{10^3 \cdot 10^8} = \dfrac{10^{20}}{10^{3+8}} = 10^{20-11} = 10^9$

95. $(2x^3)(3x^4) = 2 \cdot 3 \cdot x^3 x^4 = 6x^{3+4} = 6x^7$

97. $(5a^2b)(8a^3b^4) = 8 \cdot 5 \cdot a^2 a^3 b b^4 = 40a^5b^5$

99. $s^3 \cdot t^5 \cdot t \cdot t^{10} \cdot s^6 = s^3 \cdot s^6 \cdot t^5 \cdot t \cdot t^{10} = s^9 t^{16}$

101. $(-2v^2)(3v)(5v^5) = -2 \cdot 3 \cdot 5v^2 v v^5 = -30v^8$

103. $\left(\dfrac{2}{3}m^{13}n^8\right)(24m^7n^2) = \dfrac{2}{3} \cdot 24m^{13}m^7 n^8 n^2$
$= 16m^{20}n^{10}$

105. $\dfrac{14c^4d^5}{7c^3d} = 2c^{4-3}d^{5-1} = 2cd^4$

107. $\dfrac{z^3 z^{11}}{z^4 z^6} = \dfrac{z^{3+11}}{z^{4+6}} = \dfrac{z^{14}}{z^{10}} = z^{14-10} = z^4$

109. $\dfrac{25h^3 jk^5}{12h^2k} = \dfrac{25}{12} \cdot \dfrac{h^3}{h^2} \cdot j \cdot \dfrac{k^5}{k}$
$= \dfrac{25h^{3-2}jk^{5-1}}{12} = \dfrac{25hjk^4}{12}$

111. $(-4p^6q^8r^4)(2pqr^2) = -4 \cdot 2p^6 pq^8 qr^4 r^2$
$= -8p^{6+1}q^{8+1}r^{4+2}$
$= -8p^7q^9r^6$

113. $\dfrac{-12s^2tu^3}{4su^2} = \dfrac{-12}{4} \cdot \dfrac{s^2}{s} \cdot t \cdot \dfrac{u^3}{u^2}$
$= -3s^{2-1}tu^{3-2} = -3stu$

115. Substitute $P = 5000$, $t = 2$, and $r = 7\% = 0.07$ into the formula.
$A = P(1+r)^t$
$= \$5000(1+0.07)^2$
$= 5000(1.07)^2$
$= \$5724.50$

117. Substitute $P = 4000$, $t = 3$, and $r = 6\% = 0.06$ into the formula.

$$A = P(1 + r)^t$$
$$= \$4000(1 + 0.06)^3$$
$$= 4000(1.06)^3$$
$$= \$4764.06$$

119. Substitute $r = 16$ and $\pi = 3.14$ into the formula.

$$A = \pi r^2$$
$$= (3.14)(8 \text{ in.})^2$$
$$= 3.14(64)$$
$$= 200.96$$
$$\approx 201 \text{ in.}^2 \qquad \text{Round to the nearest in}^2$$

121. Since the radius of a sphere is half of its diameter, $r = \dfrac{d}{2} = \dfrac{8}{2} = 4$ in. Substitute $r = 4$ and $\pi = 3.14$ into the formula.

$$V = \frac{4}{3}\pi r^3$$
$$= \frac{4}{3}(3.14)(4)^3$$
$$= 267.9466667$$
$$\approx 268 \text{ in.}^3 \qquad \text{Round to the nearest in.}^3$$

123. $x^n x^{n+1} = x^{n+n+1} = x^{2n+1}$

125. $p^{3m+5} p^{-m-2} = p^{3m+5-m-2} = p^{2m+3}$

127. $\dfrac{z^{b+1}}{z^b} = z^{b+1-b} = z$

129. $\dfrac{r^{3a+3}}{r^{3a}} = r^{3a+3-3a} = r^3$

Section 5.2 Practice Exercises

1. $4^2 \cdot 4^7 = 4^{2+7} = 4^9$

3. $a^{13} \cdot a \cdot a^6 = a^{13+1+6} = a^{20}$

5. $\dfrac{d^{13}d}{d^5} = d^{14-5} = d^9$

7. $\dfrac{7^{11}}{7^5} = 7^{11-5} = 7^6$

9. When multiplying expressions with the same base, add the exponents. When dividing expression with the same base, subtract the exponents.

11. $(5^3)^4 = 5^{3 \cdot 4} = 5^{12}$

13. $(12^3)^2 = 12^{3 \cdot 2} = 12^6$

15. $(y^7)^2 = y^{7 \cdot 2} = y^{14}$

17. $(w^5)^5 = w^{5 \cdot 5} = w^{25}$

19. $(a^2 a^4)^6 = (a^6)^6 = a^{6 \cdot 6} = a^{36}$

21. $(y^3 y^4)^2 = (y^7)^2 = y^{7 \cdot 2} = y^{14}$

23. $(2^2)^3 = 2^6$; $(2^3)^2 = 2^6$, they are both equal to 2^6.

25. $2^{(2^4)} = 2^{16}$ and $(2^2)^4 = 2^8$; the expression $2^{(2^4)}$ is greater than $(2^2)^4$.

27. $(5w)^2 = 5^2 w^2 = 25w^2$

29. $(srt)^4 = s^4 r^4 t^4$

31. $\left(\dfrac{2}{r}\right)^4 = \dfrac{2^4}{r^4} = \dfrac{16}{r^4}$

33. $\left(\dfrac{x}{y}\right)^5 = \dfrac{x^5}{y^5}$

35. $(-3a)^4 = (-3)^4 a^4 = 81a^4$

37. $(-3abc)^3 = (-3)^3 a^3 b^3 c^3 = -27a^3 b^3 c^3$

39. $\left(-\dfrac{4}{x}\right)^3 = (-1)^3 \dfrac{4^3}{x^3} = -\dfrac{64}{x^3}$

41. $\left(-\dfrac{a}{b}\right)^2 = (-1)^2 \dfrac{a^2}{b^2} = \dfrac{a^2}{b^2}$

43. $(6u^2v^4)^3 = 6^3(u^2)^3(v^4)^3$
$= 6^3u^6v^{12}$ or $216u^6v^{12}$

45. $5(x^2y)^4 = 5(x^2)^4(y)^4 = 5x^8y^4$

47. $(-h^4)^7 = (-1)^7h^{28} = -h^{28}$

49. $(-m^2)^6 = (-1)^6m^{12} = m^{12}$

51. $\left(\dfrac{4}{rs^4}\right)^5 = \dfrac{4^5}{(rs^4)^5}$
$= \dfrac{4^5}{(r)^5(s^4)^5}$
$= \dfrac{4^5}{r^5s^{20}}$ or $\dfrac{1024}{r^5s^{20}}$

53. $\left(\dfrac{3p}{q^3}\right)^5 = \dfrac{(3p)^5}{(q^3)^5}$
$= \dfrac{3^5p^5}{q^{3\cdot5}}$
$= \dfrac{3^5p^5}{q^{15}}$ or $\dfrac{243p^5}{q^{15}}$

55. $\dfrac{y^8(y^3)^4}{(y^2)^3} = \dfrac{y^8y^{12}}{y^6} = \dfrac{y^{20}}{y^6} = y^{14}$

57. $(x^2)^5(x^3)^7 = x^{10}x^{21} = x^{31}$

59. $(2a^2b)^3(5a^4b^3)^2$
$= 2^3(a^2)^3b^3 \cdot 5^2(a^4)^2(b^3)^2$
$= 8a^6b^3 \cdot 25a^8b^6$
$= 8 \cdot 25a^{6+8}b^{3+6}$
$= 200a^{14}b^9$

61. $(-2p^2q^4)^4 = (-2)^4(p^2)^4(q^4)^4 = 16p^8q^{16}$

63. $(-m^7n^3)^5 = (-1)^5(m^7)^5(n^3)^5 = -m^{35}n^{15}$

65. $\dfrac{(5a^3b)^4(a^2b)^4}{(5ab)^2} = \dfrac{5^4a^{12}b^4a^8b^4}{5^2a^2b^2}$
$= \dfrac{5^4a^{20}b^8}{5^2a^2b^2}$
$= 5^2a^{18}b^6$
$= 25a^{18}b^6$

67. $\left(\dfrac{2c^3d^4}{3c^2d}\right)^2 = \dfrac{2^2c^6d^8}{3^2c^4d^2} = \dfrac{4c^2d^6}{9}$

69. $(2c^3d^2)^5\left(\dfrac{c^6d^8}{4c^2d}\right)^3 = (2^5c^{15}d^{10})\left(\dfrac{c^{18}d^{24}}{64c^6d^3}\right)$
$= \dfrac{32c^{15}d^{10} \cdot c^{18}d^{24}}{64c^6d^3}$
$= \dfrac{32c^{33}d^{34}}{64c^6d^3}$
$= \dfrac{c^{27}d^{31}}{2}$

71. $\left(\dfrac{-3a^3b}{c^2}\right)^3 = \dfrac{(-3)^3a^9b^3}{c^6} = -\dfrac{27a^9b^3}{c^6}$

73. $\dfrac{(-8b^6)^2(b^3)^5}{4b} = \dfrac{(-8)^2b^{12}b^{15}}{4b}$
$= \dfrac{64b^{27}}{4b}$
$= 16b^{27-1}$
$= 16b^{26}$

75. $(x^m)^2 = x^{2m}$

77. $(5a^{2n})^3 = 5^3(a^{2n})^3 = 125a^{6n}$

79. $\left(\dfrac{m^2}{n^3}\right)^b = \dfrac{(m^2)^b}{(n^3)^b} = \dfrac{m^{2b}}{n^{3b}}$

81. $\left(\dfrac{3a^3}{5b^4}\right)^n = \dfrac{3^n(a^3)^n}{5^n(b^4)^n} = \dfrac{3^na^{3n}}{5^nb^{4n}}$

Section 5.3 Practice Exercises

1. (a) 1

(b) $\left(\dfrac{1}{b}\right)^n$ or $\dfrac{1}{b^n}$

3. $c^7c^2 = c^{7+2} = c^9$

5. $\dfrac{y^9}{y^8} = y^{9-8} = y^1 = y$

7. $\dfrac{3^{14}}{3^3 \cdot 3^5} = \dfrac{3^{14}}{3^8} = 3^6$ or 729

9. $(7w^7 z^2)^4 = 7^4 (w^7)^4 (z^2)^4$
$ = 7^4 w^{28} z^8$ or $2401w^{28} z^8$

11. (a) $d^0 = 1$

    (b) $\dfrac{d^3}{d^3} = d^{3-3} = d^0 = 1$

13. $p^0 = 1$

15. $5^0 = 1$

17. $-4^0 = -(4^0) = -1$

19. $(-6)^0 = 1$

21. $(8x)^0 = 8^0 x^0 = 1$

23. $-7x^0 = -7 \cdot 1 = -7$

25. (a) $t^{-5} = \dfrac{1}{t^5}$

    (b) $\dfrac{t^3}{t^8} = t^{3-8} = t^{-5} = \dfrac{1}{t^5}$

27. $\left(\dfrac{2}{7}\right)^{-3} = \left(\dfrac{7}{2}\right)^3 = \dfrac{7^3}{2^3} = \dfrac{343}{8}$

29. $\left(-\dfrac{1}{5}\right)^{-2} = (-5)^2 = 25$

31. $a^{-3} = \dfrac{1}{a^3}$

33. $12^{-1} = \dfrac{1}{12}$

35. $(4b)^{-2} = \dfrac{1}{(4b)^2} = \dfrac{1}{16b^2}$

37. $6x^{-2} = 6 \cdot \dfrac{1}{x^2} = \dfrac{6}{x^2}$

39. $(-8)^{-2} = \dfrac{1}{(-8)^2} = \dfrac{1}{64}$

41. $-3y^{-4} = -3 \cdot \dfrac{1}{y^4} = \dfrac{-3}{y^4}$

43. $(-t)^{-3} = \dfrac{1}{(-t)^3} = -\dfrac{1}{t^3}$

45. $\dfrac{1}{a^{-5}} = a^5$

47. $\dfrac{x^4}{x^{-6}} = x^{4-(-6)} = x^{10}$

49. $2a^{-3} = 2 \cdot \dfrac{1}{a^3} = \dfrac{2}{a^3}$

51. $x^{-8} x^4 = x^{-8+4} = x^{-4} = \dfrac{1}{x^4}$

53. $a^{-8} a^8 = a^{-8+8} = a^0 = 1$

55. $y^{17} y^{-13} = y^{17+(-13)} = y^4$

57. $(m^{-6} n^9)^3 = (m^{-6})^3 (n^9)^3 = m^{-18} n^{27} = \dfrac{n^{27}}{m^{18}}$

59. $(-3j^{-5} k^6)^4 = (-3)^4 (j^{-5})^4 (k^6)^4$
$\phantom{(-3j^{-5} k^6)^4} = 81 j^{-20} k^{24}$
$\phantom{(-3j^{-5} k^6)^4} = \dfrac{81 k^{24}}{j^{20}}$

61. $\dfrac{p^3}{p^9} = p^{3-9} = p^{-6} = \dfrac{1}{p^6}$

63. $\dfrac{r^{-5}}{r^{-2}} = r^{-5-(-2)} = r^{-3} = \dfrac{1}{r^3}$

65. $\dfrac{a^2}{a^{-6}} = a^{2-(-6)} = a^8$

67. $\dfrac{y^{-2}}{y^6} = y^{-2-6} = y^{-8} = \dfrac{1}{y^8}$

69. $\dfrac{7^3}{7^2 \cdot 7^8} = \dfrac{7^3}{7^{10}} = 7^{3-10} = 7^{-7} = \dfrac{1}{7^7}$

71. $\dfrac{a^2 a}{a^3} = \dfrac{a^{2+1}}{a^3} = \dfrac{a^3}{a^3} = 1$

73. $\dfrac{a^{-1}b^2}{a^3 b^8} = a^{-1-3}b^{2-8} = a^{-4}b^{-6} = \dfrac{1}{a^4 b^6}$

75. $\dfrac{w^{-8}(w^2)^{-5}}{w^3} = \dfrac{w^{-8}w^{-10}}{w^3}$

$= \dfrac{w^{-18}}{w^3}$

$= w^{-21}$

$= \dfrac{1}{w^{21}}$

77. $\dfrac{3^{-2}}{3} = 3^{-2-1} = 3^{-3} = \dfrac{1}{3^3} = \dfrac{1}{27}$

79. $\left(\dfrac{p^{-1}q^5}{p^{-6}}\right)^0 = 1$

81. $(8x^3 y^0)^{-2} = 8^{-2}x^{-6}y^0 = \dfrac{1}{8^2}\cdot\dfrac{1}{x^6}\cdot 1 = \dfrac{1}{64x^6}$

83. $(-8y^{-12})(2y^{16}z^{-2}) = (-8)(2)y^{-12+16}z^{-2}$

$= -16y^4 z^{-2}$

$= \dfrac{-16y^4}{z^2}$

85. $\dfrac{-18a^{10}b^6}{108a^{-2}b^6} = \dfrac{-18}{108}a^{10-(-2)}b^{6-6}$

$= -\dfrac{1}{6}a^{12}b^0$

$= -\dfrac{a^{12}}{6}$

87. $\dfrac{(-4c^{12}d^7)^2}{(5c^{-3}d^{10})^{-1}} = \dfrac{(-4)^2(c^{12})^2(d^7)^2}{(5)^{-1}(c^{-3})^{-1}(d^{10})^{-1}}$

$= \dfrac{16c^{24}d^{14}}{5^{-1}c^3 d^{-10}}$

$= 16(5)c^{24-3}d^{14-(-10)}$

$= 80c^{21}d^{24}$

89. $\dfrac{(2x^3 y^2)^{-3}}{(3x^2 y^4)^{-2}} = \dfrac{(2)^{-3}(x^3)^{-3}(y^2)^{-3}}{(3)^{-2}(x^2)^{-2}(y^4)^{-2}}$

$= \dfrac{2^{-3}x^{-9}y^{-6}}{3^{-2}x^{-4}y^{-8}}$

$= \dfrac{3^2 x^{-9}y^{-6}}{2^3 x^{-4}y^{-8}}$

$= \dfrac{9}{8}x^{-9-(-4)}y^{-6-(-8)}$

$= \dfrac{9}{8}x^{-5}y^2$

$= \dfrac{9y^2}{8x^5}$

91. $\left(\dfrac{5cd^{-3}}{10d^5}\right)^{-2} = \left(\dfrac{1}{2}cd^{-3-5}\right)^{-2}$

$= \left(\dfrac{1}{2}\right)^{-2}c^{-2}(d^{-8})^{-2}$

$= 2^2 \cdot \dfrac{1}{c^2}d^{16}$

$= \dfrac{4d^{16}}{c^2}$

93. $(2xy^3)\left(\dfrac{9xy}{4x^3 y^2}\right) = \dfrac{18x^2 y^4}{4x^3 y^2}$

$= \dfrac{18}{4}x^{2-3}y^{4-2}$

$= \dfrac{9}{2}x^{-1}y^2$

$= \dfrac{9y^2}{2x}$

95. $5^{-1} + 2^{-2} = \dfrac{1}{5} + \dfrac{1}{2^2} = \dfrac{1}{5} + \dfrac{1}{4} = \dfrac{4}{20} + \dfrac{5}{20} = \dfrac{9}{20}$

97. $10^0 - 10^{-1} = 1 - \dfrac{1}{10} = \dfrac{10}{10} - \dfrac{1}{10} = \dfrac{9}{10}$

99. $2^{-2} + 1^{-2} = \dfrac{1}{2^2} + \dfrac{1}{1^2} = \dfrac{1}{4} + 1 = \dfrac{5}{4}$

101. $4 \cdot 5^0 - 2 \cdot 3^{-1} = 4 \cdot 1 - 2 \cdot \dfrac{1}{3}$

$$= 4 - \dfrac{2}{3}$$

$$= \dfrac{12}{3} - \dfrac{2}{3}$$

$$= \dfrac{10}{3}$$

103. $\dfrac{y^4 y^{\square}}{y^{-2}} = y^8$

$$\dfrac{y^{4+\square}}{y^{-2}} = y^8$$

$$y^{4+\square-(-2)} = y^8$$

$$y^{6+\square} = y^8$$

$$6 + \square = 8$$

$$\square = 2$$

105. $\dfrac{w^{-9}}{w^{\square}} = w^2$

$$w^{-9-\square} = w^2$$

$$-9 - \square = 2$$

$$-\square = 11$$

$$\square = -11$$

Chapter 5 Problem Recognition Exercises

1. (a) $t^3 t^5 = t^{3+5} = t^8$

(b) $\dfrac{t^3}{t^5} = t^{3-5} = t^{-2} = \dfrac{1}{t^2}$

(c) $(t^3)^5 = t^{3 \cdot 5} = t^{15}$

3. (a) $\dfrac{y^7}{y^2} = y^{7-2} = y^5$

(b) $(y^7)^2 = y^{7 \cdot 2} = y^{14}$

(c) $y^7 y^2 = y^{7+2} = y^9$

5. (a) $(w^4)^{-2} = w^{4 \cdot (-2)} = w^{-8} = \dfrac{1}{w^8}$

(b) $w^4 w^{-2} = w^{4-2} = w^2$

(c) $\dfrac{w^4}{w^{-2}} = w^{4-(-2)} = w^6$

7. (a) $\left(r^2 s^4\right)^2 = (r^2)^2 (s^4)^2 = r^{2 \cdot 2} s^{4 \cdot 2} = r^4 s^8$

(b) $\left(\dfrac{r^2}{s^4}\right)^2 = \dfrac{(r^2)^2}{(s^4)^2} = \dfrac{r^{2 \cdot 2}}{s^{4 \cdot 2}} = \dfrac{r^4}{s^8}$

(c) $\dfrac{(r^2)^2}{s^4} = \dfrac{r^{2 \cdot 2}}{s^4} = \dfrac{r^4}{s^4}$

9. (a) $\dfrac{x^{-7} x^4}{x^{-3}} = \dfrac{x^{-7+4}}{x^{-3}} = \dfrac{x^{-3}}{x^{-3}} = 1$

(b) $\dfrac{x^{-7} y^4}{y^{-3}} = x^{-7} y^{4-(-3)} = \dfrac{y^7}{x^7}$

(c) $\dfrac{x^4 y^{-7}}{z^{-3}} = \dfrac{x^4 z^3}{y^7}$

11. (a) $\left(2^5 b^{-3}\right)^{-3} = \dfrac{1}{\left(2^5 b^{-3}\right)^3} = \dfrac{1}{2^{5 \cdot 3} b^{-3 \cdot 3}}$

$$= \dfrac{1}{2^{15} b^{-9}} = \dfrac{b^9}{2^{15}}$$

(b) $\left(\dfrac{2^5}{b^{-3}}\right)^{-3} = \dfrac{\left(2^5\right)^{-3}}{\left(b^{-3}\right)^{-3}} = \dfrac{2^{5 \cdot (-3)}}{b^{-3 \cdot (-3)}}$

$$= \dfrac{2^{-15}}{b^9} = \dfrac{1}{2^{15} b^9}$$

(c) $\left(\dfrac{2b^2}{2^5 b^{-3}}\right)^{-3} = \left(2^{1-5} b^{2-(-3)}\right)^{-3}$

$$= \left(2^{-4} b^5\right)^{-3} = 2^{-4 \cdot (-3)} b^{5 \cdot (-3)}$$

$$= 2^{12} b^{-15} = \dfrac{2^{12}}{b^{15}}$$

13. (a) $\dfrac{\left(t^{-2}\right)^3}{t^{-4}} = \dfrac{t^{-2\cdot(3)}}{t^{-4}} = \dfrac{t^{-6}}{t^{-4}} = t^{-6-(-4)}$

$= t^{-2} = \dfrac{1}{t^2}$

(b) $\dfrac{t^{-2}t^3}{t^{-4}} = \dfrac{t^{-2+3}}{t^{-4}} = \dfrac{t^1}{t^{-4}} = t^{1-(-4)} = t^5$

(c) $\left(\dfrac{t^{-2}t}{t^{-4}}\right)^3 = \left(\dfrac{t^{-2+1}}{t^{-4}}\right)^3 = \left(\dfrac{t^{-1}}{t^{-4}}\right)^3$

$= \left(t^{-1-(-4)}\right)^3 = \left(t^3\right)^3 = t^9$

15. (a) $\left(5h^{-2}k^0\right)^3 \left(5k^{-2}\right)^{-4}$

$= 5^3 h^{-2\cdot3} k^{0\cdot3} \cdot 5^{-4} k^{-2\cdot(-4)}$

$= 5^3 h^{-6} k^0 \cdot 5^{-4} k^8$

$= 5^{3+(-4)} h^{-6} k^{0+8}$

$= 5^{-1} h^{-6} k^8 = \dfrac{k^8}{5h^6}$

(b) $\left(\dfrac{5h^{-2}k^0}{5hk^2}\right)^3 = \left(5^{1-1} h^{-2-1} k^{0-2}\right)^3$

$= \left(5^0 h^{-3} k^{-2}\right)^3 = h^{-3\cdot3} k^{-2\cdot3}$

$= h^{-9} k^{-6} = \dfrac{1}{h^9 k^6}$

(c) $\dfrac{\left(5h^{-2}k^0\right)^3}{\left(5k^{-2}\right)^{-4}} = \dfrac{5^3 h^{-6} k^0}{5^{-4} k^8}$

$= \dfrac{5^7 h^{-6}}{k^8} = \dfrac{5^7}{h^6 k^8}$

Calculator Exercises

1. `(5.2E6)*(4.6E-3)`
   `                23920`

3. `(4.76E-5)/(2.38E`
   `9)`
   `              2E-14`

5. `(9.6E7)*(4.0E-3)`
   `/(2.0E-2)`
   `         19200000`

Section 5.4 Practice Exercises

1. scientific notation

3. $b^5 b^8 = b^{5+8} = b^{13}$

5. $10^5 \cdot 10^8 = 10^{5+8} = 10^{13}$

7. $\dfrac{y^2}{y^7} = y^{2-7} = y^{-5} = \dfrac{1}{y^5}$

9. $(x^5 y^{-3})^4 = (x^5)^4 (y^{-3})^4 = x^{20} y^{-12} = \dfrac{x^{20}}{y^{12}}$

11. $\dfrac{w^{-2} w^5}{w^{-1}} = \dfrac{w^3}{w^{-1}} = w^{3-(-1)} = w^4$

13. $\dfrac{10^{-2} \cdot 10^5}{10^{-1}} = \dfrac{10^3}{10^{-1}} = 10^{3-(-1)} = 10^4$

15. Move the decimal point between 2 and 3 and multiply by 10^{-10}; 2.3×10^{-10}

17. Move the decimal point between 5 and 0 and multiply by 10^4; $50,000 = 5 \times 10^4$

19. Move the decimal point between 2 and 0 and multiply by 10^5; $208,000 = 2.08 \times 10^5$

21. Move the decimal point between 6 and 0 and multiply by 10^6; $6,010,000 = 6.01 \times 10^6$

23. Move the decimal point between 8 and 0 and multiply by 10^{-6}; $0.000008 = 8 \times 10^{-6}$

25. Move the decimal point between 1 and 2 and multiply by 10^{-4}; $0.000125 = 1.25 \times 10^{-4}$

27. Move the decimal point between 6 and 7 and multiply by 10^{-3}; $0.006708 = 6.708 \times 10^{-3}$

29. Move the decimal point between 1 and 7 and multiply by 10^{-24}; 1.7×10^{-24} g

31. Move the decimal point between 2 and 7 and multiply by 10^{10}; $\$2.7 \times 10^{10}$

33. Move the decimal point between 6 and 8 and multiply by 10^{7}; 6.8×10^{7} gal
Move the decimal point between 1 and 0 and multiply by 10^{2}; 1.0×10^{2} mi

35. Move the decimal point nine places to the left; 0.0000000031

37. Move the decimal point five places to the left; 0.00005

39. Move the decimal point three places to the right; 2800

41. Move the decimal point four places to the left; 0.000603

43. Move the decimal point six places to the right; 2,400,000

45. Move the decimal point two places to the left; 0.019

47. Move the decimal point three places to the right; 7032

49. Move the decimal point twelve places to the left; 0.000000000001 g

51. Move the decimal point three places to the right; 1600 Cal; move the decimal point three places to the right; 2800 Cal

53. $(2.5 \times 10^{6})(2 \times 10^{-2}) = 2.5(2) \times 10^{6} \cdot 10^{-2}$
$\qquad = 5 \times 10^{4}$

55. $(1.2 \times 10^{4})(3 \times 10^{7}) = 1.2(3) \times 10^{4} \cdot 10^{7}$
$\qquad = 3.6 \times 10^{11}$

57. $\dfrac{7.7 \times 10^{6}}{3.5 \times 10^{2}} = \dfrac{7.7}{3.5} \times 10^{6-2} = 2.2 \times 10^{4}$

59. $\dfrac{9 \times 10^{-6}}{4 \times 10^{7}} = \dfrac{9}{4} \times 10^{-6-7} = 2.25 \times 10^{-13}$

61. $(8 \times 10^{10})(4 \times 10^{3}) = 8(4) \times 10^{10} \cdot 10^{3}$
$\qquad = 32 \times 10^{13}$
$\qquad = 3.2 \times 10^{1} \times 10^{3}$
$\qquad = 3.2 \times 10^{14}$

63. $(3.2 \times 10^{-4})(7.6 \times 10^{-7})$
$\qquad = 3.2(7.6) \times 10^{-4} \cdot 10^{-7}$
$\qquad = 24.32 \times 10^{-11}$
$\qquad = 2.432 \times 10^{1} \times 10^{-11}$
$\qquad = 2.432 \times 10^{-10}$

65. $\dfrac{2.1 \times 10^{11}}{7 \times 10^{-3}} = \dfrac{2.1}{7} \times 10^{11-(-3)}$
$\qquad = 0.3 \times 10^{14}$
$\qquad = 3 \times 10^{-1} \times 10^{3}$
$\qquad = 3 \times 10^{13}$

67. $\dfrac{5.7 \times 10^{-2}}{9.5 \times 10^{-8}} = \dfrac{5.7}{9.5} \times 10^{-2-(-8)}$
$\qquad = 0.6 \times 10^{6}$
$\qquad = 6 \times 10^{-1} \times 10^{6}$
$\qquad = 6 \times 10^{5}$

69. $6,000,000,000 \times 0.0000000023$
$\qquad = (6 \times 10^{9})(2.3 \times 10^{-9})$
$\qquad = 13.8 \times 10^{9+(-9)}$
$\qquad = 1.38 \times 10^{1} \times 10^{0}$
$\qquad = 1.38 \times 10^{1}$

71. $\dfrac{0.0000000003}{6000} = \dfrac{3 \times 10^{-10}}{6 \times 10^{3}}$
$\qquad = 0.5 \times 10^{-10-3}$
$\qquad = 5 \times 10^{-1} \times 10^{-13}$
$\qquad = 5 \times 10^{-14}$

73. (thickness of paper)(no. of pieces)

$= (3 \times 10^{-3})(1.25 \times 10^{3})$

$= 3(1.25) \times 10^{-3+3}$

$= 3.75 \times 10^{0}$

$= 3.75$ in.

75. (number of shares)(price per share)

$= (1,100,000,000)(27)$

$= (1.1 \times 10^{9})(27)$

$= 29.7 \times 10^{9}$

$= 2.97 \times 10^{1} \times 10^{9}$

$= \$2.97 \times 10^{10}$ or $\$29,700,000,000$

77. (a) 65 million $= 65,000,000 = 6.5 \times 10^{7}$

(b) 6.5×10^{7} years

$= (6.5 \times 10^{7}$ years$)(365$ days$)$

$= (6.5 \times 10^{7}$ years$)(3.65 \times 10^{2}$ days$)$

$= 23.725 \times 10^{9}$

$= 2.3725 \times 10^{1} \times 10^{9}$

$= 2.3725 \times 10^{10}$ days

(c) 6.5×10^{7} years

$= (2.3725 \times 10^{10}$ days$)(24$ hours$)$

$= (2.3725 \times 10^{10}$ days$)(2.4 \times 10^{1}$ hours$)$

$= 5.694 \times 10^{11}$ hours

(d)

6.5×10^{7} years

$= (5.694 \times 10^{11}$ hours$)(3600$ seconds$)$

$= (5.694 \times 10^{11}$ hours$)(3.6 \times 10^{3}$ seconds$)$

$= 20.4984 \times 10^{14}$

$= 2.04984 \times 10^{1} \times 10^{14}$

$= 2.04984 \times 10^{15}$ seconds

Section 5.5 Practice Exercises

1. (a) polynomial

(b) coefficient; degree

(c) 1

(d) one

(e) binomial

(f) trinomial

(g) leading; leading coefficient

(h) greatest

(i) zero

3. $(3x)^{2}(5x^{-4}) = 9 \cdot 5 \cdot x^{2} \cdot x^{-4} = 45x^{-2} = \dfrac{45}{x^{2}}$

5. $\dfrac{8t^{-6}}{4t^{-2}} = 2t^{-6-(-2)} = 2t^{-4} = \dfrac{2}{t^{4}}$

7. $\dfrac{3^{4} \cdot 3^{-8}}{3^{12} \cdot 3^{-4}} = \dfrac{3^{-4}}{3^{8}} = 3^{-4-8} = 3^{-12} = \dfrac{1}{3^{12}}$

9. 4×10^{-2} is in scientific notation in which 10 is raised to the -2 power. 4^{-2} is not in scientific notation and 4 is being raised to the -2 power.

11. (a) To write the polynomial in descending order, start with the term with the highest power; $-7x^{5} + 7x^{2} + 9x + 6$
(b) The leading coefficient is -7 since it is the coefficient of the term with the highest power.
(c) The degree of the polynomial is 5 since it is the degree of the term with the highest power.

13. Binomial;
$10(-3)^{2} + 5(-3) = 10(9) - 15 = 75$

15. Monomial; $6(-3)^{2} = 6(9) = 54$

17. Binomial; $2(2) - (2)^{4} = 4 - 16 = -12$

19. Trinomial;
$2(2)^{4} - 3(2) + 1 = 2(16) - 6 + 1 = 32 - 5 = 27$

21. Monomial;
$-32(-3)(2)(-1) = (96)(-2) = -192$

23. Like terms are terms with the same exponents on the same variable factors.

The exponents on the x-factors are different here so $3x$ and $3x^2$ are not like terms.

25. $23x^2y + 12x^2y = 35x^2y$

27. $3b^5d^2 + (5b^5d^2 - 9d) = 3b^5d^2 + 5b^5d^2 - 9d$
$$= 8b^5d^2 - 9d$$

29. $(7y^2 + 2y - 9) + (-3y^2 - y)$
$$= 7y^2 - 3y^2 - y + 2y - 9$$
$$= 4y^2 + y - 9$$

31. $(5x + 3x^2 - x^3) + (2x^2 + 4x - 10)$
$$= -x^3 + 3x^2 + 2x^2 + 5x + 4x - 10$$
$$= -x^3 + 5x^2 + 9x - 10$$

33. $(6.1y + 3.2x) + (4.8y - 3.2x)$
$$= 6.1y + 4.8y + 3.2x - 3.2x$$
$$= 10.9y$$

35.
$$
\begin{array}{r}
6a + 2b - 5c \\
+\ \underline{-2a - 2b - 3c} \\
4a \qquad -8c
\end{array}
$$

37. $\left(\dfrac{2}{5}a + \dfrac{1}{4}b - \dfrac{5}{6}\right) + \left(\dfrac{3}{5}a - \dfrac{3}{4}b - \dfrac{7}{6}\right)$
$$= \frac{2}{5}a + \frac{3}{5}a + \frac{1}{4}b - \frac{3}{4}b - \frac{5}{6} - \frac{7}{6}$$
$$= a - \frac{1}{2}b - 2$$

39. $\left(z - \dfrac{8}{3}\right) + \left(\dfrac{4}{3}z^2 - z + 1\right) = \dfrac{4}{3}z^2 + z - z - \dfrac{8}{3} + 1$
$$= \frac{4}{3}z^2 - \frac{5}{3}$$

41.
$$
\begin{array}{r}
7.9t^3 \qquad\quad + 2.6t - 1.1 \\
+\ \underline{\qquad -3.4t^2 + 3.4t - 3.1} \\
7.9t^3 - 3.4t^2 + 6t - 4.2
\end{array}
$$

43. $-(4h - 5) = -4h + 5$

45. $-(-2m^2 + 3m - 15) = 2m^2 - 3m + 15$

47. $-(3v^3 + 5v^2 + 10v + 22)$
$$= -3v^3 - 5v^2 - 10v - 22$$

49. $4a^3b^2 - 12a^3b^2 = -8a^3b^2$

51. $-32x^3 - 21x^3 = -53x^3$

53. $(7a - 7) - (12a - 4) = 7a - 7 - 12a + 4$
$$= -5a - 3$$

55.
$$
\begin{array}{ccc}
4k + 3 & = & 4k + 3 \\
-\ \dfrac{(-12k - 6)}{16k + 9} & + & \dfrac{12k + 6}{16k + 9}
\end{array}
$$

57. $25s - (23s + 14) = 25s - 23s - 14 = 2s - 14$

59. $(5t^2 - 3t - 2) - (2t^2 + t + 1)$
$$= 5t^2 - 2t^2 - 3t - t - 2 - 1$$
$$= 3t^2 - 4t - 3$$

61. To subtract the polynomials vertically, add the opposite of the second polynomial to the first.
$$
\begin{array}{r}
10r - 6s + 2t \\
+\ \underline{-12r + 3s + t} \\
-2r - 3s + 3t
\end{array}
$$

63. $\left(\dfrac{7}{8}x + \dfrac{2}{3}y - \dfrac{3}{10}\right) - \left(\dfrac{1}{8}x + \dfrac{1}{3}y\right)$
$$= \frac{7}{8}x - \frac{1}{8}x + \frac{2}{3}y - \frac{1}{3}y - \frac{3}{10}$$
$$= \frac{3}{4}x + \frac{1}{3}y - \frac{3}{10}$$

65. $\left(\dfrac{2}{3}h^2 - \dfrac{1}{5}h - \dfrac{3}{4}\right) - \left(\dfrac{4}{3}h^2 - \dfrac{4}{5}h + \dfrac{7}{4}\right)$
$$= \frac{2}{3}h^2 - \frac{4}{3}h^2 - \frac{1}{5}h + \frac{4}{5}h - \frac{7}{4} - \frac{3}{4}$$
$$= -\frac{2}{3}h^2 + \frac{3}{5}h - \frac{5}{2}$$

67. To subtract the polynomials vertically, add the opposite of the second polynomial to the first.

$$4.5x^4 - 3.1x^2 \qquad -6.7$$
$$+ \quad \underline{-2.1x^4 \qquad\qquad -4.4x}$$
$$2.4x^4 - 3.1x^2 - 4.4x - 6.7$$

69. $(4b^3 + 6b - 7) - (-12b^2 + 11b + 5)$
$$= 4b^3 + 12b^2 + 6b - 11b - 7 - 5$$
$$= 4b^3 + 12b^2 - 5b - 12$$

71. $(-2x^2 + 6x - 21) - (3x^3 - 5x + 10)$
$$= -3x^3 - 2x^2 + 6x + 5x - 21 - 10$$
$$= -3x^3 - 2x^2 + 11x - 31$$

73. $(y^2 + 3) + (3y^3 - y^2 - 1) + (y^3 + 2y^2)$
$$= 3y^3 + y^3 + y^2 - y^2 + 2y^2 + 3 - 1$$
$$= 4y^3 + 2y^2 + 2$$

75. $P = 5a^2 - 2a + 1$

$$\begin{pmatrix} \text{missing} \\ \text{side} \end{pmatrix} + (a - 3) + (2a^2 - 1) = P$$

$$\begin{pmatrix} \text{missing} \\ \text{side} \end{pmatrix} = 5a^2 - 2a + 1 - (a - 3) - (2a^2 - 1)$$

$$\begin{pmatrix} \text{missing} \\ \text{side} \end{pmatrix} = 3a^2 - 3a + 5$$

77. $(2ab^2 + 9a^2b) + (7ab^2 - 3ab + 7a^2b)$
$$= 2ab^2 + 7ab^2 + 9a^2b + 7a^2b - 3ab$$
$$= 9ab^2 - 3ab + 16a^2b$$

79. To subtract the polynomials vertically, add the opposite of the second polynomial to the first.

$$4z^5 \qquad + \ z^3 - 3z + 13$$
$$+ \quad \underline{z^4 + 8z^3 \qquad -15}$$
$$4z^5 + z^4 + 9z^3 - 3z - 2$$

81. $(9x^4 + 2x^3 - x + 5) + (9x^3 - 3x^2 + 8x + 3) - (7x^4 - x + 12)$
$$= 9x^4 - 7x^4 + 2x^3 + 9x^3 - 3x^2 - x + x + 8x + 5 + 3 - 12$$
$$= 2x^4 + 11x^3 - 3x^2 + 8x - 4$$

83. $(0.2w^2 + 3w + 1.3) - (w^3 - 0.7w + 2)$
$$= -w^3 + 0.2w^2 + 3w + 0.7w + 1.3 - 2$$
$$= -w^3 + 0.2w^2 + 3.7w - 0.7$$

85. $(7p^2q - 3pq^2) - (8p^2q + pq) + (4pq - pq^2)$
$$= 7p^2q - 3pq^2 - 8p^2q - pq + 4pq - pq^2$$
$$= -p^2q - 4pq^2 + 3pq$$

87. $(5x - 2x^3) + (2x^3 - 5x) = 5x - 5x - 2x^3 + 2x^3 = 0$

89. To subtract the polynomials vertically, add the opposite of the second polynomial to the first.

$$2a^2b - 4ab + ab^2$$
$$+ \quad \underline{-2a^2b - ab + 5ab^2}$$
$$-5ab + 6ab^2$$

91. $[(3y^2 - 5y) - (2y^2 + y - 1)] + (10y^2 - 4y - 5)$
$= (3y^2 - 2y^2 - 5y - y + 1) + (10y^2 - 4y - 5)$
$= y^2 - 6y + 1 + 10y^2 - 4y - 5$
$= y^2 + 10y^2 - 6y - 4y + 1 - 5$
$= 11y^2 - 10y - 4$

93. Answers will vary; $x^3 + 6$

95. Answers will vary; $8x^5$

97. Answers will vary; $-6x^2 + 2x + 5$

Section 5.6 Practice Exercises

1. (a) $5 - 2x$

  (b) squares; $a^2 - b^2$

  (c) perfect; $a^2 + 2ab + b^2$

3. $2y^2 - 4y^2 = -2y^2$

5. $(2y^2)(-4y^2) = -8y^4$

7. $7uvw^2 + uvw^2 = 8uvw^2$

9. $(7uvw^2)(uvw^2) = 7u^2v^2w^4$

11. $-2(6y) = -12y$

13. $7(3p) = 21p$

15. $(a^{13}b^4)(12ab^4) = 12a^{13+1}b^{4+4} = 12a^{14}b^8$

17. $(2c^7d)(-c^3d^{11}) = -2c^{7+3}d^{1+11} = -2c^{10}d^{12}$

19. $8pq(2pq - 3p + 5q)$
$= 8pq(2pq) - 8pq(3p) + 8pq(5q)$
$= 16p^2q^2 - 24p^2q + 40pq^2$

21. $(k^2 - 13k - 6)(-4k)$
$= k^2(-4k) - 13k(-4k) - 6(-4k)$
$= -4k^3 + 52k^2 + 24k$

23. $-15pq(3p^2 + p^3q^2 - 2q)$
$= -15pq(3p^2) - 15pq(p^3q^2) - 15pq(-2q)$
$= -45p^3q - 15p^4q^3 + 30pq^2$

25. $(y + 10)(y + 9) = y^2 + 9y + 10y + 90$
$= y^2 + 19y + 90$

27. $(m - 12)(m - 2) = m^2 - 2m - 12m + 24$
$= m^2 - 14m + 24$

29. $(3p - 2)(4p + 1) = 12p^2 + 3p - 8p - 2$
$= 12p^2 - 5p - 2$

31. $(-4w + 8)(-3w + 2) = 12w^2 - 8w - 24w + 16$
$= 12w^2 - 32w + 16$

33. $(p - 3w)(p - 11w)$
$= p^2 - 11pw - 3pw + 33w^2$
$= p^2 - 14pw + 33w^2$

35. $(6x - 1)(2x + 5) = 6x(2x) + 5(6x) - 2x - 5$
$= 12x^2 + 28x - 5$

37. $(4a - 9)(1.5a - 2) = 4a(1.5a) - 8a - 13.5a + 18$
$= 6a^2 - 21.5a + 18$

39. $(3t - 7)(3t^2 + 1) = 3t(3t^2) + 3t - 21t^2 - 7$
$= 9t^3 - 21t^2 + 3t - 7$

41 $3(3m + 4n)(m + 2n)$
$= 3(3m(m) + 6mn + 4mn + 4n(2n))$
$= 3(3m^2 + 10mn + 8n^2)$
$= 9m^2 + 30mn + 24n^2$

43. $(5s + 3)(s^2 + s - 2)$
$= 5s(s^2) + 5s(s) - 5s(2) + 3s^2 + 3s - 6$
$= 5s^3 + 5s^2 - 10s + 3s^2 + 3s - 6$
$= 5s^3 + 8s^2 - 7s - 6$

45. $(3w-2)(9w^2+6w+4)$

$=3w(9w^2)+3w(6w)+12w-2(9w^2)-12w-8$

$=27w^3+18w^2+12w-18w^2-12w-8$

$=27w^3-8$

47. $(p^2+p-5)(p^2+4p-1)$

$=p^4+4p^3-p^2+p^3+4p^2$

$\quad -p-5p^2-20p+5$

$=p^4+5p^3-2p^2-21p+5$

49. $\left(\dfrac{1}{2}x+\dfrac{3}{4}\right)\left(\dfrac{3}{2}x-\dfrac{1}{4}\right)$

$=\left(\dfrac{1}{2}\cdot\dfrac{3}{2}\right)x^2-\dfrac{1}{2}\cdot\dfrac{1}{4}x+\dfrac{3}{4}\cdot\dfrac{3}{2}x-\dfrac{3}{4}\cdot\dfrac{1}{4}$

$=\dfrac{3}{4}x^2-\dfrac{1}{8}x+\dfrac{9}{8}x-\dfrac{3}{16}$

$=\dfrac{3}{4}x^2+x-\dfrac{3}{16}$

51.

$$
\begin{array}{r}
3a^2-4a+9 \\
\times \quad 2a-5 \\
\hline
6a^3-\ 8a^2+18a \\
+\quad -15a^2+20a-45 \\
\hline
6a^3-23a^2+38a-45
\end{array}
$$

53.

$$
\begin{array}{r}
4x^2-12xy+9y^2 \\
\times \quad 2x-3y \\
\hline
8x^3-24x^2y+18xy^2 \\
+\quad -12x^2y+36xy^2-27y^3 \\
\hline
8x^3-36x^2y+54xy^2-27y^3
\end{array}
$$

55. $(3a-4b)(3a+4b)=(3a)^2-(4b)^2$

$\qquad\qquad\qquad\quad =9a^2-16b^2$

57. $(9k+6)(9k-6)=(9k)^2-6^2=81k^2-36$

59. $\left(\dfrac{1}{2}-t\right)\left(\dfrac{1}{2}+t\right)=\left(\dfrac{1}{2}\right)^2-t^2=\dfrac{1}{4}-t^2$

61. $(u^3+5v)(u^3-5v)=(u^3)^2-(5v)^2$

$\qquad\qquad\qquad\qquad =u^6-25v^2$

63. $(2-3a)(2+3a)=2^2-(3a)^2=4-9a^2$

65. $\left(\dfrac{2}{3}-p\right)\left(\dfrac{2}{3}+p\right)=\left(\dfrac{2}{3}\right)^2-p^2=\dfrac{4}{9}-p^2$

67. $(a+5)^2=a^2+2a(5)+(5)^2=a^2+10a+25$

69. $(x-y)^2=x^2-2xy+y^2$

71. $(2c+5)^2=(2c)^2+2(2c)(5)+5^2$

$\qquad\qquad =4c^2+20c+25$

73. $(3t^2-4s)^2=(3t^2)^2-2(3t^2)(4s)+(4s)^2$

$\qquad\qquad\quad =9t^4-24st^2+16s^2$

75. $(7-t)^2=7^2-2(7)(t)+t^2$

$\qquad\quad =49-14t+t^2$

$\qquad\quad =t^2-14t+49$

77. $(3+4q)^2=3^2+2(3)(4q)+(4q)^2$

$\qquad\quad =9+24q+16q^2$

$\qquad\quad =16q^2+24q+9$

79. (a) $(2+4)^2=(6)^2=36$

(b) $2^2+4^2=4+16=20$

(c) $(a+b)^2\neq a^2+b^2$ in general

81. $A=(2x+5)(2x-5)=(2x)^2-5^2=4x^2-25$

83. $A=(4p+5)^2$

$\quad =(4p)^2+2(4p)(5)+5^2$

$\quad =16p^2+40p+25$

85. V

$=s^3$

$=(3p-5)^3$

$=(3p-5)(3p-5)^2$

$=(3p-5)(9p^2-30p+25)$

$=27p^3-90p^2+75p-45p^2+150p-125$

$=27p^3-135p^2+225p-125$

149

87. $A = \dfrac{1}{2}bh$

$\quad = \dfrac{1}{2}(5a^3 - 2)(6a^2)$

$\quad = \dfrac{1}{2}(30a^5 - 12a^2)$

$\quad = 15a^5 - 6a^2$

89. $(7x + y)(7x - y) = (7x)^2 - y^2 = 49x^2 - y^2$

91. $(5s + 3t)^2 = (5s)^2 + 2(5s)(3t) + (3t)^2$

$\quad\quad\quad\quad = 25s^2 + 30st + 9t^2$

93. $(7x - 3y)(3x - 8y)$

$\quad = 7x(3x) - 7x(8y) - 3y(3x) + 3y(8y)$

$\quad = 21x^2 - 65xy + 24y^2$

95. $\left(\dfrac{2}{3}t + 2\right)(3t + 4)$

$\quad = \dfrac{2}{3}t(3t) + \dfrac{2}{3}t(4) + 2(3t) + 2(4)$

$\quad = 2t^2 + \dfrac{8}{3}t + 6t + 8$

$\quad = 2t^2 + \dfrac{26}{3}t + 8$

97. $-5(3x + 5)(2x - 1)$

$\quad = -5\big(3x \cdot 2x + 10x - 3x - 5 \cdot 1\big)$

$\quad = -5(6x^2 + 7x - 5)$

$\quad = -30x^2 - 35x + 25$

99. $(3a - 2)(5a + 1 + 2a^2)$

$\quad = 15a^2 + 3a + 6a^3 - 10a - 2 - 4a^2$

$\quad = 6a^3 + 11a^2 - 7a - 2$

101. $(y^2 + 2y + 4)(y - 5)$

$\quad = y^3 - 5y^2 + 2y^2 - 10y + 4y - 20$

$\quad = y^3 - 3y^2 - 6y - 20$

103. $\left(\dfrac{1}{3}m - n\right)^2 = \left(\dfrac{1}{3}m\right)^2 - 2\left(\dfrac{1}{3}m\right)(n) + n^2$

$\quad\quad\quad\quad\quad = \dfrac{1}{9}m^2 - \dfrac{2}{3}mn + n^2$

105. $6w^2(7w - 14) = 6w^2(7w) - 6w^2(14)$

$\quad\quad\quad\quad\quad\quad = 42w^3 - 84w^2$

107. $(4y - 8.1)(4y + 8.1)$

$\quad (4y)^2 - (8.1)^2 = 16y^2 - 65.61$

109. $(3c^2 + 4)(7c^2 - 8)$

$\quad = 3c^2(7c^2) - (3c^2)(8) + 4(7c^2) - 4(8)$

$\quad = 21c^4 + 4c^2 - 32$

111. $(3.1x + 4.5)^2$

$\quad = (3.1x)^2 + 2(3.1x)(4.5) + (4.5)^2$

$\quad = 9.61x^2 + 27.9x + 20.25$

113. $(k - 4)^3 = (k - 4)(k - 4)^2$

$\quad\quad\quad\quad = (k - 4)(k^2 - 8k + 16)$

$\quad\quad\quad\quad = k^3 - 8k^2 + 16k - 4k^2 + 32k - 64$

$\quad\quad\quad\quad = k^3 - 12k^2 + 48k - 64$

115. $(5x + 3)^3$

$\quad = (5x + 3)^2(5x + 3)$

$\quad = (25x^2 + 30x + 9)(5x + 3)$

$\quad = 125x^3 + 75x^2 + 150x^2 + 90x + 45x + 27$

$\quad = 125x^3 + 225x^2 + 135x + 27$

117. $(y^2 + 2y + 1)(2y^2 - y + 3)$

$\quad = 2y^4 - y^3 + 3y^2 + 4y^3 - 2y^2 + 6y$

$\quad\quad\quad + 2y^2 - y + 3$

$\quad = 2y^4 + 3y^3 + 3y^2 + 5y + 3$

119. $2a(3a - 4)(a + 5) = 2a(3a^2 + 11a - 20)$

$\quad\quad\quad\quad\quad\quad\quad = 6a^3 + 22a^2 - 40a$

121. $(x - 3)(2x + 1)(x - 4)$

$\quad = (x - 3)(2x^2 - 7x - 4)$

$\quad = 2x^3 - 7x^2 - 4x - 6x^2 + 21x + 12$

$\quad = 2x^3 - 13x^2 + 17x + 12$

123. $(3x + 5)(a + b) = 6x^2 - 11x - 35$

$\quad\quad 3ax = 6x^2 \quad$ and $\quad 5b = -35$

$\quad\quad a = 2x \quad$ and $\quad b = -7$

So, $a = 2$, $b = -7$, and the binomial is
$y = 2x - 7$.

125. $x^2 + kx + 25 = x^2 + kx + 5^2$
$k = 2(5) = 10$
or
$x^2 + kx + 25 = x^2 + kx + (-5)^2$
$k = 2(-5) = -10$

127. $a^2 + ka + 16 = a^2 + ka + 4^2$
$k = 2(4) = 8$
or
$a^2 + ka + 16 = a^2 + ka + (-4)^2$
$k = 2(-4) = -8$

Section 5.7 Practice Exercises

1. division; quotient; remainder

3. $(6z^5 - 2z^3 + z - 6) - (10z^4 + 2z^3 + z^2 + z)$
$= 6z^5 - 2z^3 + z - 6 - 10z^4 - 2z^3 - z^2 - z$
$= 6z^5 - 10z^4 - 4z^3 - z^2 - 6$

5. $(10x + y)(x - 3y) = 10x^2 - 30xy + xy - 3y^2$
$= 10x^2 - 29xy - 3y^2$

7. $(10x + y) + (x - 3y) = 10x + y + x - 3y$
$= 11x - 2y$

9. $\left(\frac{4}{3}y^2 - \frac{1}{2}y + \frac{3}{8}\right) - \left(\frac{1}{3}y^2 + \frac{1}{4}y - \frac{1}{8}\right)$
$= \frac{4}{3}y^2 - \frac{1}{2}y + \frac{3}{8} - \frac{1}{3}y^2 - \frac{1}{4}y + \frac{1}{8}$
$= \frac{4}{3}y^2 - \frac{1}{3}y^2 - \frac{1}{2}y - \frac{1}{4}y + \frac{3}{8} + \frac{1}{8}$
$= y^2 - \frac{3}{4}y + \frac{1}{2}$

11. $(a + 3)(a^2 - 3a + 9)$
$= a^3 - 3a^2 + 9a + 3a^2 - 9a + 27$
$= a^3 + 27$

13. (a) $\frac{15t^3 + 18t^2}{3t} = \frac{15t^3}{3t} + \frac{18t^2}{3t} = 5t^2 + 6t$

(b) $3t(5t^2 + 6t) = 15t^3 + 18t^2$

15. $(6a^2 + 4a - 14) \div 2 = \frac{6a^2}{2} + \frac{4a}{2} - \frac{14}{2}$
$= 3a^2 + 2a - 7$

17. $\frac{-5x^2 - 20x + 5}{-5} = \frac{-5x^2}{-5} - \frac{20x}{-5} + \frac{5}{-5}$
$= x^2 + 4x - 1$

19. $\frac{3p^3 - p^2}{p} = \frac{3p^3}{p} - \frac{p^2}{p} = 3p^2 - p$

21. $(4m^2 + 8m) \div 4m^2 = \frac{4m^2}{4m^2} + \frac{8m}{4m^2} = 1 + \frac{2}{m}$

23. $\frac{14y^4 - 7y^3 + 21y^2}{-7y^2} = \frac{14y^4}{-7y^2} - \frac{7y^3}{-7y^2} + \frac{21y^2}{-7y^2}$
$= -2y^2 + y - 3$

25. $(4x^3 - 24x^2 - x + 8) \div (4x)$
$= \frac{4x^3}{4x} - \frac{24x^2}{4x} - \frac{x}{4x} + \frac{8}{4x}$
$= x^2 - 6x - \frac{1}{4} + \frac{2}{x}$

27. $\frac{-a^3 b^2 + a^2 b^2 - ab^3}{-a^2 b^2}$
$= \frac{-a^3 b^2}{-a^2 b^2} + \frac{a^2 b^2}{-a^2 b^2} - \frac{ab^3}{-a^2 b^2}$
$= a - 1 + \frac{b}{a}$

29. $(6t^4 - 2t^3 + 3t^2 - t + 4) \div (2t^3)$
$= \frac{6t^4}{2t^3} - \frac{2t^3}{2t^3} + \frac{3t^2}{2t^3} - \frac{t}{2t^3} + \frac{4}{2t^3}$
$= 3t - 1 + \frac{3}{2t} - \frac{1}{2t^2} + \frac{2}{t^3}$

31. (a)

$$z+5 \overline{\smash{\big)}\ z^2+7z+11} \quad \begin{array}{r} z+2 \end{array}$$

$$\underline{-(z^2+5z)}$$
$$2z+11$$
$$\underline{-(2z+10)}$$
$$1$$

$$z+2+\frac{1}{z+5}$$

(b) $(z+5)(z+2)+1 = z^2+2z+5z+10+1$
$$= z^2+7z+11$$

33.

$$t+1 \overline{\smash{\big)}\ t^2+4t+5} \quad \begin{array}{r} t+3 \end{array}$$

$$\underline{-(t^2+\ t)}$$
$$3t+5$$
$$\underline{-(3t+3)}$$
$$2$$

$$t+3+\frac{2}{t+1}$$

35.

$$b-1 \overline{\smash{\big)}\ 7b^2-3b-4} \quad \begin{array}{r} 7b+4 \end{array}$$

$$\underline{-(7b^2-7b)}$$
$$4b-4$$
$$\underline{-(4b-4)}$$
$$7b+4$$

37.

$$5k+1 \overline{\smash{\big)}\ 5k^2-29k-6} \quad \begin{array}{r} k-6 \end{array}$$

$$\underline{-(5k^2+k)}$$
$$-30k-6$$
$$\underline{-(-30k-6)}$$
$$k-6$$

39.

$$2p+3 \overline{\smash{\big)}\ 4p^3+12p^2+p-12} \quad \begin{array}{r} 2p^2+3p-4 \end{array}$$

$$\underline{-(4p^3+6p^2)}$$
$$6p^2+p$$
$$\underline{-(6p^2+9p)}$$
$$-8p-12$$
$$\underline{-(-8p-12)}$$
$$2p^2+3p-4$$

41. Arrange both the dividend and divisor in descending order.

$$k+1 \overline{\smash{\big)}\ k^2-k-6} \quad \begin{array}{r} k-2 \end{array}$$

$$\underline{-(k^2+k)}$$
$$-2k-6$$
$$\underline{-(-2k-2)}$$
$$-4$$

$$k-2+\frac{-4}{k+1}$$

43.

$$2x-3 \overline{\smash{\big)}\ 4x^3-8x^2+15x-16} \quad \begin{array}{r} 2x^2-x+6 \end{array}$$

$$\underline{-(4x^3-6x^2)}$$
$$-2x^2+15x$$
$$\underline{-(-2x^2+\ 3x)}$$
$$12x-16$$
$$\underline{-(12x-18)}$$
$$2$$

$$2x^2-x+6+\frac{2}{2x-3}$$

45.

$$
\begin{array}{r}
y^2+2y+1 \\
3y-1{\overline{\smash{\big)}\,3y^3+5y^2+\ y+1}} \\
\underline{-(3y^3-y^2)} \\
6y^2+\ \ y \\
\underline{-(6y^2-2y)} \\
3y+1 \\
\underline{-(3y-1)} \\
2
\end{array}
$$

$$y^2+2y+1+\frac{2}{3y-1}$$

47. Arrange the dividend in descending order. The term $0a$ is a placeholder for the missing term.

$$
\begin{array}{r}
a-3 \\
a+3{\overline{\smash{\big)}\,a^2+0a+9}} \\
\underline{-(a^2+3a)} \\
-3a+9 \\
\underline{-(-3a-9)} \\
18
\end{array}
$$

$$a-3+\frac{18}{a+3}$$

49. The term $0x^2$ is a placeholder for the missing term in the dividend.

$$
\begin{array}{r}
4x^2+8x+13 \\
x-2{\overline{\smash{\big)}\,4x^3+0x^2\ -3x-26}} \\
\underline{-(4x^3-8x^2)} \\
8x^2\ -3x \\
\underline{-(8x^2-16x)} \\
13x-26 \\
\underline{-(13x-26)} \\
\end{array}
$$

$$4x^2+8x+13$$

51. The term $0w$ is a placeholder for the missing term in the divisor.

$$
\begin{array}{r}
w^2+5w-2 \\
w^2+0w-3{\overline{\smash{\big)}\,w^4+5w^3-5w^2-15w+7}} \\
\underline{-(w^4+0w^3-3w^2)} \\
5w^3-2w^2-15w \\
\underline{-(5w^3+0w^2-15w)} \\
-2w^2\ +0w+7 \\
\underline{-(-2w^2\ +0w+6)} \\
1
\end{array}
$$

$$w^2+5w-2+\frac{1}{w^2-3}$$

53.

$$
\begin{array}{r}
n^2+n-6 \\
2n^2+3n-2{\overline{\smash{\big)}\,2n^4+5n^3-11n^2-20n+12}} \\
\underline{-(2n^4+3n^3\ -2n^2)} \\
2n^3-9n^2-20n \\
\underline{-(2n^3+3n^2\ -2n)} \\
-12n^2-18n+12 \\
\underline{-(-12n^2-18n+12)} \\
\end{array}
$$

$$n^2+n-6$$

55.

$$
\begin{array}{r}
x-1 \\
5x^2+5x+1{\overline{\smash{\big)}\,5x^3+0x^2-4x-9}} \\
\underline{-(5x^3+5x^2\ +x)} \\
-5x^2-5x-9 \\
\underline{-(-5x^2-5x-1)} \\
-8
\end{array}
$$

$$x-1+\frac{-8}{5x^2+5x+1}$$

57.

$$
\begin{array}{r}
3y^2-3 \\
y^2+1{\overline{\smash{\big)}\,3y^4+2y+3}} \\
\underline{-(3y^4+3y^2)} \\
-3y^2+2y \\
\underline{-(-3y^2-3)} \\
2y+3+3
\end{array}
$$

$$3y^2-3+\frac{2y+6}{y^2+1}$$

59. To check, multiply the divisor $(x-2)$ by the quotient (x^2+4).

$(x-2)(x^2+4) = x^3 - 2x^2 + 4x - 8$ which does not equal $x^3 - 8$.

61. Monomial division;

$$\frac{9a^3}{3a} + \frac{12a^2}{3a} = 3a^2 + 4a$$

63. Long division;

$$
\begin{array}{r}
p+2 \\
p^2 - p - 2 \overline{)\ p^3 + p^2 - 4p - 4} \\
\underline{-(p^3 - p^2 - 2p)} \\
2p^2 - 2p - 4 \\
\underline{-(2p^2 - 2p - 4)} \\
\end{array}
$$
$p+2$

65. Long division; the terms $0t^3$ and $0t$ are placeholders for the missing terms in the dividend.

$$
\begin{array}{r}
t^3 - 2t^2 + 5t - 10 \\
t+2 \overline{)\ t^4 + 0t^3 + t^2 + 0t - 16} \\
\underline{-(t^4 + 2t^3)} \\
-2t^3 + t^2 \\
\underline{-(-2t^3 - 4t^2)} \\
5t^2 + 0t \\
\underline{-(5t^2 + 10t)} \\
-10t - 16 \\
\underline{-(-10t - 20)} \\
4 \\
\end{array}
$$
$$t^3 - 2t^2 + 5t - 10 + \frac{4}{t+2}$$

67. Long division; the terms $0w^3$ and $0w$ are placeholders for the missing terms in the dividend and $0w$ in the divisor.

$$
\begin{array}{r}
w^2 + 3 \\
w^2 + 0w - 2 \overline{)\ w^4 + 0w^3 + w^2 + 0w - 5} \\
\underline{-(w^4 + 0w^3 - 2w^2)} \\
3w^2 + 0w - 5 \\
\underline{-(3w^2 + 0w - 6)} \\
1 \\
\end{array}
$$
$$w^2 + 3 + \frac{1}{w^2 - 2}$$

69. Long division; the terms $0n^2$ and $0n$ are placeholders for the missing terms in the dividend.

$$
\begin{array}{r}
n^2 + 4n + 16 \\
n-4 \overline{)\ n^3 + 0n^2 + 0n - 64} \\
\underline{-(n^3 - 4n^2)} \\
4n^2 + 0n \\
\underline{-(4n^2 - 16n)} \\
16n - 64 \\
\underline{-(16n - 64)} \\
\end{array}
$$
$n^2 + 4n + 16$

71. Monomial division;

$$\frac{9r^3}{-3r^2} + \frac{-12r^2}{-3r^2} + \frac{9}{-3r^2} = -3r + 4 - \frac{3}{r^2}$$

73. Insert placeholders for missing terms in the dividend.

$$
\begin{array}{r}
x+1 \\
x-1 \overline{)\ x^2 + 0x - 1} \\
\underline{-(x^2 - x)} \\
x-1 \\
\underline{-(x-1)} \\
\end{array}
$$
$$\frac{x^2 - 1}{x-1} = x+1$$

75. Insert placeholders for missing terms in the dividend.

$$\begin{array}{r} x^3 + x^2 + x + 1 \\ x-1\overline{\smash{\big)}\, x^4 + 0x^3 + 0x^2 + 0x - 1} \\ \underline{-(x^4 - x^3)} \\ x^3 + 0x^2 \\ \underline{-(x^3 - x^2)} \\ x^2 + 0x \\ \underline{-(x^2 - x)} \\ x - 1 \\ \underline{-(x-1)} \end{array}$$

$$\frac{x^4 - 1}{x - 1} = x^3 + x^2 + x + 1$$

77. Insert placeholders for missing terms in the dividend.

$$\begin{array}{r} x + 1 \\ x-1\overline{\smash{\big)}\, x^2 + 0x + 0} \\ \underline{-(x^2 - x)} \\ x + 0 \\ \underline{-(x-1)} \\ 1 \end{array}$$

$$\frac{x^2}{x - 1} = x + 1 + \frac{1}{x - 1}$$

79. Insert placeholders for missing terms in the dividend.

$$\begin{array}{r} x^3 + x^2 + x + 1 \\ x-1\overline{\smash{\big)}\, x^4 + 0x^3 + 0x^2 + 0x + 0} \\ \underline{-(x^4 - x^3)} \\ x^3 + 0x^2 \\ \underline{-(x^3 - x^2)} \\ x^2 + 0x \\ \underline{-(x^2 - x)} \\ x + 0 \\ \underline{-(x-1)} \\ 1 \end{array}$$

$$\frac{x^4}{x - 1} = x^3 + x^2 + x + 1 + \frac{1}{x - 1}$$

Problem Recognition Exercises

1. (a) $6x^2 + 2x^2 = (6 + 2)x^2 = 8x^2$

   (b) $(6x^2)(2x^2) = (6 \cdot 2)x^2 \cdot x^2 = 12x^4$

3. (a) $(4x + y)^2 = (4x)^2 + 2(4x)y + y^2$
   $$= 16x^2 + 8xy + y^2$$

   (b) $(4xy)^2 = 4^2 x^2 y^2 = 16x^2 y^2$

5. (a) $(2x+3) + (4x-2) = 2x + 4x + 3 - 2 = 6x + 1$

   (b) $(2x+3)(4x-2) = 2 \cdot 4 \cdot x \cdot x + 12x - 4x - 6$
   $$= 8x^2 + 8x - 6$$

7. (a) $(3z + 2)^2 = (3z)^2 + 2(3z)(2) + (2)^2$
   $$= 9z^2 + 12z + 4$$

   (b) $(3z + 2)(3z - 2) = (3z)^2 - (2)^2 = 9z^2 - 4$

9. (a) $(2x-4)(x^2 - 2x + 3)$
   $$= 2x^3 - 4x^2 + 6x - 4x^2 + 8x - 12$$
   $$= 2x^3 - 8x^2 + 14x - 12$$

   (b) $(2x - 4) + (x^2 - 2x + 3)$
   $$= x^2 + 2x - 2x - 4 + 3$$
   $$= x^2 - 1$$

11. (a) $x + x = 2x$

   (b) $x \cdot x = x^2$

13. $(4xy)^2 = 4^2 x^2 y^2 = 16x^2 y^2$

15. $(-2x^4 - 6x^3 + 8x^2) \div (2x^2)$
   $$= \frac{-2x^4}{2x^2} - \frac{6x^3}{2x^2} + \frac{8x^2}{2x^2}$$
   $$= -x^2 - 3x + 4$$

17. $(m^3 - 4m^2 - 6) - (3m^2 + 7m) + (-m^3 - 9m + 6)$
   $$= m^3 - 4m^2 - 6 - 3m^2 - 7m - m^3 - 9m + 6$$
   $$= -7m^2 - 16m$$

19. Insert a placeholder for the missing term in the dividend.

$$\begin{array}{r} 8x^2 + 16x + 34 \\ x-2 \overline{\smash{\big)}\ 8x^3 + 0x^2 + 2x + 6} \\ \underline{-(8x^3 - 16x^2)} \\ 16x^2 + 2x \\ \underline{-(16x^2 - 32x)} \\ 34x + 6 \\ \underline{-(34x - 68)} \\ 74 \end{array}$$

$$8x^2 + 16x + 34 + \frac{74}{x-2}$$

21. $(2x - y)(3x^2 + 4xy - y^2)$

$= 6x^3 + 8x^2y - 2xy^2 - 3x^2y - 4xy^2 + y^3$

$= 6x^3 + 5x^2y - 6xy^2 + y^3$

23. $(x + y^2)(x^2 - xy^2 + y^4)$

$= x^3 - x^2y^2 + xy^4 + x^2y^2 - xy^4 + y^6$

$= x^3 + y^6$

25. $(a^2 + 2b) - (a^2 - 2b) = a^2 + 2b - a^2 + 2b$

$= 4b$

27. $(a^3 + 2b)(a^3 - 2b) = (a^3)^2 - (2b)^2$

$= a^6 - 4b^2$

29. $\begin{array}{r} 4p + 4 \\ 2p-1 \overline{\smash{\big)}\ 8p^2 + 4p - 6} \\ \underline{-(8p^2 - 4p)} \\ 8p - 6 \\ \underline{-(8p - 4)} \\ -2 \end{array}$

$4p + 4 + \dfrac{-2}{2p-1}$

31. $\dfrac{12x^3y^7}{3xy^5} = \dfrac{12}{3}x^{3-1}y^{7-5} = 4x^2y^2$

33. $\left(\dfrac{3}{7}x - \dfrac{1}{2}\right)\left(\dfrac{3}{7}x + \dfrac{1}{2}\right)$

$= \left(\dfrac{3}{7}x\right)^2 - \left(\dfrac{1}{2}\right)^2$

$= \dfrac{9}{49}x^2 - \dfrac{1}{4}$

35. $\left(\dfrac{1}{9}x^3 + \dfrac{2}{3}x^2 + \dfrac{1}{6}x - 3\right) -$

$\qquad\qquad \left(\dfrac{4}{3}x^3 + \dfrac{1}{9}x^2 + \dfrac{2}{3}x + 1\right)$

$= \dfrac{1}{9}x^3 + \dfrac{2}{3}x^2 + \dfrac{1}{6}x - 3$

$\qquad - \dfrac{4}{3}x^3 - \dfrac{1}{9}x^2 - \dfrac{2}{3}x - 1$

$= \dfrac{1}{9}x^3 - \dfrac{4}{3}x^3 + \dfrac{2}{3}x^2 - \dfrac{1}{9}x^2$

$\qquad + \dfrac{1}{6}x - \dfrac{2}{3}x - 3 - 1$

$= \dfrac{1}{9}x^3 - \dfrac{12}{9}x^3 + \dfrac{6}{9}x^2 - \dfrac{1}{9}x^2$

$\qquad + \dfrac{1}{6}x - \dfrac{4}{6}x - 3 - 1$

$= -\dfrac{11}{9}x^3 + \dfrac{5}{9}x^2 - \dfrac{3}{6}x - 4$

$= -\dfrac{11}{9}x^3 + \dfrac{5}{9}x^2 - \dfrac{1}{2}x - 4$

37. $(0.05x^2 - 0.16x - 0.75)$

$\qquad\qquad + (1.25x^2 - 0.14x + 0.25)$

$= 0.05x^2 + 1.25x^2 - 0.16x - 0.14x$

$\qquad\qquad - 0.75 + 0.25$

$= 1.3x^2 - 0.3x - 0.5$

39. $(3x^2y)(-2xy^5) = 3(-2)x^2 \cdot x \cdot y \cdot y^5 = -6x^3y^6$

Group Activity

1. $a = 5$
$b = 12$
$c = 13$

$$a^2 + b^2 = c^2$$

$$(5)^2 + (12)^2 \overset{?}{=} (13)^2$$

$$25 + 144 = 169$$

3. $(a+b)^2 = c^2 + 4 \cdot (\tfrac{1}{2}ab)$

$$a^2 + 2ab + b^2 = c^2 + 2ab$$

$$a^2 + b^2 = c^2$$

Chapter 5 Review Exercises

Section 5.1

1. Base 5; exponent 3

3. Base (-2); exponent 0

5. **(a)** $6^2 = 6 \cdot 6 = 36$

   (b) $(-6)^2 = (-6)(-6) = 36$

   (c) $-6^2 = -(6 \cdot 6) = -36$

   (b) $(-4)^3 = (-4)(-4)(-4) = -64$

   (c) $-4^3 = -(4 \cdot 4 \cdot 4) = -64$

7. $5^3 \cdot 5^{10} = 5^{3+10} = 5^{13}$

9. $x \cdot x^6 \cdot x^2 = x^{1+6+2} = x^9$

11. $\dfrac{10^7}{10^4} = 10^{7-4} = 10^3$

13. $\dfrac{b^9}{b} = b^{9-1} = b^8$

15. $\dfrac{k^2 k^3}{k^4} = k^{5-4} = k^1 = k$

17. $\dfrac{2^8 \cdot 2^{10}}{2^3 \cdot 2^7} = \dfrac{2^{18}}{2^{10}} = 2^{18-10} = 2^8$

19. Exponents are added only when multiplying factors with the same base. In such a case, the base does not change.

21. Substitute $P = 6000$, $t = 3$, and $r = 0.06$ into the formula.
$A = P(1+r)^t = 6000(1+0.06)^3 = \$7146.10$

Section 5.2

23. $(7^3)^4 = 7^{3 \cdot 4} = 7^{12}$

25. $(p^4 p^2)^3 = (p^6)^3 = p^{6 \cdot 3} = p^{18}$

27. $\left(\dfrac{a}{b}\right)^2 = \dfrac{a^2}{b^2}$

29. $\left(\dfrac{5}{c^2 d^5}\right)^2 = \dfrac{5^2}{(c^2 d^5)^2} = \dfrac{5^2}{c^4 d^{10}}$

31. $(2ab^2)^4 = 2^4 a^4 (b^2)^4 = 2^4 a^4 b^8$

33. $\left(\dfrac{-3x^3}{5y^2 z}\right)^3 = \dfrac{(-3x^3)^3}{(5y^2 z)^3}$

$$= \dfrac{-3^3 (x^3)^3}{5^3 (y^2)^3 z^3}$$

$$= -\dfrac{3^3 x^9}{5^3 y^6 z^3}$$

35. $\dfrac{a^4 (a^2)^8}{(a^3)^3} = \dfrac{a^4 a^{16}}{a^9} = \dfrac{a^{20}}{a^9} = a^{11}$

37. $\dfrac{(4h^2 k)^2 (h^3 k)^4}{(2hk^3)^2} = \dfrac{4^2 h^4 k^2 h^{12} k^4}{2^2 h^2 k^6}$

$$= \dfrac{16 h^{16} k^6}{4 h^2 k^6}$$

$$= 4h^{14}$$

39. $\left(\dfrac{2x^4 y^3}{4xy^2}\right)^2 = \dfrac{(2x^4 y^3)^2}{(4xy^2)^2} = \dfrac{2^2 x^8 y^6}{4^2 x^2 y^4} = \dfrac{x^6 y^2}{4}$

Section 5.3

41. $8^0 = 1$

43. $-x^0 = -1$

157

45. $2y^0 = 2(1) = 2$

47. $z^{-5} = \dfrac{1}{z^5}$

49. $(6a)^{-2} = \dfrac{1}{(6a)^2} = \dfrac{1}{36a^2}$

51. $4^0 + 4^{-2} = 1 + \dfrac{1}{4^2} = \dfrac{16}{16} + \dfrac{1}{16} = \dfrac{17}{16}$

53. $t^{-6}t^{-2} = t^{-8} = \dfrac{1}{t^8}$

55. $\dfrac{12x^{-2}y^3}{6x^4 y^{-4}} = 2x^{-2-4}y^{3-(-4)} = 2x^{-6}y^7 = \dfrac{2y^7}{x^6}$

57. $(-2m^2 n^{-4})^{-4} = (-2)^{-4} m^{-8} n^{16}$
$$= \dfrac{n^{16}}{2^4 m^8}$$
$$= \dfrac{n^{16}}{16m^8}$$

59. $\dfrac{(k^{-6})^{-2}(k^3)}{5k^{-6}k^0} = \dfrac{k^{12}k^3}{5k^{-6}} = \dfrac{k^{15-(-6)}}{5} = \dfrac{k^{21}}{5}$

61. $2 \cdot 3^{-1} - 6^{-1} = 2 \cdot \dfrac{1}{3} - \dfrac{1}{6}$
$$= \dfrac{2}{3} - \dfrac{1}{6}$$
$$= \dfrac{4}{6} - \dfrac{1}{6}$$
$$= \dfrac{3}{6}$$
$$= \dfrac{1}{2}$$

Section 5.4

63. **(a)** 9.74×10^7

   (b) 4.2×10^{-3} in.

65. $(4.1 \times 10^{-6})(2.3 \times 10^{11}) = (4.1)(2.3) \times 10^{-6+11}$
$$= 9.43 \times 10^5$$

67. $\dfrac{2000}{0.000008} = \dfrac{2.0 \times 10^3}{8.0 \times 10^{-6}}$
$$= \dfrac{2.0}{8.0} \times 10^{3-(-6)}$$
$$= 0.25 \times 10^9$$
$$= 2.5 \times 10^{-1} \times 10^9$$
$$= 2.5 \times 10^8$$

69. $5^{20} \approx 9.5367 \times 10^{13}$
This number has too many digits to fit on most calculator displays.

71. **(a)** $C = 2\pi r$
$$= 2(3.14)(9.3 \times 10^7)$$
$$\approx 58.4 \times 10^7$$
$$= 5.84 \times 10^1 \times 10^7$$
$$= 5.84 \times 10^8 \text{ miles}$$

   (b) $\dfrac{5.84 \times 10^8 \text{ miles}}{8.76 \times 10^3 \text{ hours}} \approx 0.667 \times 10^5$
$$= 6.67 \times 10^{-1} \times 10^5$$
$$= 6.67 \times 10^4 \text{ mph}$$

Section 5.5

73. **(a)** Trinomial

   (b) degree 4

   (c) leading coefficient 7

75. $(4x + 2) + (3x - 5) = 7x - 3$

77. $(9a^2 - 6) - (-5a^2 + 2a)$
$$= 9a^2 + 5a^2 - 2a - 6$$
$$= 14a^2 - 2a - 6$$

79.
$$\begin{array}{r} 8w^4 \qquad\quad -6w + 3 \\ +\ \underline{\ 2w^4 + 2w^3 - \ w + 1\ } \\ 10w^4 + 2w^3 - 7w + 4 \end{array}$$

81. $(7x^2 - 5x) - (9x^2 + 4x + 6)$
$$= 7x^2 - 5x - 9x^2 - 4x - 6$$
$$= -2x^2 - 9x - 6$$

83. Answers will vary; $-5x^2 + 2x - 4$

85. $P = 2w + 2l$

$\quad = 2(w) + 2(2w + 3)$

$\quad = 2w + 4w + 6$

$\quad = 6w + 6$

Section 5.6

87. $(9a^6)(2a^2b^4) = 18a^{6+2}b^4 = 18a^8b^4$

89. $(x^2 + 5x - 3)(-2x)$

$\quad = x^2(-2x) + 5x(-2x) - 3(-2x)$

$\quad = -2x^3 - 10x^2 + 6x$

91. $(4t - 1)(5t + 2) = 4t(5t) + 4t(2) - 1(5t) - 1(2)$

$\quad\quad\quad = 20t^2 + 8t - 5t - 2$

$\quad\quad\quad = 20t^2 + 3t - 2$

93. $(2a - 6)(a + 5) = 2a(a) + 2a(5) - 6a - 6(5)$

$\quad\quad\quad = 2a^2 + 4a - 30$

95. $(b - 4)^2 = b^2 - 2(b)(4) + 4^2 = b^2 - 8b + 16$

97. $(2w - 1)(-w^2 - 3w - 4)$

$\quad = -2w(w^2) - 2w(3w) - 2w(4) + w^2 + 3w + 4$

$\quad = -2w^3 - 6w^2 - 8w + w^2 + 3w + 4$

$\quad = -2w^3 - 5w^2 - 5w + 4$

99.

$$\begin{array}{r} 4a^2 + a - 5 \\ \times \quad\quad\quad 3a + 2 \\ \hline 12a^3 + 3a^2 - 15a \\ + \quad\quad 8a^2 + 2a - 10 \\ \hline 12a^3 + 11a^2 - 13a - 10 \end{array}$$

101. $\left(\dfrac{1}{3}r^4 - s^2\right)\left(\dfrac{1}{3}r^4 + s^2\right)$

$\quad = \left(\dfrac{1}{3}r^4\right)^2 - (s^2)^2$

$\quad = \dfrac{1}{9}r^8 - s^4$

103. $\quad (2h + 3)(h^4 - h^3 + h^2 - h + 1)$

$\quad = 2h^5 - 2h^4 + 2h^3 - 2h^2 + 2h$

$\quad\quad\quad + 3h^4 - 3h^3 + 3h^2 - 3h + 3$

$\quad = 2h^5 + h^4 - h^3 + h^2 - h + 3$

Section 5.7

105. $\dfrac{20y^3 - 10y^2}{5y} = \dfrac{20y^3}{5y} - \dfrac{10y^2}{5y}$

$\quad\quad\quad\quad = 4y^2 - 2y$

107. $(12x^4 - 8x^3 + 4x^2) \div (-4x^2)$

$\quad = \dfrac{12x^4}{-4x^2} + \dfrac{-8x^3}{-4x^2} + \dfrac{4x^2}{-4x^2}$

$\quad = -3x^2 + 2x - 1$

109.

$$\begin{array}{r} x + 2 \\ x + 5 \overline{)\,x^2 + 7x + 10} \\ \underline{-(x^2 + 5x)} \\ 2x + 10 \\ \underline{-(2x + 10)} \\ x + 2 \end{array}$$

111.

$$\begin{array}{r} p - 3 \\ 2p + 7 \overline{)\,2p^2 + p - 16} \\ \underline{-(2p^2 + 7p)} \\ -6p - 16 \\ \underline{-(-6p - 21)} \\ 5 \end{array}$$

$\quad p - 3 + \dfrac{5}{2p + 7}$

113.

$$\begin{array}{r} b^2 + 5b + 25 \\ b - 5 \overline{)\,b^3 + 0b^2 + 0b - 125} \\ \underline{-(b^3 - 5b^2)} \\ 5b^2 + 0b \\ \underline{-(5b^2 - 25b)} \\ 25b - 125 \\ \underline{-(25b - 125)} \\ 0 \end{array}$$

$\quad b^2 + 5b + 25$

115.

$$y^2 + 0y + 3 \overline{)\ y^4 - 4y^3 + 5y^2\ -3y + 2}$$

$$\underline{-(y^4 + 0y^3 + 3y^2)}$$

$$-4y^3 + 2y^2\ -3y$$

$$\underline{-(-4y^3 + 0y^2 - 12y)}$$

$$2y^2\ +9y + 2$$

$$\underline{-(2y^2\ +0y + 6)}$$

$$9y - 4$$

$$y^2 - 4y + 2 + \frac{9y - 4}{y^2 + 3}$$

117.

$$2w^2 - w + 3 \overline{)\ 2w^4 + w^3 + 0w^2 + 4w - 3}$$

$$\underline{-(2w^4 - w^3 + 3w^2)}$$

$$2w^3 - 3w^2 + 4w$$

$$\underline{-(2w^3\ - w^2 + 3w)}$$

$$-2w^2\ +w - 3$$

$$\underline{-(2w^2\ +w - 3)}$$

$$w^2 + w - 1$$

Chapter 5 Test

1. $\dfrac{3^4 \cdot 3^3}{3^6} = \dfrac{(3 \cdot 3 \cdot 3 \cdot 3)(3 \cdot 3 \cdot 3)}{3 \cdot 3 \cdot 3 \cdot 3 \cdot 3 \cdot 3} = 3$

3. $\dfrac{q^{10}}{q^2} = q^{10-2} = q^8$

5. $c^{-3} = \dfrac{1}{c^3}$

7. $4 \cdot 8^{-1} + 16^0 = 4 \cdot \dfrac{1}{8} + 1 = \dfrac{1}{2} + 1 = \dfrac{3}{2}$

9. $\left(\dfrac{2x}{y^3}\right)^4 = \dfrac{(2x)^4}{(y^3)^4} = \dfrac{2^4 x^4}{y^{12}} = \dfrac{16x^4}{y^{12}}$

11. $\dfrac{(s^2 t)^3 (7s^4 t)^4}{(7s^2 t^3)^2} = \dfrac{s^6 t^3 \, 7^4 s^{16} t^4}{7^2 s^4 t^6}$

$$= \dfrac{7^4 s^{22} t^7}{7^2 s^4 t^6}$$

$$= 7^2 s^{18} t$$

$$= 49 s^{18} t$$

13. $\left(\dfrac{6a^{-5}b}{8ab^{-2}}\right)^{-2} = \left(\dfrac{6a^{-5-1}b^{1-(-2)}}{8}\right)^{-2}$

$$= \left(\dfrac{6a^{-6}b^3}{8}\right)^{-2}$$

$$= \left(\dfrac{3b^3}{4a^6}\right)^{-2}$$

$$= \left(\dfrac{4a^6}{3b^3}\right)^{2}$$

$$= \dfrac{16a^{12}}{9b^6}$$

15. $\left(1.2 \times 10^6\right)\left(7 \times 10^{-15}\right) = (1.2)(7) \times 10^{6-15}$

$$= 8.4 \times 10^{-9}$$

17. (a) 1440 minutes in one day;

1.68×10^5 m^3/min(1440 min/day)

$$= 1.68 \times 10^5 \times 1.44 \times 10^3$$

$$= 2.4192 \times 10^8 \text{ m}^3 \text{ in one day}$$

(b) 2.4192×10^8 m^3 in one day

2.4192×10^8 m^3/day(365 days/year)

$$= 2.4192 \times 10^8 \times 3.65 \times 10^2$$

$$= 8.83008 \times 10^{10} \text{ m}^3/\text{year}$$

19. $(5t^4 - 2t^2 - 17) + (12t^3 + 2t^2 + 7t - 2)$

$$= 5t^4 + 12t^3 - 2t^2 + 2t^2 + 7t - 17 - 2$$

$$= 5t^4 + 12t^3 + 7t - 19$$

21. $-2x^3(5x^2 + x - 15)$

$$= -2x^3(5x^2) + (-2x^3)(x) + 2x^3(15)$$

$$= -10x^5 - 2x^4 + 30x^3$$

23. $(4y-5)(y^2-5y+3)$
$=4y^3-20y^2+12y-5y^2+25y-15$
$=4y^3-25y^2+37y-15$

25. $(5z-6)^2=(5z)^2-2(5z)(6)+6^2$
$=25z^2-60z+36$

27. $\left(2x^2+5x\right)-\left(6x^2-7\right)$
$=2x^2-6x^2+5x+7=-4x^2+5x+7$

29. $(10x^3-4x^2+1)-(3x^2-5x^3+2x)$
$=10x^3+5x^3-4x^2-3x^2-2x+1$
$=15x^3-7x^2-2x+1$

$A=lw=(5x+2)(x-3)=5x^2-13x-6$

31. $\dfrac{-12x^8+x^6-8x^3}{4x^2}$
$=\dfrac{-12x^8}{4x^2}+\dfrac{x^6}{4x^2}+\dfrac{-8x^3}{4x^2}$
$=-3x^6+\dfrac{x^4}{4}-2x$

33. $y-3\overline{)\,2y^2-13y+21\,}$ quotient $2y-7$
$\underline{-(2y^2-6y)}$
$-7y+21$
$\underline{-(-7y+21)}$
$2y-7$

35. $x^2+4\overline{)\,3x^4+x^3+0x^2+4x-33\,}$ quotient $3x^2+x-12$
$\underline{-(3x^4+0x^3+12x^2)}$
x^3-12x^2+4x
$\underline{-(x^3+4x)}$
$-12x^2+0x-33$
$\underline{-(-12x^2-48)}$
15

$3x^2+x-12+\dfrac{15}{x^2+4}$

Cumulative Review Exercises
Chapters 1–5

1. $-5-\dfrac{1}{2}[4-3(-7)]=-5-\dfrac{1}{2}[4+21]$
$=-5-\dfrac{1}{2}(25)$
$=-5-\dfrac{25}{2}$
$=-\dfrac{35}{2}$

3. $5^2-\sqrt{4};\ 25-2=23$

5. $-2y-3=-5(y-1)+3y$
$-2y-3=-5y+5+3y$
$-2y-3=-2y+5$
$-3=5$ Contradiction

No solution

7. y-axis

9. (a) Substitute $x=4$ into the equation.
$y=\dfrac{3}{2}(4)+6=6+6=12$ in.

(b) Substitute $x=9$ into the equation.
$y=\dfrac{3}{2}(9)+6=\dfrac{27}{2}+6=19.5$ in.

(c) Substitute $y=14\dfrac{1}{4}$ into the equation and solve for x.
$14\dfrac{1}{4}=\dfrac{3}{2}x+6$
$4\left(\dfrac{57}{4}\right)=4\left(\dfrac{3}{2}x+6\right)$
$57=6x+24$
$6x=33$
$x=\dfrac{33}{6}=5.5$ hours

11. $2 - 3(2x + 4) \leq -2x - (x - 5)$

$2 - 6x - 12 \leq -2x - x + 5$

$-6x - 10 \leq -3x + 5$

$-3x \leq 15$

$x \geq -5$

$[-5, \infty)$

13. $(2y + 3z)(-y - 5z)$

$= -2y^2 - 10yz - 3yz - 15z^2$

$= -2y^2 - 13yz - 15z^2$

15. $\left(\dfrac{2}{5}a + \dfrac{1}{3}\right)\left(\dfrac{2}{5}a - \dfrac{1}{3}\right) = \left(\dfrac{2}{5}a\right)^2 - \left(\dfrac{1}{3}\right)^2$

$= \dfrac{4}{25}a^2 - \dfrac{1}{9}$

17.

$$
\begin{array}{r}
4m^2 + 8m + 11 \\
m - 2 \overline{)\, 4m^3 + 0m^2 \;\; - 5m + 2} \\
\underline{-(4m^3 - 8m^2)} \\
8m^2 \;\; - 5m \\
\underline{-(8m^2 - 16m)} \\
11m \;\; + 2 \\
\underline{-(11m - 22)} \\
24
\end{array}
$$

$4m^2 + 8m + 11 + \dfrac{24}{m - 2}$

19. $\dfrac{10a^{-2}b^{-3}}{5a^0 b^{-6}} = 2a^{-2}b^3 = \dfrac{2b^3}{a^2}$

Chapter 6

Chapter 6 opener

5	2	6	^{A}1	3	^{B}4
3	^{C}1	4	2	^{D}6	^{E}5
^{F}4	6	1	5	^{G}2	^{H}3
^{I}2	3	5	4	1	6
1	4	3	6	5	2
6	5	2	^{J}3	4	1

Section 6.1 Practice Exercises

1. **(a)** product

   (b) prime

   (c) greatest common factor

   (d) prime

   (e) greatest common factor (GCF)

   (f) grouping

3. 7

5. 6

7. y

9. $4w^2z$

11. $2xy^4z^2$

13. $(x-y)$

15. **(a)** $3(x-2y) = 3x - 3(2y) = 3x - 6y$

    (b) $3x - 6y = 3x - 3(2y) = 3(x - 2y)$

17. $4p + 12 = 4p + 4 \cdot 3 = 4(p+3)$

19. $5c^2 - 10c + 15 = 5 \cdot c^2 - 5 \cdot 2c + 5 \cdot 3$
    $= 5(c^2 - 2c + 3)$

21. $x^5 + x^3 = x^3 x^2 + x^3 = x^3(x^2 + 1)$

23. $t^4 - 4t + 8t^2 = tt^3 - 4t + 8tt = t(t^3 - 4 + 8t)$

25. $2ab + 4a^3b = 2ab + 2ab(2a^2)$
    $= 2ab(1 + 2a^2)$

27. $38x^2y - 19x^2y^4 = 19x^2y(2) - 19x^2y(y^3)$
    $= 19x^2y(2 - y^3)$

29. $6x^3y^5 - 18xy^9z = 6xy^5(x^2 - 3y^4z)$

31. $5 + 7y^3$ is prime because it is not factorable.

33. $42p^3q^2 + 14pq^2 - 7p^4q^4$
    $= 7pq^2(6p^2 + 2 - p^3q^2)$

35. $t^5 + 2rt^3 - 3t^4 + 4r^2t^2$
    $= t^2(t^3 + 2rt - 3t^2 + 4r^2)$

37. **(a)** $-2x^3 - 4x^2 + 8x = -2x(x^2 + 2x - 4)$

    (b) $-2x^3 - 4x^2 + 8x = 2x(-x^2 - 2x + 4)$

39. $-8t^2 - 9t - 2 = -1(8t^2 + 9t + 2)$

41. $-15p^3 - 30p^2 = -15p^2(p + 2)$

43. $-q^4 + 2q^2 - 9q = -q(q^3 - 2q + 9)$

45. $-7x - 6y - 2z = -1(7x + 6y + 2z)$

47. $13(a+6) - 4b(a+6) = (a+6)(13 - 4b)$

49. $8v(w^2 - 2) + (w^2 - 2)$
    $= 8v(w^2 - 2) + 1(w^2 - 2)$
    $= (w^2 - 2)(8v + 1)$

51. $21x(x+3) + 7x^2(x+3) = 7x(x+3)(3+x)$
    $= 7x(x+3)^2$

53. $8a^2 - 4ab + 6ac - 3bc$
    $= 4a(2a - b) + 3c(2a - b)$
    $= (2a - b)(4a + 3c)$

55. $3q + 3p + qr + pr = 3(q + p) + r(q + p)$
$\quad\quad\quad = (q + p)(3 + r)$

57. $6x^2 + 3x + 4x + 2 = 3x(2x + 1) + 2(2x + 1)$
$\quad\quad\quad = (2x + 1)(3x + 2)$

59. $2t^2 + 6t - t - 3 = 2t(t + 3) - 1(t + 3)$
$\quad\quad\quad = (t + 3)(2t - 1)$

61. $6y^2 - 2y - 9y + 3 = 2y(3y - 1) + (-3)(3y - 1)$
$\quad\quad\quad = (3y - 1)(2y - 3)$

63. $b^4 + b^3 - 4b - 4 = b^3(b + 1) + (-4)(b + 1)$
$\quad\quad\quad = (b + 1)(b^3 - 4)$

65. $3j^2k + 15k + j^2 + 5 = 3k(j^2 + 5) + 1(j^2 + 5)$
$\quad\quad\quad = (j^2 + 5)(3k + 1)$

67. $14w^6x^6 + 7w^6 - 2x^6 - 1$
$\quad = 7w^6(2x^6 + 1) + (-1)(2x^6 + 1)$
$\quad = (2x^6 + 1)(7w^6 - 1)$

69. $ay + bx + by + ax = ay + ax + by + bx$
$\quad\quad\quad = a(y + x) + b(y + x)$
$\quad\quad\quad = (a + b)(y + x)$

71. $vw^2 - 3 + w - 3wv = vw^2 - 3vw + w - 3$
$\quad\quad\quad = vw(w - 3) + 1(w - 3)$
$\quad\quad\quad = (vw + 1)(w - 3)$

73. $15x^4 + 15x^2y^2 + 10x^3y + 10xy^3$
$\quad = 5x(3x^3 + 3xy^2 + 2x^2y + 2y^3)$
$\quad = 5x(3x(x^2 + y^2) + 2y(x^2 + y^2))$
$\quad = 5x(x^2 + y^2)(3x + 2y)$

75. $4abx - 4b^2x - 4ab + 4b^2$
$\quad = 4b(ax - bx - a + b)$
$\quad = 4b(x(a - b) - 1(a - b))$
$\quad = 4b(a - b)(x - 1)$

77. $6st^2 - 18st - 6t^4 + 18t^3$
$\quad = 6t(st - 3s - t^3 + 3t^2)$
$\quad = 6t(s(t - 3) - t^2(t - 3))$
$\quad = 6t(t - 3)(s - t^2)$

79. $P = 2l + 2w$
$\quad P = 2(l + w)$

81. $S = 2\pi r^2 + 2\pi rh$
$\quad S = 2\pi r(r + h)$

83. $\dfrac{1}{7}x^2 + \dfrac{3}{7}x - \dfrac{5}{7} = \dfrac{1}{7}(x^2 + 3x - 5)$

85. $\dfrac{5}{4}w^2 + \dfrac{3}{4}w + \dfrac{9}{4} = \dfrac{1}{4}(5w^2 + 3w + 9)$

87. Answers may vary. For example: $6x^2 + 9x$

89. Answers may vary. For example:
$\quad 16p^4q^2 + 8p^3q - 4p^2q$

Section 6.2 Practice Exercises

1. (a) positive

(b) different

(c) Both are correct.

(d) $3(x + 6)(x + 2)$

3. $3t(t - 5) - 6(t - 5) = 3(t - 5)(t - 2)$

5. $ax + 2bx - 5a - 10b$
$\quad = x(a + 2b) - 5(a + 2b)$
$\quad = (a + 2b)(x - 5)$

7. $x^2 + 10x + 16 = (x + 8)(x + 2)$

9. $z^2 - 11z + 18 = (z - 9)(z - 2)$

11. $z^2 - 3z - 18 = (z - 6)(z + 3)$

13. $p^2 - 3p - 40 = (p - 8)(p + 5)$

15. $t^2 + 6t - 40 = (t + 10)(t - 4)$

17. $x^2 - 3x + 20$ is prime

19. $n^2 + 8n + 16 = (n + 4)(n + 4) = (n + 4)^2$

21. a

23. c

25. They are both correct because multiplication of polynomials is a commutative operation.

27. The expressions are equal and both are correct

29. Descending order

31. $-13x + x^2 - 30 = x^2 - 13x - 30$
$= (x - 15)(x + 2)$

33. $-18w + 65 + w^2 = w^2 - 18w + 65$
$= (w - 13)(w - 5)$

35. $22t + t^2 + 72 = t^2 + 22t + 72$
$= (t + 18)(t + 4)$

37. $3x^2 - 30x - 72 = 3(x^2 - 10x - 24)$
$= 3(x - 12)(x + 2)$

39. $8p^3 - 40p^2 + 32p = 8p(p^2 - 5p + 4)$
$= 8p(p - 1)(p - 4)$

41. $y^4 z^2 - 12y^3 z^2 + 36y^2 z^2$
$= y^2 z^2 (y^2 - 12y + 36)$
$= y^2 z^2 (y - 6)(y - 6)$ or $y^2 z^2 (y - 6)^2$

43. $-x^2 + 10x - 24 = -(x^2 - 10x + 24)$
$= -(x - 4)(x - 6)$

45. $-m^2 + m + 6 = -(m^2 - m - 6)$
$= -(m - 3)(m + 2)$

47. $-4 - 2c^2 - 6c = -2(c^2 + 3c + 2)$
$= -2(c + 2)(c + 1)$

49. $x^2 y^3 - 19x^2 y^3 + 60xy^3$
$= xy^3 (x^2 - 19x + 60)$
$= xy^3 (x - 15)(x - 4)$

51. $12p^2 - 96p + 84 = 12(p^2 - 8p + 7)$
$= 12(p - 7)(p - 1)$

53. $-2m^2 + 22m - 20 = -2(m^2 - 11m + 10)$
$= -2(m - 10)(m - 1)$

55. $c^2 + 6cd + 5d^2 = (c + 5d)(c + d)$

57. $a^2 - 9ab + 14b^2 = (a - 2b)(a - 7b)$

59. $a^2 + 4a + 18$ is Prime

61. $2q + q^2 - 63 = q^2 + 2q - 63$
$= (q - 7)(q + 9)$

63. $x^2 + 20x + 100 = (x + 10)(x + 10)$
$= (x + 10)^2$

65. $t^2 + 18t - 40 = (t + 20)(t - 2)$

67. The student forgot to factor out the GCF before factoring the trinomial further. The polynomial is not factored completely, because $(2x - 4)$ has a common factor of 2.

69. $(x - 4)(x + 13) = x^2 + 9x - 52$

71. **(a)** The perimeter can be found by adding the lengths of the sides.
$(5x) + (2x - 3) + (3x^2 - 5) + (2x - 4)$
$= 3x^2 + (5 + 2 + 2)x + (-3 - 5 - 4)$
$= 3x^2 + 9x - 12$

(b) $3x^2 + 9x - 12 = 3(x^2 + 3x - 4)$
$= 3(x + 4)(x - 1)$

73. $x^4 + 10x^2 + 9 = (x^2 + 1)(x^2 + 9)$

75. $w^4 + 2w^2 - 15 = (w^2 + 5)(w^2 - 3)$

77. $7, 5, -7, -5$

79. For example, $c = -16$

Section 6.3 Practice Exercises

1. (a) Both are correct.

   (b) $2(3x - 5)(x + 1)$

3. $mn - m - 2n + 2 = m(n - 1) - 2(n - 1)$
$$= (n - 1)(m - 2)$$

5. $6a^2 - 30a - 84 = 6(a^2 - 5a - 12)$
$$= 6(a - 7)(a + 2)$$

7. a

9. b

11. $3n^2 + 13n + 4 = (3n + 1)(n + 4)$

13. $2y^2 - 3y - 2 = (2y + 1)(y - 2)$

15. $5x^2 - 14x - 3 = (5x + 1)(x - 3)$

17. $12c^2 - 5c - 2 = (4c + 1)(3c - 2)$

19. $-12 + 10w^2 + 37w = 10w^2 + 37w - 12$
$$= (10w - 3)(w + 4)$$

21. $-5q - 6 + 6q^2 = 6q^2 - 5q - 6$
$$= (3q + 2)(2q - 3)$$

23. $6b - 23 + 4b^2 = 4b^2 + 6b - 23$ is prime

25. $-8 + 25m^2 - 10m = 25m^2 - 10m - 8$
$$= (5m + 2)(5m - 4)$$

27. $6y^2 + 19xy - 20x^2 = (6y - 5x)(y + 4x)$

29. $2m^2 - 12m - 80 = 2(m^2 - 6m - 40)$
$$= 2(m - 10)(m + 4)$$

31. $2y^5 + 13y^4 + 6y^3 = y^3(2y^2 + 13y + 6)$
$$= y^3(2y + 1)(y + 6)$$

33. $-a^2 - 15a + 34 = -(a^2 + 15a - 34)$
$$= -(a + 17)(a - 2)$$

35. $80m^2 - 100mp - 30p^2$
$$= 10(8m^2 - 10mp - 3p^2)$$
$$= 10(4m + p)(2m - 3p)$$

37. $x^4 + 10x^2 + 9 = (x^2 + 1)(x^2 + 9)$

39. $w^4 + 2w^2 - 15 = (w^2 + 5)(w^2 - 3)$

41. $2x^4 - 7x^2 - 15 = (2x^2 + 3)(x^2 - 5)$

43. $-2z^2 + 20z - 18 = -2(z^2 - 10z + 9)$
$$= -2(z - 9)(z - 1)$$

45. $q^2 - 13q + 42 = (q - 7)(q - 6)$

47. $6t^2 + 7t - 3 = (2t + 3)(3t - 1)$

49. $4m^2 - 20m + 25 = (2m - 5)(2m - 5)$
$$= (2m - 5)^2$$

51. $5c^2 - c + 2$ is prime

53. $6x^2 - 19xy + 10y^2 = (2x - 5y)(3x - 2y)$

55. $12m^2 + 11mn - 5n^2 = (4m + 5n)(3m - n)$

57. $30r^2 + 5r - 10 = 5(6r^2 + r - 2)$
$$= 5(3r + 2)(2r - 1)$$

59. $4s^2 - 8st + t^2$ is prime

61. $10t^2 - 23t - 5 = (2t - 5)(5t + 1)$

63. $14w^2 + 13w - 12 = (7w - 4)(2w + 3)$

65. $a^2 - 10a - 24 = (a - 12)(a + 2)$

67. $x^2 + 9xy + 20y^2 = (x + 5y)(x + 4y)$

69. $a^2 + 21ab + 20b^2 = (a + 20b)(a + b)$

71. $t^2 - 10t + 21 = (t - 7)(t - 3)$

73. $5d^6 + 3d^2 - 10d = d(5d^5 + 3d - 10)$

75. $4b^3 - 4b^2 - 80b = 4b(b^2 - b - 20)$
$$= 4b(b - 5)(b + 4)$$

77. $x^2y^2 - 13xy^2 + 30y^2 = y^2(x^2 - 13x + 30)$
$$= y^2(x - 3)(x - 10)$$

79. $p^2q^2 - 14pq^2 + 33q^2 = q^2(p^2 - 14p + 33)$
$$= q^2(p - 3)(p - 11)$$

81. $8x^4 + 14x^2 + 3 = (2x^2 + 3)(4x^2 + 1)$

83. (a) Let $t = 1$ in the polynomial
$-16t^2 + 12t + 40$.
$$-16(1)^2 + 12(1) + 40 = -16(1) + 12 + 40$$
$$= -16 + 12 + 40$$
$$= 36$$

The height of the rock after 1 second is 36 ft.

(b) $-16t^2 + 12t + 40 = -4(4t^2 + 3t + 10)$
$$= -4(4t + 5)(t - 2)$$
$$-4(4(1) + 5)(1 - 2) = -4(4 + 5)(1 - 2)$$
$$= -4(9)(-1)$$
$$= 36$$
Yes, the result is the same as part (a).

85. (a) $x^2 - 10x - 24 = (x - 12)(x + 2)$

(b) $x^2 - 10x + 24 = (x - 6)(x - 4)$

87. (a) $x^2 - 5x - 6 = (x - 6)(x + 1)$

(b) $x^2 - 5x + 6 = (x - 2)(x - 3)$

Section 6.4 Practice Exercises

1. (a) Both are correct.

(b) $3(4x + 3)(x - 2)$

3. $8(y + 5) + 9y(y + 5) = (y + 5)(8 + 9y)$

5. 12 and 1

7. -8 and -1

9. 5 and -4

11. 9 and -2

13. $3x^2 + 13x + 4 = 3x^2 + 12x + x + 4$
$$= 3x(x + 4) + 1(x + 4)$$
$$= (x + 4)(3x + 1)$$

15. $4w^2 - 9w + 2 = 4w^2 - 8w - w + 2$
$$= 4w(w - 2) - 1(w - 2)$$
$$= (w - 2)(4w - 1)$$

17. $x^2 + 7x - 18 = x^2 + 9x - 2x - 18$
$$= x(x + 9) - 2(x + 9)$$
$$= (x + 9)(x - 2)$$

19. $2m^2 + 5m - 3 = 2m^2 + 6m - m - 3$
$$= 2m(m + 3) - 1(m + 3)$$
$$= (m + 3)(2m - 1)$$

21. $8k^2 - 6k - 9 = 8k^2 - 12k + 6k - 9$
$$= 4k(2k - 3) + 3(2k - 3)$$
$$= (4k - 3)(2k + 3)$$

23. $4k^2 - 20k + 25 = 4k^2 - 10k - 10k + 25$
$$= 2k(2k - 5) - 5(2k - 5)$$
$$= (2k - 5)(2k - 5)$$
$$= (2k - 5)^2$$

25. Prime

27. $9z^2 - 21z + 10 = 9z^2 - 15z - 6z + 10$
$$= 3z(3z - 5) - 2(3z - 5)$$
$$= (3z - 5)(3z - 2)$$

29. $12y^2 + 8yz - 15z^2$
$$= 12y^2 + 18yz - 10yz - 15z^2$$
$$= 6y(2y + 3z) - 5z(2y + 3z)$$
$$= (6y - 5z)(2y + 3z)$$

31. $50y + 24 + 14y^2 = 2(7y^2 + 25y + 12)$
$$= 2(7y^2 + 21y + 4y + 12)$$
$$= 2[7y(y + 3) + 4(y + 3)]$$
$$= 2(7y + 4)(y + 3)$$

33. $-15w^2 + 22w + 5 = -(15w^2 - 22w - 5)$

$= -(3w - 5)(5w + 1)$

35. $-12x^2 + 20xy - 8y^2 = -4(3x^2 - 5xy + 2y^2)$

$= -4(3x - 2y)(x - y)$

37. $18y^3 + 60y^2 + 42y = 6y(3y^2 + 10y + 7)$

$= 6y(3y + 7)(y + 1)$

39. $a^4 + 5a^2 + 6 = (a^2 + 3)(a^2 + 2)$

41. $6x^4 - x^2 - 15 = (3x^2 - 5)(2x^2 + 3)$

43. $8p^4 + 37p^2 - 15 = (8p^2 - 3)(p^2 + 5)$

45. $20p^2 - 19p + 3 = 20p^2 - 15p - 4p + 3$

$= 5p(4p - 3) - 1(4p - 3)$

$= (4p - 3)(5p - 1)$

47. $6u^2 - 19uv + 10v^2$

$= 6u^2 - 15uv - 4uv + 10v^2$

$= 3u(2u - 5v) - 2v(2u - 5v)$

$= (2u - 5v)(3u - 2v)$

49. $12a^2 + 11ab - 5b^2$

$= 12a^2 + 15ab - 4ab - 5b^2$

$= 3a(4a + 5b) - b(4a + 5b)$

$= (4a + 5b)(3a - b)$

51. $3h^2 + 19hk - 14k^2$

$= 3h^2 + 21hk - 2hk - 14k^2$

$= 3h(h + 7k) - 2k(h + 7k)$

$= (h + 7k)(3h - 2k)$

53. Prime

55. $16z^2 - 14z + 3 = 16z^2 - 8z - 6z + 3$

$= 8z(2z - 1) - 3(2z - 1)$

$= (2z - 1)(8z - 3)$

57. $b^2 - 8b + 16 = b^2 - 4b - 4b + 16$

$= b(b - 4) - 4(b - 4)$

$= (b - 4)(b - 4)$

$= (b - 4)^2$

59. $-5x^2 + 25x - 30 = -5(x^2 - 5x + 6)$

$= -5(x^2 - 3x - 2x + 6)$

$= -5(x(x - 3) - 2(x - 3))$

$= -5(x - 2)(x - 3)$

61. $t^2 - t - 6 = t^2 - 3t + 2t - 6$

$= t(t - 3) + 2(t - 3)$

$= (t - 3)(t + 2)$

63. Prime

65. $72x^2 + 18x - 2 = 2(36x^2 + 9x - 1)$

$= 2(36x^2 + 12x - 3x - 1)$

$= 2(12x(3x + 1) - 1(3x + 1))$

$= 2(3x + 1)(12x - 1)$

67. $p^3 - 6p^2 - 27p = p(p^2 - 6p - 27)$

$= p(p^2 - 9p + 3p - 27)$

$= p(p(p - 9) + 3(p - 9))$

$= p(p + 3)(p - 9)$

69. $3x^3 + 10x^2 + 7x = x(3x^2 + 10x + 7)$

$= x(3x^2 + 3x + 7x + 7)$

$= x(3x + 7)(x + 1)$

71. $2p^3 - 38p^2 + 120p$

$= 2p(p^2 - 19p + 60)$

$= 2p(p^2 - 4p - 15p + 60)$

$= 2p(p(p - 4) - 15(p - 4))$

$= 2p(p - 15)(p - 4)$

73. $x^2y^2 + 14x^2y + 33x^2$

$= x^2(y^2 + 14y + 33)$

$= x^2(y^2 + 11y + 3y + 33)$

$= x^2(y(y + 11) + 3(y + 11))$

$= x^2(y + 3)(y + 11)$

75. $-k^2 - 7k - 10 = -1(k^2 + 7k + 10)$
$= -1(k^2 + 5k + 2k + 10)$
$= -1(k(k+5) + 2(k+5))$
$= -1(k+2)(k+5)$

77. $-3n^2 - 3n + 90 = -3(n^2 + n - 30)$
$= -3(n^2 + 6n - 5n - 30)$
$= -3(n(n+6) - 5(n+6))$
$= -3(n+6)(n-5)$

79. $x^4 - 7x^2 + 10 = (x^2 - 5)(x^2 - 2)$

81. No. $(2x + 4)$ contains a common factor of 2.

83. **(a)** Let $x = 10$ in the polynomial $-2x^2 + 40x - 72$.
$-2(10)^2 + 40(10) - 72 = -2(100) + 40(10) - 72$
$= -200 + 400 - 72$
$= 128$
There are 128 customers when 10 tables are set up.

(b)
$-2x^2 + 40x - 72 = -2(x^2 - 20x + 36)$
$= -2(x^2 - 18x - 2x + 36)$
$= -2[x(x-18) - 2(x-18)]$
$= -2(x-18)(x-2)$
$-2(10-18)(10-2) = -2(-8)(8) = 128$;
Yes, the result is the same as part (a).

85. **(a)** Let $n = 6$ in the polynomial $n^2 + n$.
$6^2 + 6 = 36 + 6 = 42$

(b) $n^2 + n = n(n+1)$ $6(6+1) = 6(7) = 42$

Yes, the result is the same as part (a).

Section 6.5 Practice Exercises

1. **(a)** difference; $(a+b)(a-b)$.

(b) sum

(c) is not

(d) square

(e) $(a+b)^2$; . $(a-b)^2$.

3. $6x^2 - 17x + 5 = (3x-1)(2x-5)$

5. $2x^2 - x - 1 = (2x+1)(x-1)$

7. $ax + ab - 6x - 6b = a(x+b) - 6(x+b)$
$= (x+b)(a-6)$

9. $6y - 10 + y^2 = y^2 + 6y - 40 = (y+10)(y-4)$

11. $x^2 - 5^2 = x^2 - 25$

13. $(2p - 3q)(2p + 3q) = 4p^2 - 9q^2$

15. $x^2 - 36 = (x+6)(x-6)$

17. $3w^2 - 300 = 3(w^2 - 100) = 3(w+10)(w-10)$

19. $4a^2 - 121b^2 = (2a)^2 - (11b)^2$
$= (2a + 11b)(2a - 11b)$

21. $49m^2 - 16n^2 = (7m)^2 - (4n)^2$
$= (7m + 4n)(7m - 4n)$

23. $9q^2 + 16$ is prime

25. $y^2 - 4z^2 = (y + 2z)(y - 2z)$

27. $a^2 - b^4 = (a + b^2)(a - b^2)$

29. $25p^2q^2 - 1 = (5pq - 1)(5pq + 1)$

31. $c^2 - \dfrac{1}{25} = (c)^2 - \left(\dfrac{1}{5}\right)^2 = \left(c - \dfrac{1}{5}\right)\left(c + \dfrac{1}{5}\right)$

33. $50 - 32t^2 = 2(25 - 16t^2)$
$= 2[5^2 - (4t)^2]$
$= 2(5 + 4t)(5 - 4t)$

35. $z^4 - 16 = (z^2 + 4)(z^2 - 4)$
$= (z^2 + 4)(z + 2)(z - 2)$

37. $16 - z^4 = \left(4\right)^2 - \left(z^2\right)^2$

$\qquad = \left(4 - z^2\right)\left(4 + z^2\right)$

$\qquad = \left(2 - z\right)\left(2 + z\right)\left(4 + z^2\right)$

39. $x^3 + 5x^2 - 9x - 45 = x^2(x+5) - 9(x+5)$

$\qquad = (x+5)(x^2 - 9)$

$\qquad = (x+5)(x+3)(x-3)$

41. $c^3 - c^2 - 25c + 25 = c^2(c-1) - 25(c-1)$

$\qquad = (c-1)(c^2 - 25)$

$\qquad = (c-1)(c+5)(c-5)$

43. $2x^2 - 18 + x^2 y - 9y = 2(x^2 - 9) + y(x^2 - 9)$

$\qquad = (2+y)(x^2 - 9)$

$\qquad = (2+y)(x+3)(x-3)$

45. $x^2 y^2 - 9x^2 - 4y^2 + 36$

$= x^2(y^2 - 9) - 4(y^2 - 9)$

$= (x^2 - 4)(y^2 - 9)$

$= (x+2)(x-2)(y+3)(y-3)$

47. $(3x+5)^2 = 9x^2 + 30x + 25$

49. (a) $x^2 + 4x + 4$ is a perfect square trinomial

(b) $x^2 + 4x + 4 = (x+2)^2$;

$x^2 + 5x + 4 = (x+1)(x+4)$

51. $x^2 + 18x + 81 = x^2 + 2(9)(x) + 9^2 = (x+9)^2$

53. $25z^2 - 20z + 4 = (5z)^2 - 2(5z)(2) + 2^2$

$\qquad = (5z - 2)^2$

55. $49a^2 + 42ab + 9b^2$

$= (7a)^2 + 2(7a)(3b) + (3b)^2$

$= (7a + 3b)^2$

57. $-2y + y^2 + 1 = y^2 - 2y + 1 = (y-1)^2$

59. $80z^2 + 120zw + 45w^2$

$= 5(16z^2 + 24zw + 9w^2)$

$= 5((4z)^2 + 2(4z)(3w) + (3w)^2)$

$= 5(4z + 3w)^2$

61. $9y^2 + 78y + 25 = (3y + 25)(3y + 1)$

63. $2a^2 - 20a + 50 = 2(a^2 - 10a + 25)$

$\qquad = 2(a - 5)^2$

65. $4x^2 + x + 9$ is prime

67. $4x^2 + 4xy + y^2 = (2x)^2 + 2(2x)y + y^2$

$\qquad = (2x + y)^2$

69. $3x^3 - 6x^2 + 3x = 3x\left(x^2 - 2x + 1\right)$

$\qquad = 3x\left(x - 1\right)\left(x - 1\right)$

$\qquad = 3x\left(x - 1\right)^2$

71. $(y-3)^2 - 9 = ((y-3) + 3)((y-3) - 3)$

$\qquad = y(y - 6)$

73. $(2p+1)^2 - 36 = ((2p+1) + 6)((2p+1) - 6)$

$\qquad = (2p + 7)(2p - 5)$

75. $16 - (t+2)^2$

$= (4 + (t+2))(4 - (t+2))$

$= (4 + t + 2)(4 - t - 2)$

$= (t+6)(-t+2)$ or $-(t+6)(t-2)$

77. $(2a-5)^2 - 100b^2$

$= ((2a-5) - 10b)((2a-5) + 10b)$

$= (2a - 5 - 10b)(2a - 5 + 10b)$

79. (a) a^2 (area of outer square)

b^2 (area of inner square)

Area of shaded region $= a^2 - b^2$

(b) $a^2 - b^2 = (a+b)(a-b)$

Section 6.6 Practice Exercises

1. **(a)** sum; cubes

   (b) difference; cubes

   (c) $(a-b)(a^2+ab+b^2)$

   (d) $(a+b)(a^2-ab+b^2)$

3. $20-5t^2 = 5(4-t^2) = 5(2-t)(2+t)$

5. $2t+2u+st+su = 2(t+u)+s(t+u)$
   $= (t+u)(2+s)$

7. $3v^2+5v-12 = (3v-4)(v+3)$

9. $-c^2-10c-25 = -(c^2+10c+25) = -(c+5)^2$

11. x^3, 8, y^6, $27q^3$, w^{12}, $r^3 s^6$

13. $t^3, -1, 27, a^3 b^6, 125, y^6$

15. $y^3-8 = y^3-2^3 = (y-2)(y^2+2y+4)$

17. $1-p^3 = (1-p)(1+p+p^2)$

19. $w^3+64 = w^3+4^3 = (w+4)(w^2-4w+16)$

21. $x^3-1000y^3 = x^3-(10y)^3$
    $= (x-10y)(x^2+10xy+100y^2)$

23. $64t^3+1 = (4t)^3+1^3 = (4t+1)(16t^2-4t+1)$

25. $1000a^3+27 = (10a)^3+3^3$
    $= (10a+3)(100a^2-30a+9)$

27. $n^3-\dfrac{1}{8} = n^3-\left(\dfrac{1}{2}\right)^3 = \left(n-\dfrac{1}{2}\right)\left(n^2+\dfrac{1}{2}n+\dfrac{1}{4}\right)$

29. $125x^3+8y^3 = (5x)^3+(2y)^3$
    $= (5x+2y)(25x^2-10xy+4y^2)$

31. $x^4-4 = (x^2)^2-2^2 = (x^2+2)(x^2-2)$

33. a^2+9 is prime.

35. $t^3+64 = t^3+4^3 = (t+4)(t^2-4t+16)$

37. g^3-4 is prime.

39. $4b^3+108 = 4(b^3+27)$
    $= 4(b+3)(b^2-3b+9)$

41. $5p^2-125 = 5(p^2-25) = 5(p+5)(p-5)$

43. $\dfrac{1}{64}-8h^3 = \left(\dfrac{1}{4}\right)^3-(2h)^3$
    $= \left(\dfrac{1}{4}-2h\right)\left(\dfrac{1}{16}+\dfrac{1}{2}h+4h^2\right)$

45. $x^4-16 = (x^2)^2-4^2$
    $= (x^2+4)(x^2-4)$
    $= (x^2+4)(x+2)(x-2)$

47. $\dfrac{4}{9}x^2-w^2 = \left(\dfrac{2}{3}x\right)^2-(w)^2$
    $= \left(\dfrac{2}{3}x-w\right)\left(\dfrac{2}{3}x+w\right)$

49. q^6-64
    $= (q^3)^2-8^2$
    $= (q^3+8)(q^3-8)$
    $= (q+2)(q^2-2q+4)(q-2)(q^2+2q+4)$

51. $x^9+64y^3 = \left(x^3\right)^3+(4y)^3$
    $= \left(x^3+4y\right)\left(x^6-4x^3y+16y^2\right)$

53. $2x^3+3x^2-2x-3 = x^2(2x+3)-1(2x+3)$
    $= (2x+3)(x^2-1)$
    $= (2x+3)(x+1)(x-1)$

55. $16x^4-y^4 = (4x^2+y^2)(4x^2-y^2)$
    $= (4x^2+y^2)(2x+y)(2x-y)$

57. $81y^4-16 = (9y^2+4)(9y^2-4)$
    $= (9y^2+4)(3y+2)(3y-2)$

59. $a^3+b^6 = a^3+(b^2)^3$
    $= (a+b^2)(a^2-ab^2+b^4)$

61. $x^4 - y^4 = (x^2 + y^2)(x^2 - y^2)$
$= (x^2 + y^2)(x + y)(x - y)$

63. $k^3 + 4k^2 - 9k - 36 = k^2(k + 4) - 9(k + 4)$
$= (k + 4)(k^2 - 9)$
$= (k + 4)(k + 3)(k - 3)$

65. $2t^3 - 10t^2 - 2t + 10 = 2t^2(t - 5) - 2(t - 5)$
$= 2(t - 5)(t^2 - 1)$
$= 2(t - 5)(t + 1)(t - 1)$

67. $\dfrac{64}{125}p^3 - \dfrac{1}{8}q^3$

$= \left(\dfrac{4}{5}p\right)^3 - \left(\dfrac{1}{2}q\right)^3$

$= \left(\dfrac{4}{5}p - \dfrac{1}{2}q\right)\left(\dfrac{16}{25}p^2 + \dfrac{2}{5}pq + \dfrac{1}{4}q^2\right)$

69. $a^{12} + b^{12} = (a^4)^3 + (b^4)^3$
$= (a^4 + b^4)(a^8 - a^4 b^4 + b^8)$

71. (a)
$$\begin{array}{r} x^2 + 2x + 4 \\ x - 2 \overline{)\, x^3 + 0x^2 + 0x - 8} \\ \underline{-(x^3 - 2x^2)} \\ 2x^2 + 0x \\ \underline{-(2x^2 - 4x)} \\ 4x - 8 \\ \underline{-(4x - 8)} \end{array}$$

(b) $x^3 - 8 = (x - 2)(x^2 + 2x + 4)$

73. $x^2 + 4x + 16$

75. $2x + 1$

Problem Recognition Exercises

1. A prime polynomial cannot be factored further.

3. Look for the difference of squares: $a^2 - b^2$, a difference of cubes: $a^3 - b^3$, or a sum of cubes: $a^3 + b^3$

5. (a) Difference of squares

(b) $2a^2 - 162 = 2(a^2 - 81) = 2(a + 9)(a - 9)$

7. (a) None of these

(b) $6w^2 - 6w = 6w(w - 1)$

9. (a) Non-perfect square trinomial

(b) $3t^2 + 13t + 4 = (3t + 1)(t + 4)$

11. (a) Four terms-grouping

(b) $\quad 3ac + ad - 3bc - bd$
$= a(3c + d) - b(3c + d)$
$= (3c + d)(a - b)$

13. (a) Sum of cubes

(b) $y^3 + 8 = (y + 2)(y^2 - 2y + 4)$

15. (a) Non-perfect square trinomial

(b) $3q^2 - 9q - 12 = 3(q^2 - 3q - 4)$
$= 3(q - 4)(q + 1)$

17. (a) None of these

(b) $18a^2 + 12a = 6a(3a + 2)$

19. (a) Difference of squares

(b) $4t^2 - 100 = 4(t^2 - 25) = 4(t + 5)(t - 5)$

21. (a) Non-perfect square trinomial

(b) $10c^2 + 10c + 10 = 10(c^2 + c + 1)$

23. (a) Sum of cubes

(b) $x^3 + 0.001 = (x + 0.1)(x^2 - 0.1x + 0.01)$

25. (a) Perfect square trinomial

(b) $64 + 16k + k^2 = 8^2 + 2(8)(k) + k^2$
$= (8 + k)^2$

27. (a) Four terms-grouping

(b) $2x^2 + 2x - xy - y$
$= 2x(x+1) - y(x+1)$
$= (x+1)(2x-y)$

29. (a) Difference of cubes

(b) $a^3 - c^3 = (a-c)(a^2 + ac + c^2)$

31. (a) Non-perfect square trinomial

(b) $c^2 + 8c + 9$ is Prime

33. (a) Perfect square trinomial

(b) $b^2 + 10b + 25 = (b+5)^2$

35. (a) Non-perfect square trinomial

(b) $-p^3 - 5p^2 - 4p = -p(p^2 + 5p + 4)$
$= -p(p+1)(p+4)$

37. (a) Non-perfect square trinomial

39. (a) None of these

(b) $5a^2bc^3 - 7abc^2 = abc^2(5ac - 7)$

41. (a) Non-perfect square trinomial

(b) $t^2 + 2t - 63 = (t-7)(t+9)$

43. (a) Four terms-grouping

(b) $ab + ay - b^2 - by$
$= a(b+y) - b(b+y)$
$= (b+y)(a-b)$

45. (a) Non-perfect square trinomial

(b) $14u^2 - 11uv + 2v^2 = (7u - 2v)(2u - v)$

47. (a) Non-perfect square trinomial

(b) $4q^2 - 8q - 6 = 2(2q^2 - 4q - 3)$

49. (a) Sum of squares

(b) $9m^2 + 16n^2$ is prime

51. (a) Non-perfect square trinomial

(b) $6r^2 + 11r + 3 = (3r+1)(2r+3)$

53. (a) Difference of squares

(b) $16a^4 - 1 = (4a^2 + 1)(4a^2 - 1)$
$= (4a^2 + 1)(2a+1)(2a-1)$

55. (a) Perfect square trinomial

(b) $81u^2 - 90uv + 25v^2$
$= (9u)^2 - 2(9u)(5v) + (5v)^2$
$= (9u - 5v)^2$

57. (a) Non-perfect square trinomial

(b) $x^2 - 5x - 6 = (x-6)(x+1)$

59. (a) Four terms-grouping

(b) $2ax - 6ay + 4bx - 12by$
$= 2a(x-y) + 4b(x-y)$
$= 2(x - 3y)(a + 2b)$

61. (a) Non-perfect square trinomial

(b) $21x^4y + 41x^3y + 10x^2y$
$= x^2y(21x^2 + 41x + 10)$
$= x^2y(7x + 2)(3x + 5)$

63. (a) Four terms-grouping

(b) $8uv - 6u + 12v - 9$
$= 2u(4v - 3) + 3(4v - 3)$
$= (4v - 3)(2u + 3)$

65. (a) Perfect square trinomial

(b) $12x^2 - 12x + 3 = 3(4x^2 - 4x + 1)$
$= 3([2x]^2 - 2(2x)(1) + 1^2)$
$= 3(2x - 1)^2$

67. (a) Non-perfect square trinomial

(b) $6n^3 + 5n^2 - 4n$

$= n(6n^2 + 5n - 4)$

$= n(6n^2 - 3n + 8n - 4)$

$= n[3n(2n-1) + 4(2n-1)]$

$= n[(2n-1)(3n+4)]$

$= n(2n-1)(3n+4)$

69. (a) Difference of squares

(b) $64 - y^2 = 8^2 - y^2 = (8-y)(8+y)$

71. (a) Non-perfect square trinomial

(b) $b^2 - 4b + 10$ is prime.

73. (a) Non-perfect square trinomial

(b) $c^4 - 12c^2 + 20 = (c^2 - 10)(c^2 - 2)$

Section 6.7 Practice Exercises

1. (a) quadratic

(b) 0; 0

3. $4b^2 - 44b + 120 = 4(b^2 - 11b + 30)$

$= 4(b-6)(b-5)$

5. $3x^2 + 10x - 8 = (3x-2)(x+4)$

7. $4x^2 + 16y^2 = 4(x^2 + 4y^2)$

9. Neither

11. Quadratic

13. Linear

15. $(x+3)(x-1) = 0$

$x + 3 = 0$ or $x - 1 = 0$

$x = -3$ $\qquad x = 1$

17. $(2x-7)(2x+7) = 0$

$2x - 7 = 0$ or $2x + 7 = 0$

$2x = 7$ $\qquad 2x = -7$

$x = \dfrac{7}{2}$ $\qquad x = -\dfrac{7}{2}$

19. $3(x+5)(x+5) = 0$

$x + 5 = 0$

$x = -5$

21. $x(5x-1) = 0$

$x = 0$ or $5x - 1 = 0$

$x = 0$ or $\qquad 5x = 1$

$x = 0$ or $\qquad x = \dfrac{1}{5}$

23. The polynomial must be factored completely before applying the zero product rule.

25. $p^2 - 2p - 15 = 0$

$(p+3)(p-5) = 0$

$p + 3 = 0$ or $p - 5 = 0$

$p = -3$ $\qquad p = 5$

27. $z^2 + 10z - 24 = 0$

$(z+12)(z-2) = 0$

$z + 12 = 0$ or $z - 2 = 0$

$z = -12$ $\qquad z = 2$

29. $2q^2 - 7q - 4 = 0$

$(2q+1)(q-4) = 0$

$2q + 1 = 0$ or $q - 4 = 0$

$q = -\dfrac{1}{2}$ $\qquad q = 4$

31. $0 = 9x^2 - 4$

$0 = (3x+2)(3x-2)$

$3x + 2 = 0$ or $3x - 2 = 0$

$x = -\dfrac{2}{3}$ $\qquad x = \dfrac{2}{3}$

33. $2k^2 - 28k + 96 = 0$

$2(k^2 - 14k + 48) = 0$

$2(k-6)(k-8) = 0$

$k - 6 = 0$ or $k - 8 = 0$

$k = 6$ $\qquad k = 8$

35. $0 = 2m^3 - 5m^2 - 12m$

$m(2m^2 - 5m - 12) = 0$

$m(2m + 3)(m - 4) = 0$

$m = 0$ or $2m + 3 = 0$ or $m - 4 = 0$

$m = 0$ $\qquad m = -\dfrac{3}{2}$ $\qquad m = 4$

37. $(3p + 1)(p - 3)(p + 6) = 0$

$3p + 1 = 0$ or $p - 3 = 0$ or $p + 6 = 0$

$p = -\dfrac{1}{3}$ or $p = 3$ or $p = -6$

39. $x(x - 4)(2x + 3) = 0$

$x = 0$ or $x - 4 = 0$ or $2x + 3 = 0$

$x = 0$ or $x = 4$ or $2x = -3$

$x = 0$ or $x = 4$ or $x = -\dfrac{3}{2}$

41. $-5x(2x + 9)(x - 11) = 0$

$-5x = 0$ or $2x + 9 = 0$ or $x - 11 = 0$

$x = \dfrac{0}{-5}$ or $2x = -9$ or $x = 11$

$x = 0$ or $x = -\dfrac{9}{2}$ or $x = 11$

43. $x^3 - 16x = 0$

$x(x^2 - 16) = 0$

$x(x + 4)(x - 4) = 0$

$x = 0$ or $x + 4 = 0$ or $x - 4 = 0$

$x = 0$ or $x = -4$ or $x = 4$

45. $3x^2 + 18x = 0$

$3x(x + 6) = 0$

$x = 0$ or $x + 6 = 0$

$x = 0$ or $x = -6$

47. $16m^2 = 9$

$16m^2 - 9 = 0$

$(4m + 3)(4m - 3) = 0$

$4m + 3 = 0$ or $4m - 3 = 0$

$m = -\dfrac{3}{4}$ or $m = \dfrac{3}{4}$

49. $2y^3 + 14y^2 = -20y$

$2y^3 + 14y^2 + 20y = 0$

$2y(y^2 + 7y + 10) = 0$

$2y(y + 5)(y + 2) = 0$

$2y = 0$ or $y + 5 = 0$ or $y + 2 = 0$

$y = 0$ or $y = -5$ or $y = -2$

51. $5t - 2(t - 7) = 0$

$5t - 2t + 14 = 0$

$3t + 14 = 0$

$t = -\dfrac{14}{3}$

53. $2c(c - 8) = -30$

$2c^2 - 16c + 30 = 0$

$2(c^2 - 8c + 15) = 0$

$2(c - 5)(c - 3) = 0$

$c - 5 = 0$ or $c - 3 = 0$

$c = 5$ or $c = 3$

55. $b^3 = -4b^2 - 4b$

$b^3 + 4b^2 + 4b = 0$

$b(b^2 + 4b + 4) = 0$

$b(b + 2)(b + 2) = 0$

$b = 0$ or $b + 2 = 0$ or $b + 2 = 0$

$b = 0$ or $b = -2$ or $b = -2$

57. $3(a^2 + 2a) = 2a^2 - 9$

$3a^2 + 6a = 2a^2 - 9$

$a^2 + 6a + 9 = 0$

$(a + 3)(a + 3) = 0$

$a + 3 = 0$

$a = -3$

59. $2n(n + 2) = 6$

$2n^2 + 4n - 6 = 0$

$2(n^2 + 2n - 3) = 0$

$2(n + 3)(n - 1) = 0$

$n + 3 = 0$ or $n - 1 = 0$

$n = -3$ $\qquad n = 1$

175

61. $x(2x+5)-1=2x^2+3x+2$

$2x^2+5x-1=2x^2+3x+2$

$2x-3=0$

$x=\dfrac{3}{2}$

63. $27q^2=9q$

$27q^2-9q=0$

$9q(3q-1)=0$

$9q=0$ or $3q-1=0$

$q=0$ $\qquad q=\dfrac{1}{3}$

65. $3(c^2-2c)=0$

$3c(c-2)=0$

$3c=0$ or $c-2=0$

$c=0$ $\qquad c=2$

67. $y^3-3y^2-4y+12=0$

$y^2(y-3)-4(y-3)=0$

$(y-3)(y+2)(y-2)=0$

$y-3=0$ or $y+2=0$ or $y-2=0$

$y=3$ $\qquad y=-2$ $\qquad y=2$

69. $(x-1)(x+2)=18$

$x^2+x-2-18=0$

$x^2+x+20=0$

$(x+5)(x-4)=0$

$x+5=0$ or $x-4=0$

$x=-5$ $\qquad x=4$

71. $(p+2)(p+3)=1-p$

$p^2+5p+6=1-p$

$p^2+6p+5=0$

$(p+5)(p+1)=0$

$p+5=0$ or $p+1=0$

$p=-5$ $\qquad p=-1$

Problem Recognition Exercises

1. (a) $x^2+6x-7=(x+7)(x-1)$

(b) $(x+7)(x-1)=0$

$x+7=0$ or $x-1=0$

$x=-7$ $\qquad x=1$

3. (a) $2y^2+7y+3=(2y+1)(y+3)$

(b) $(2y+1)(y+3)=0$

$2y+1=0$ or $y+3=0$

$y=-\dfrac{1}{2}$ $\qquad y=-3$

5. (a) $5q^2+q-4=(5q-4)(q+1)=0$

$5q-4=0$ or $q+1=0$

$q=\dfrac{4}{5}$ $\qquad q=-1$

(b) $5q^2+q-4=(5q-4)(q+1)$

7. (a) $a^2-64=(a+8)(a-8)=0$

$a+8=0$ or $a-8=0$

$a=-8$ $\qquad a=8$

(b) $a^2-64=(a+8)(a-8)$

9. (a) $4b^2-81=(2b+9)(2b-9)$

(b) $(2b+9)(2b-9)=0$

$2b+9=0$ or $2b-9=0$

$b=-\dfrac{9}{2}$ $\qquad b=\dfrac{9}{2}$

11. (a) $8x^2+16x+6=2(4x^2+8x+3)$

$\qquad\qquad\qquad =2(2x+3)(2x+1)=0$

$2x+3=0$ or $2x+1=0$

$x=-\dfrac{3}{2}$ $\qquad x=-\dfrac{1}{2}$

(b) $8x^2+16x+6=2(2x+3)(2x+1)$

13. (a) $x^3-8x^2-20x=x(x^2-8x-20)$

$\qquad\qquad\qquad\quad =x(x-10)(x+2)$

(b) $x(x-10)(x+8)=0$

$x=0$ or $x-10=0$ or $x+2=0$

$x=0$ $\qquad x=10$ $\qquad x=-2$

15. **(a)** $b^3 + b^2 - 9b - 9$

$\qquad = b^2(b+1) - 9(b+1)$

$\qquad = (b+1)(b^2 - 9)$

$\qquad = (b+1)(b+3)(b-3) = 0$

$b+1=0$ or $b+3=0$ or $b-3=0$

$\quad b=-1 \qquad\quad b=-3 \qquad\quad b=3$

(b) $b^3 + b^2 - 9b - 9 = (b+1)(b-3)(b+3)$

17. $2s^2 - 6s + rs - 3r = 2s(s-3) + r(s-3)$

$\qquad\qquad\qquad\qquad = (s-3)(2s+r)$

19. $8x^3 - 2x = 2x(4x^2 - 1)$

$\qquad\qquad = 2x(2x-1)(2x+1) = 0$

$2x=0$ or $2x-1=0$ or $2x+1=0$

$\qquad\qquad\quad x=\dfrac{1}{2} \qquad\quad x=-\dfrac{1}{2}$

$x=0$

21. $2x^3 - 4x^2 + 2x = 2x(x^2 - 2x + 1)$

$\qquad\qquad\qquad = 2x(x-1)^2 = 0$

$2x=0$ or $x-1=0$ or $x-1=0$

$x=0 \qquad\quad x=1 \qquad\quad x=1$

23. $7c^2 - 2c + 3 = 7(c^2 + c)$

$7c^2 - 2c + 3 = 7c^2 + 7c$

$\qquad\qquad 3 = 9c$

$\qquad\qquad c = \dfrac{3}{9}$

$\qquad\qquad c = \dfrac{1}{3}$

25. $8w^3 + 27 = (2w)^3 + (3)^3$

$\qquad\qquad = (2w+3)(4w^2 - 6w + 9)$

27. $5z^2 + 2z = 7$

$5z^2 + 2z - 7 = 0$

$(5z+7)(z-2) = 0$

$5z+7=0$ or $z-2=0$

$\quad z=-\dfrac{7}{5} \qquad\quad z=2$

29. $3b(b+6) = b-10$

$3b^2 + 18b = b-10$

$3b^2 + 17b + 10 = 0$

$(3b+2)(b+5) = 0$

$3b+2=0$ or $b+5=0$

$b=-\dfrac{2}{3} \qquad\quad b=-5$

31. $5(2x-3) - 2(3x+1) = 4 - 3x$

$10x - 15 - 6x - 2 = 4 - 3x$

$\qquad\quad 4x - 17 = 4 - 3x$

$\qquad\qquad\quad 7x = 21$

$\qquad\qquad\quad x = 3$

33. $4s^2 = 64$

$4s^2 - 64 = 0$

$4(s^2 - 16) = 0$

$4(s+4)(s-4) = 0$

$s+4=0$ or $s-4=0$

$s=-4 \qquad\quad s=4$

35. $(x-3)(x-4) = 6$

$x^2 - 7x + 12 = 6$

$x^2 - 7x + 6 = 0$

$(x-6)(x-1) = 0$

$x-6=0$ or $x-1=0$

$x=6 \qquad\quad x=1$

Section 6.8 Practice Exercises

1. **(a)** $x+1$

(b) $x+2$

(c) $x+2$

(d) lw

(e) $\dfrac{1}{2}bh$

(f) $a^2 + b^2 = c^2$

3. $9x(3x + 2) = 0$

$9x = 0 \quad \text{or} \quad 3x + 2 = 0$

$x = 0 \qquad\qquad x = -\dfrac{2}{3}$

5. $x^2 - 5x = 6$

$x^2 - 5x - 6 = 0$

$(x - 6)(x + 1) = 0$

$x - 6 = 0 \quad \text{or} \quad x + 1 = 0$

$x = 6 \qquad\qquad x = -1$

7. $6x^2 - 7x - 10 = 0$

$(6x + 5)(x - 2) = 0$

$6x + 5 = 0 \quad \text{or} \quad x - 2 = 0$

$x = -\dfrac{5}{6} \qquad\qquad x = 2$

9. Let x = the number. Then,

$x^2 + 11 = 60$

$x^2 - 49 = 0$

$(x + 7)(x - 7) = 0$

$x + 7 = 0 \quad \text{or} \quad x - 7 = 0$

$x = -7 \qquad\qquad x = 7$

The numbers are 7 and –7.

11. Let x = a number. Then,

$12 + 6x = x^2 - 28$

$x^2 - 6x - 40 = 0$

$(x + 4)(x - 10) = 0$

$x + 4 = 0 \quad \text{or} \quad x - 10 = 0$

$x = -4 \qquad\qquad x = 10$

The numbers are –4 and 10.

13. Let x = first integer. Then the next consecutive odd integer is $(x + 2)$.

$x(x + 2) = 63$

$x^2 + 2x - 63 = 0$

$(x + 9)(x - 7) = 0$

$x + 9 = 0 \quad \text{or} \quad x - 7 = 0$

$x = -9 \qquad\qquad x = 7$

The numbers are –9 and –7, or 7 and 9.

15. Let x = first integer. Then the next consecutive integer is $(x + 1)$.

$x^2 + (x + 1)^2 = 10(x + 1) + 1$

$x^2 + x^2 + 2x + 1 = 10x + 11$

$2x^2 - 8x - 10 = 0$

$2(x + 1)(x - 5) = 0$

$x + 1 = 0 \quad \text{or} \quad x - 5 = 0$

$x = -1 \qquad\qquad x = 5$

The numbers are –1 and 0, or 5 and 6.

17. Let x = the width of the painting and $x + 2$ is the height (or length) of the painting.

$A = lw$

$99 = (x + 2)(x)$

$99 = x^2 + 2x$

$x^2 + 2x - 99 = 0$

$(x + 11)(x - 9) = 0$

$x + 11 = 0 \quad \textbf{or} \quad x - 9 = 0$

$x = \cancel{-11} \qquad\qquad x = 9$

Since x represents the width, it must be positive. The width of the painting is 9 ft, and the height of the painting is 11 ft.

19. Let x = length and $x - 3$ be the width.

(a) $\qquad A = lw$

$28 = x(x - 3)$

$x^2 - 3x - 28 = 0$

$(x + 4)(x - 7) = 0$

$x = -4 \text{ or } x = 7$

The dimensions are 7 m by 4 m.

(b) $P = 2w + 2l$

$P = 2(4) + 2(7)$

$P = 8 + 14 = 22$ m

21. Let x = height and $x + 3$ = the base.

$A = \dfrac{1}{2}bh$

$14 = \dfrac{1}{2}(x)(x + 3)$

$28 = x^2 + 3x$

$x^2 + 3x - 28 = 0$

$(x + 7)(x - 4) = 0$

$x + 7 = 0$ or $x - 4 = 0$

$x = \cancel{-7}$ $x = 4$

The height is 4 ft and the base is 7 ft.

23. Let $x =$ the base of the triangle and $3x - 7$ is the height of the triangle.

$A = \dfrac{1}{2}bh$

$20 = \dfrac{1}{2}(x)(3x - 7)$

$40 = (x)(3x - 7)$

$40 = 3x^2 - 7x$

$3x^2 - 7x - 40 = 0$

$(3x + 8)(x - 5) = 0$

$3x + 8 = 0$ or $x - 5 = 0$

$x = \cancel{-\dfrac{8}{3}}$ $x = 5$

Since x is the base of the triangle, it must be positive. The base is 5 cm and the height is $3(5) - 7$ or 8 cm.

25. If you let $h = 0$, then,

$0 = -16t^2 + 144$

$0 = -16(t^2 - 9)$

$-16(t + 3)(t - 3) = 0$

$t + 3 = 0$ or $t - 3 = 0$

$t = -3$ $t = 3$

It will take 3 seconds to hit the ground.

27. Ground level is when $h = 0$. Then,

$0 = -16t^2 + 24t$

$0 = -8t(2t - 3)$

$-8t = 0$ or $2t - 3 = 0$

$t = 0$ $t = \dfrac{3}{2} = 1.5$

The times are 0 seconds and 1.5 seconds.

29. Pictures may vary.

31.

$c^2 = a^2 + b^2$

$c^2 = 24^2 + 7^2$

$c^2 = 625$

$c^2 - 625 = 0$

$(c + 25)(c - 25) = 0$

$c = \cancel{-25}$ or $c = 25$

$c = 25$ cm

33. $c^2 = a^2 + b^2$

$17^2 = a^2 + 8^2$

$289 = a^2 + 64$

$a^2 - 225 = 0$

$(a + 15)(a - 15) = 0$

$a = \cancel{-15}$ or $a = 15$

$a = 15$ in.

35. Let $x = c$.

$c^2 = a^2 + b^2$

$x^2 = 16^2 + 12^2$

$x^2 = 400$

$x^2 - 400 = 0$

$(x + 20)(x - 20) = 0$

$x = \cancel{-20}$ or $x = 20$

The brace is 20 in. long.

37. Let $h =$ height of the kite. Then, $a = h - 1$.

$c^2 = a^2 + b^2$

$30^2 = (h - 1)^2 + 24^2$

$900 = h^2 - 2h + 1 + 576$

$h^2 - 2h - 323 = 0$

$(h + 17)(h - 19) = 0$

$h = \cancel{-17}$ or $h = 19$

The kit is 19 yd high.

39. Let $x =$ distance between the base of the ladder and the bottom of the house. Then, the distance between the top of the ladder and the ground is $x + 7$ ft.

179

$$c^2 = a^2 + b^2$$
$$17^2 = x^2 + (x+7)^2$$
$$289 = x^2 + x^2 + 14x + 49$$
$$2x^2 + 14x - 240 = 0$$
$$2(x+15)(x-8) = 0$$
$$x = \cancel{-15} \text{ or } x = 8$$

The bottom of the ladder is 8 ft from the house. The distance from the top of the ladder to the ground is 15 ft.

41. Let x = hypotenuse. Then, one leg is $x - 4$ m and the other leg is $x - 2$ m.
$$x^2 = (x-4)^2 + (x-2)^2$$
$$x^2 = x^2 - 8x + 16 + x^2 - 4x + 4$$
$$x^2 = 2x^2 - 12x + 20$$
$$x^2 - 12x + 20 = 0$$
$$(x-10)(x-2) = 0$$
$$x = 10 \text{ or } x = \cancel{2}$$
The hypotenuse is 10 m.

Group Activity

1. Answers will vary. For example:
$$2x^3 - 10x^2 + 2x = 2x(x^2 - 5x + 1)$$

3. Answers will vary. For example:
$$x^2 - 5x - 14 = (x+2)(x-7)$$

5. Answers will vary. For example:
$$7x^3 - 14x^2 - 21x = 7x(x-3)(x+1)$$

7. Answers will vary. For example:
$$216 - 125b^3 = (6 - 5b)(36 + 30b + 25b^2)$$

9. Answers will vary. For example:
$$(x-4)(x+7) = 0$$
$$x^2 + 3x - 28 = 0$$

Chapter 6 Review Exercises

Section 6.1

1. GCF: $3a^2b$

3. GCF: $2c(3c - 5)$

5. $6x^2 + 2x^4 - 8x = 2x(3x + x^3 - 4)$

7. $-t^2 + 5t = -t(t-5)$ or $t(-t+5)$

9. $3b(b+2) - 7(b+2) = (b+2)(3b-7)$

11. $7w^2 + 14w + wb + 2b = 7w(w+2) + b(w+2)$
$$= (w+2)(7w+b)$$

13. $60y^2 - 45y - 12y + 9$
$$= 3(20y^2 - 15y - 4y + 3)$$
$$= 3(5y(4y - 3) - 1(4y - 3))$$
$$= 3(4y - 3)(5y - 1)$$

Section 6.2

15. $x^2 - 10x + 21 = x^2 - 7x - 3x + 21$
$$= x(x - 7) - 3(x - 7)$$
$$= (x - 3)(x - 7)$$

17. $-6z + z^2 - 72 = z^2 - 6z - 72$
$$= z^2 + 6z - 12z - 72$$
$$= z(z + 6) - 12(z + 6)$$
$$= (z - 12)(z + 6)$$

19. $3p^2w + 36pw + 60w$
$$= 3w(p^2 + 12p + 20)$$
$$= 3w(p^2 + 2p + 10p + 20)$$
$$= 3w(p(p + 2) + 10(p + 2))$$
$$= 3w(p + 10)(p + 2)$$

21. $-t^2 + 10t - 16 = -(t^2 - 10t + 16)$
$$= -(t^2 - 2t - 8t + 16)$$
$$= -(t(t - 2) - 8(t - 2))$$
$$= -(t - 8)(t - 2)$$

23. $a^2 + 12ab + 11b^2 = a^2 + 11ab + ab + 11b^2$
$$= a(a + 11b) + b(a + 11b)$$
$$= (a + b)(a + 11b)$$

Section 6.3

25. Different

27. Both positive

29. $2y^2 - 5y - 12 = 2y^2 - 8y + 3y - 12$
$$= 2y(y-4) + 3(y-4)$$
$$= (2y+3)(y-4)$$

31. $10z^2 + 29z + 10 = 10z^2 + 4z + 25z + 10$
$$= 2z(5z+2) + 5(5z+2)$$
$$= (2z+5)(5z+2)$$

33. $2p^2 - 5p + 1$ is prime.

35. $10w^2 - 60w - 270 = 10(w^2 - 6w - 27)$
$$= 10(w^2 - 9w + 3w - 27)$$
$$= 10(w(w-9) + 3(w-9))$$
$$= 10(w+3)(w-9)$$

37. $9c^2 - 30cd + 25d^2$
$$= 9c^2 - 15cd - 15cd + 25d^2$$
$$= 3c(3c-5d) - 5d(3c-5d)$$
$$= (3c-5d)(3c-5d)$$
$$= (3c-5d)^2$$

39. $6g^2 + 7gh + 2h^2 = 6g^2 + 3gh + 4gh + 2h^2$
$$= 3g(2g+h) + 2h(2g+h)$$
$$= (2g+h)(3g+2h)$$

41. $v^4 - 2v^2 - 3 = (v^2)^2 - 2v^2 - 3$
$$= (v^2)^2 + v^2 - 3v^2 - 3$$
$$= v^2(v^2+1) - 3(v^2+1)$$
$$= (v^2-3)(v^2+1)$$

Section 6.4

43. $5, -1$

45. $3c^2 - 5c - 2 = 3c^2 - 6c + c - 2$
$$= 3c(c-2) + 1(c-2)$$
$$= (c-2)(3c+1)$$

47. $t^2 + 13t + 12 = t^2 + 12t + t + 12$
$$= t(t+12) + 1(t+12)$$
$$= (t+1)(t+12)$$

49. $w^4 + 7w^2 + 10 = w^4 + 2w^2 + 5w^2 + 10$
$$= w^2(w^2+2) + 5(w^2+2)$$
$$= (w^2+2)(w^2+5)$$

51. $-40v^2 - 22v + 6 = -2(20v^2 + 11v - 3)$
$$= -2(20v^2 + 15v - 4v - 3)$$
$$= -2(5v(4v+3) - 1(4v+3))$$
$$= -2(4v+3)(5v-1)$$

53. $a^3b - 10a^2b^2 + 24ab^3$
$$= ab(a^2 - 10ab + 24b^2)$$
$$= ab(a^2 - 6ab - 4ab + 24b^2)$$
$$= ab(a(a-6b) - 4b(a-6b))$$
$$= ab(a-6b)(a-4b)$$

55. $m + 9m^2 - 2$ is prime.

57. $49x^2 + 140x + 100$
$$= (7x)^2 + 2(7x)(10) + 10^2$$
$$= (7x+10)^2$$

Section 6.5

59. $a^2 - b^2 = a(a+b) - b(a+b)$
$$= (a-b)(a+b)$$

61. $a^2 - 49 = a^2 - 7^2 = (a+7)(a-7)$

63. $100 - 81t^2 = 10^2 - (9t)^2 = (10+9t)(10-9t)$

65. $x^2 + 16$; this is a sum of squares, not a difference.

67. $y^2 + 12y + 36 = (y+6)^2$

69. $9a^2 - 12a + 4 = (3a-2)^2$

71. $-3v^2 - 12v - 12 = -3(v^2 + 4v + 4)$
$$= -3(v+2)^2$$

73. $2c^4 - 18 = 2(c^4 - 9) = 2(c^2+3)(c^2-3)$

75. $p^3 + 3p^2 - 16p - 48$
$= p^3 + 3p^2 - 16(p+3)$
$= p^2(p+3) - 16(p+3)$
$= (p^2 - 16)(p+3)$
$= (p-4)(p+4)(p+3)$

Section 6.6

77. $a^3 + b^3 = (a+b)(a^2 - ab + b^2)$

79. $64 + a^3 = 4^3 + a^3$
$= (4+a)(4^2 - 4a + a^2)$
$= (4+a)(16 - 4a + a^2)$

81. $p^6 + 8 = (p^2)^3 + 2^3$
$= (p^2 + 2)((p^2)^2 - p^2 2 + 2^2)$
$= (p^2 + 2)(p^4 - 2p^2 + 4)$

83. $6x^3 - 48 = 6(x^3 - 8)$
$= 6(x^3 - 2^3)$
$= 6(x-2)(x^2 + 2x + 4)$

85. $x^3 - 36x = x(x^2 - 36) = x(x+6)(x-6)$

87. $8h^2 + 20 = 4(2h^2 + 5)$

89. $x^3 + 4x^2 - x - 4 = x^2(x+4) - 1(x+4)$
$= (x+4)(x^2 - 1)$
$= (x+4)(x+1)(x-1)$

91. $8n + n^4 = n(8 + n^3) = n(2+n)(4 - 2n + n^2)$

Section 6.7

93. The equation $(x-3)(2x+1) = 0$ can be solved directly by the zero product rule because it is a product of factors set equal to zero.

95. $(a-9)(2a-1) = 0$
$a - 9 = 0$ or $2a - 1 = 0$
$a = 9$ $a = \dfrac{1}{2}$

97. $6u(u-7)(4u-9) = 0$
$6u = 0$ or $u - 7 = 0$ or $4u - 9 = 0$
$u = 0$ $u = 7$ $u = \dfrac{9}{4}$

99. $4h^2 - 23h - 6 = 0$
$(4h+1)(h-6) = 0$
$4h + 1 = 0$ or $h - 6 = 0$
$h = -\dfrac{1}{4}$ $h = 6$

101. $\qquad r^2 = 25$
$r^2 - 25 = 0$
$(r+5)(r-5) = 0$
$r + 5 = 0$ or $r - 5 = 0$
$r = -5$ $r = 5$

103. $\qquad x(x-6) = -8$
$x^2 - 6x + 8 = 0$
$(x-4)(x-2) = 0$
$x - 4 = 0$ or $x - 2 = 0$
$x = 4$ $x = 2$

105. $\qquad 9s^2 + 12s = -4$
$9s^2 + 12s + 4 = 0$
$(3s+2)(3s+2) = 0$
$3s + 2 = 0$
$s = -\dfrac{2}{3}$

107. $\qquad 2(p^2 - 66) = -13p$
$2p^2 + 13p - 132 = 0$
$(2p-11)(p+12) = 0$
$2p - 11 = 0$ or $p + 12 = 0$
$p = \dfrac{11}{2}$ $p = -12$

109. $\qquad x^3 - 4x = 0$
$x(x^2 - 4) = 0$
$x(x+2)(x-2) = 0$
$x = 0$ or $x + 2 = 0$ or $x - 2 = 0$
$x = 0$ or $x = -2$ or $x = 2$

Section 6.8

111. At $h = 0$, the ball is ground level.
$$0 = -16x^2 + 16x$$
$$0 = -16x(x - 1)$$
$x = 0$ or $x = 1$
The ball is at ground level at 0 and 1 second.

113. Let x = shorter leg. Then, the other leg is $x + 2$ and the hypotenuse is $2x - 2$.
$$x^2 + (x + 2)^2 = (2x - 2)^2$$
$$x^2 + x^2 + 4x + 4 = 4x^2 - 8x + 4$$
$$2x^2 - 12x = 0$$
$$2x(x - 6) = 0$$
$x = 0$ or $x = 6$
The legs are 6 ft and 8 ft; the hypotenuse is 10 ft.

115. x = first integer and $x + 1$ is the next integer.
$$x(x + 1) = 14(x + x + 1) + 44$$
$$x^2 + x = 28x + 14 + 44$$
$$x^2 - 27x - 58 = 0$$
$$(x + 2)(x - 29) = 0$$
$x = -2$ and $x = 29$
The numbers are -2 and -1 or 29 and 30.

Chapter 6 Test

1. $15x^4 - 3x + 6x^3 = 3x(5x^3 - 1 + 2x^2)$

3. $6w^2 - 43w + 7 = (6w - 1)(w - 7)$

5. $q^2 - 16q + 64 = (q - 8)^2$

7. $a^2 + 12a + 32 = (a + 4)(a + 8)$

9. $2y^2 - 17y + 8 = (2y - 1)(y - 8)$

11. $9t^2 - 100 = (3t + 10)(3t - 10)$

13. $3a^2 + 27ab + 54b^2 = 3(a^2 + 9ab + 18b^2)$
$$= 3(a + 6b)(a + 3b)$$

15. $xy - 7x + 3y - 21 = x(y - 7) + 3(y - 7)$
$$= (y - 7)(x + 3)$$

17. $-10u^2 + 30u - 20 = -10(u^2 - 3u + 2)$
$$= -10(u - 2)(u - 1)$$

19. $5y^2 - 50y + 125 = 5(y^2 - 10y + 25)$
$$= 5(y - 5)(y - 5)$$
$$= 5(y - 5)^2$$

21. $2x^3 + x^2 - 8x - 4 = x^2(2x + 1) - 4(2x + 1)$
$$= (2x + 1)(x^2 - 4)$$
$$= (2x + 1)(x + 2)(x - 2)$$

23. $m^2n^2 - 81 = (mn + 9)(mn - 9)$

25. $64x^3 - 27y^6$
$$= (4x - 3y^2)(16x^2 + 12xy^2 + 9y^4)$$

27. $(2x - 3)(x + 5) = 0$
$2x - 3 = 0$ or $x + 5 = 0$
$$x = \frac{3}{2} \qquad x = -5$$

29. $\qquad x^2 - 6x = 16$
$$x^2 - 6x - 16 = 0$$
$$(x + 2)(x - 8) = 0$$
$x + 2 = 0$ or $x - 8 = 0$
$\qquad x = -2 \qquad x = 8$

31. $\qquad y^3 + 10y^2 - 9y - 90 = 0$
$$y^2(y + 10) - 9(y + 10) = 0$$
$$(y^2 - 9)(y + 10) = 0$$
$$(y + 3)(y - 3)(y + 10) = 0$$
$y + 3 = 0$ or $y - 3 = 0$ or $y + 10 = 0$
$\quad y = -3$ or $\quad y = 3$ or $\quad y = -10$

33. Let x represent the first odd integer. Then $x + 2$ represents the second odd integer.
$$x(x + 2) = 35$$
$$x^2 + 2x = 35$$
$$x^2 + 2x - 35 = 0$$
$$(x + 7)(x - 5) = 0$$

$x + 7 = 0$ or $x - 5 = 0$

$x = -7$ or $x = 5$

$x + 2 = -5$ or $x + 2 = 7$

The two integers are -5 and -7 or 5 and 7.

35. Let x = shorter leg, $3x - 3$ the longer leg, and $3x - 2$ the length of the hypotenuse.

$$(x)^2 + (3x - 3)^2 = (3x - 2)^2$$

$$x^2 + 9x^2 - 18x + 9 = 9x^2 - 12x + 4$$

$$x^2 - 6x + 5 = 0$$

$$(x - 1)(x - 5) = 0$$

$x = 1$ or $x = 5$

The shorter leg is 5 ft.

Cumulative Review Exercises
Chapters 1–6

1. $\dfrac{|4 - 25 \div (-5) \cdot 2|}{\sqrt{8^2 + 6^2}} = \dfrac{|4 - (-5) \cdot 2|}{\sqrt{64 + 36}}$

$= \dfrac{|4 - (-10)|}{\sqrt{100}}$

$= \dfrac{|14|}{10}$

$= \dfrac{7}{5}$

3. $3x - 2y = 8$

$-2y = 8 - 3x$

$\dfrac{-2y}{-2} = \dfrac{8 - 3x}{-2}$

$= \dfrac{3x - 8}{2}$ or $y = \dfrac{8 - 3x}{-2}$

$y = \dfrac{3}{2}x - 4$

5. $-\dfrac{5}{12}x \le \dfrac{5}{3}$

$-\dfrac{12}{5} \cdot \left(-\dfrac{5}{12}x\right) \le -\dfrac{12}{5} \cdot \dfrac{5}{3}$

$x \ge -4$

$[-4, \infty)$

7. (a) Vertical line

(b) Undefined

(c) $(5, 0)$

(d) Does not exist

9. $2x - 3y = 4$

$5x - 6y = 13$

Multiply the first equation by -2 to obtain opposite coefficients on y. Then add the equations and solve the resulting equation for x.

$-4x + 6y = -8$

$\underline{5x - 6y = 13}$

$x = 5$

$2(5) - 3y = 4$

$10 - 3y = 4$

$-3y = -6$

$y = 2$

Solution: $(5, 2)$

11. $(4p^2 - 5p - 1)(2p - 3)$

$= 8p^3 - 10p^2 - 2p - 12p^2 + 15p + 3$

$= 8p^3 - 22p^2 + 13p + 3$

13.

$$
\begin{array}{r}
r^3 + 5r^2 + 15r + 40 \\
r - 3 \overline{\smash{\big)}\, r^4 + 2r^3 + 0r^2 - 5r + 1} \\
\underline{-(r^4 - 3r^3)} \\
5r^3 + 0r^2 \\
\underline{-(5r^3 - 15r^2)} \\
15r^2 - 5r \\
\underline{-(15r^2 - 45r)} \\
40r + 1 \\
\underline{-(40r - 120)} \\
121
\end{array}
$$

$r^3 + 5r^2 + 15r + 40 + \dfrac{121}{r - 3}$

15. $\left(\dfrac{6a^2b^{-4}}{2a^4b^{-5}}\right)^{-2} = \dfrac{6^{-2}a^{-4}b^8}{2^{-2}a^{-8}b^{10}}$

$\qquad = \dfrac{6^{-2}a^{-4-(-8)}b^{8-10}}{2^{-2}}$

$\qquad = \dfrac{2^2 a^4 b^{-2}}{6^2}$

$\qquad = \dfrac{4a^4}{36b^2} = \dfrac{a^4}{9b^2}$

17. $w^4 - 16 = (w^2)^2 - 4^2$

$\qquad = (w^2 + 4)(w^2 - 4)$

$\qquad = (w^2 + 4)(w + 2)(w - 2)$

19. $4x^2 - 8x - 5 = (2x + 1)(2x - 5)$

Chapter 7

Chapter 7 opener

1. $\dfrac{7}{4}+\dfrac{5}{3}+\dfrac{1}{2}=\dfrac{7}{4}\cdot\dfrac{3}{3}+\dfrac{5}{3}\cdot\dfrac{4}{4}+\dfrac{1}{2}\cdot\dfrac{6}{6}$

$\qquad\qquad\quad =\dfrac{21}{12}+\dfrac{20}{12}+\dfrac{6}{12}$

$\qquad\qquad\quad =\dfrac{47}{12}$

3. t.

5. a.

$\dfrac{p\ r\ o\ c\ r\ a\ s\ t\ i\ n\ a\ t\ i\ o\ n}{4\ 12\quad 15\ \ 3\quad\ \ \ 53\ \ 2}$.

Section 7.1 Practice Exercises

1. (a) rational

    (b) denominator; zero

    (c) $\dfrac{p}{q}$

    (d) $1;\ -1$

3. $\dfrac{t-2}{t^2-4t+8}=\dfrac{2-2}{2^2-4(2)+8}$

$\qquad\qquad\quad =\dfrac{0}{4}$

$\qquad\qquad\quad =0$

5. $\dfrac{1}{x-6}=\dfrac{1}{-2-6}$

$\qquad\quad =\dfrac{1}{-8}$

$\qquad\quad =-\dfrac{1}{8}$

7. $\dfrac{w-4}{2w+8}=\dfrac{0-4}{2(0)+8}$

$\qquad\qquad =\dfrac{-4}{8}$

$\qquad\qquad =-\dfrac{1}{2}$

9. $\dfrac{(a-7)(a+1)}{(a-2)(a+5)}=\dfrac{(2-7)(2+1)}{(2-2)(2+5)}$

$\qquad\qquad\qquad =\dfrac{(-5)(3)}{0(7)}$

$\qquad\qquad\qquad =-\dfrac{15}{0};\ \text{undefined}$

11. (a) $t=\dfrac{24}{x}+\dfrac{24}{x+8}$

$\qquad\quad =\dfrac{24}{12}+\dfrac{24}{12+8}$

$\qquad\quad =2+\dfrac{24}{20}$

$\qquad\quad =2+1\dfrac{4}{20}$

$\qquad\quad =3\dfrac{1}{5}$

$\qquad 3\dfrac{1}{5}\ \text{hr or 3.2 hr}$

    (b) $t=\dfrac{24}{x}+\dfrac{24}{x+8}$

$\qquad\quad =\dfrac{24}{24}+\dfrac{24}{24+8}$

$\qquad\quad =1+\dfrac{24}{32}$

$\qquad\quad =1+\dfrac{3}{4}$

$\qquad\quad =1\dfrac{3}{4}$

$\qquad 1\dfrac{3}{4}\ \text{hr or 1.75 hr}$

$\$196$

$\$168$

186

13. $k = -2$

15. $x = \dfrac{5}{2}$, $x = -8$

17. $m = -2$, $m = -3$

19. There are no restricted values.

21. There are no restricted values.

23. $t = 0$

25. Answers will vary. For example: $\dfrac{1}{x-2}$

27. Answers will vary. For example:
$$\dfrac{1}{(x+3)(x-7)}$$

29. (a) $\dfrac{3x^2 - 2x - 1}{6x^2 - 7x - 3} = \dfrac{3(-1)^2 - 2(-1) - 1}{6(-1)^2 - 7(-1) - 3}$

$= \dfrac{3 + 2 - 1}{6 + 7 - 3}$

$= \dfrac{4}{10}$

$= \dfrac{2}{5}$

(b) $\dfrac{x-1}{2x-3} = \dfrac{-1-1}{2(-1)-3}$

$= \dfrac{-2}{-5}$

$= \dfrac{2}{5}$

31. (a) $\dfrac{5x+5}{x^2-1} = \dfrac{5(1)+5}{1^2-1}$

$= \dfrac{10}{0}$; undefined

(b) $\dfrac{5}{x-1} = \dfrac{5}{1-1}$

$= \dfrac{5}{0}$; undefined

33. (a) What value of y makes the
$$6y + 12 = 0$$
denominator equal 0? $6y = -12$; $y =$
$$y = -2$$
-2

(b) $\dfrac{3y+6}{6y+12} = \dfrac{3\cancel{(y+2)}}{6\cancel{(y+2)}}$

$= \dfrac{3}{6}$

$= \dfrac{1}{2}$

35. (a) What value of t makes the
denominator equal 0? $\begin{array}{l} t + 1 = 0 \\ t = -1 \end{array}$; $t =$
-1

(b) $\dfrac{t^2-1}{t+1} = \dfrac{\cancel{(t+1)}(t-1)}{\cancel{t+1}}$

$= t - 1$

37. (a) What value of w makes the
denominator equal 0?
$$21w^2 - 35w = 0$$
$$7w(3w - 5) = 0$$
$$7w = 0 \quad \text{or} \quad 3w - 5 = 0$$
$$w = 0 \quad \text{or} \quad 3w = 5$$
$$w = \dfrac{5}{3}$$

$w = 0$, $w = \dfrac{5}{3}$

(b) $\dfrac{7w}{21w^2 - 35w} = \dfrac{\cancel{7w}}{\cancel{7w}(3w-5)}$

$= \dfrac{1}{3w-5}$

39. (a) What value of x makes the

$$6x + 4 = 0$$
$$6x = -4$$

denominator equal 0? $\quad x = -\dfrac{4}{6}$ $\qquad$;

$$x = -\dfrac{2}{3}$$

$$x = -\dfrac{2}{3}$$

(b) $\dfrac{9x^2 - 4}{6x + 4} = \dfrac{\cancel{(3x+2)}\,(3x-2)}{2\,\cancel{(3x+2)}}$

$$= \dfrac{3x - 2}{2}$$

41. (a) What value of a makes the denominator equal 0?

$$a^2 + a - 6 = 0$$
$$(a - 2)(a + 3) = 0$$
$$a - 2 = 0 \quad \text{or} \quad a + 3 = 0$$
$$a = 2 \quad \text{or} \quad a = -3$$

$a = 2,\ a = -3$

(b) $\dfrac{a^2 + 3a - 10}{a^2 + a - 6} = \dfrac{\cancel{(a-2)}\,(a+5)}{\cancel{(a-2)}\,(a+3)}$

$$= \dfrac{a + 5}{a + 3}$$

43. $\dfrac{7b^2}{21b} = \dfrac{\cancel{7} \cdot \cancel{b} \cdot b}{\cancel{7} \cdot 3 \cdot \cancel{b}}$

$$= \dfrac{b}{3}$$

45. $\dfrac{-24x^2 y^5 z}{8xy^4 z^3} = \dfrac{-3 \cdot \cancel{8} \cdot \cancel{x} \cdot x \cdot \cancel{y} \cdot \cancel{y} \cdot \cancel{y} \cdot \cancel{y} \cdot y \cdot \cancel{z}}{\cancel{8} \cdot \cancel{x} \cdot \cancel{y} \cdot \cancel{y} \cdot \cancel{y} \cdot \cancel{y} \cdot \cancel{z} \cdot z \cdot z}$

$$= -\dfrac{3xy}{z^2}$$

47. $\dfrac{(p-3)\cancel{(p+5)}}{\cancel{(p+5)}(p+4)} = \dfrac{p-3}{p+4}$

49. $\dfrac{\cancel{m+11}}{4\,\cancel{(m+11)}\,(m-11)} = \dfrac{1}{4(m-11)}$

51. $\dfrac{x(2x+1)^2}{4x^3(2x+1)} = \dfrac{\cancel{x}(2x+1)\,\cancel{(2x+1)}}{4 \cdot \cancel{x} \cdot x \cdot x \cdot \cancel{(2x+1)}}$

$$= \dfrac{2x+1}{4x^2}$$

53. $\dfrac{5}{20a - 25} = \dfrac{\cancel{5}}{\cancel{5}(4a - 5)}$

$$= \dfrac{1}{4a - 5}$$

55. $\dfrac{3x^2 - 6x}{9xy + 18x} = \dfrac{\cancel{3x}(x-2)}{3 \cdot \cancel{3x}(y+2)}$

$$= \dfrac{x - 2}{3(y + 2)}$$

57. $\dfrac{2x + 4}{x^2 - 3x - 10} = \dfrac{2\,\cancel{(x+2)}}{\cancel{(x+2)}\,(x-5)}$

$$= \dfrac{2}{x - 5}$$

59. $\dfrac{a^2 - 49}{a - 7} = \dfrac{(a+7)\,\cancel{(a-7)}}{\cancel{a-7}}$

$$= a + 7$$

61. $\dfrac{q^2 + 25}{q + 5}$ Cannot simplify

63. $\dfrac{y^2 + 6y + 9}{2y^2 + y - 15} = \dfrac{\cancel{(y+3)}\,(y+3)}{\cancel{(y+3)}\,(2y-5)}$

$$= \dfrac{y + 3}{2y - 5}$$

65. $\dfrac{5q^2 + 5}{q^4 - 1} = \dfrac{5(q^2 + 1)}{(q^2 + 1)(q^2 - 1)}$

$$= \dfrac{5\,\cancel{(q^2 + 1)}}{\cancel{(q^2 + 1)}\,(q-1)(q+1)}$$

$$= \dfrac{5}{(q - 1)(q + 1)}$$

67. $\dfrac{ac-ad+2bc-2bd}{2ac+ad+4bc+2bd} = \dfrac{(ac+2bc)-(ad+2bd)}{(2ac+4bc)+(ad+2bd)}$

$\qquad\qquad = \dfrac{c(a+2b)-d(a+2b)}{2c(a+2b)+d(a+2b)}$

$\qquad\qquad = \dfrac{(c-d)\cancel{(a+2b)}}{(2c+d)\cancel{(a+2b)}}$

$\qquad\qquad = \dfrac{c-d}{2c+d}$

69. $\dfrac{5x^3+4x^2-45x-36}{x^2-9} = \dfrac{(5x^3-45x)+(4x^2-36)}{(t-4)(t+4)}$

$\qquad\qquad = \dfrac{5x(x^2-9)+4(x^2-9)}{x^2-9}$

$\qquad\qquad = \dfrac{(5x+4)\cancel{(x^2-9)}}{\cancel{x^2-9}}$

$\qquad\qquad = 5x+4$

71. $\dfrac{2x^2-xy-3y^2}{2x^2-11xy+12y^2} = \dfrac{(x+y)\cancel{(2x-3y)}}{\cancel{(2x-3y)}(x-4y)}$

$\qquad\qquad = \dfrac{x+y}{x-4y}$

73. $2-x = -(x-2)$; They are opposites.

75. $\dfrac{x-5}{5-x} = \dfrac{\cancel{x-5}}{-(\cancel{x-5})}$

$\qquad = -1$

77. $\dfrac{-4-y}{4+y} = \dfrac{-(\cancel{4+y})}{\cancel{4+y}}$

$\qquad = -1$

79. $\dfrac{3y-6}{12-6y} = \dfrac{3(y-2)}{-(6y-12)}$

$\qquad = -\dfrac{3\cancel{(y-2)}}{6\cancel{(y-2)}}$

$\qquad = -\dfrac{3}{6}$

$\qquad = -\dfrac{1}{2}$

81. $\dfrac{k+5}{5-k}$; Cannot simplify

83. $\dfrac{10x-12}{10x+12} = \dfrac{\cancel{2}(5x-6)}{\cancel{2}(5x+6)}$

$\qquad\qquad = \dfrac{5x-6}{5x+6}$

85. $\dfrac{x^2-x-12}{16-x^2} = \dfrac{(x+3)(x-4)}{(4-x)(4+x)}$

$\qquad\qquad = \dfrac{(x+3)\cancel{(x-4)}}{-\cancel{(x-4)}(4+x)}$

$\qquad\qquad = -\dfrac{x+3}{4+x}$

87. $\dfrac{3x^2+7x-6}{x^2+7x+12} = \dfrac{(3x-2)\cancel{(x+3)}}{\cancel{(x+3)}(x+4)}$

$\qquad\qquad = \dfrac{3x-2}{x+4}$

89. $\dfrac{3(m-2)}{6(2-m)} = \dfrac{-3\cancel{(2-m)}}{6\cancel{(2-m)}}$

$\qquad\qquad = -\dfrac{3}{6}$

$\qquad\qquad = -\dfrac{1}{2}$

$\qquad$ or

$\dfrac{3(m-2)}{6(2-m)} = \dfrac{3\cancel{(m-2)}}{-6\cancel{(m-2)}}$

$\qquad\qquad = -\dfrac{3}{6}$

$\qquad\qquad = -\dfrac{1}{2}$

91. $\dfrac{w^2-4}{8-4w} = \dfrac{(w-2)(w+2)}{4(2-w)}$

$\qquad\qquad = \dfrac{\cancel{(w-2)}(w+2)}{-4\cancel{(w-2)}}$

$\qquad\qquad = -\dfrac{w+2}{4}$

93. $\dfrac{18st^5}{12st^3} = \dfrac{2 \cdot 3 \cdot 3 \cdot s \cdot t \cdot t \cdot t \cdot t \cdot t}{2 \cdot 2 \cdot 3 \cdot s \cdot t \cdot t \cdot t}$

$\qquad = \dfrac{3t^2}{2}$

95. $\dfrac{4r^2 - 4rs + s^2}{s^2 - 4r^2} = \dfrac{(2r-s)(2r-s)}{(s-2r)(s+2r)}$

$\qquad = \dfrac{-(s-2r)(2r-s)}{(s-2r)(s+2r)}$

$\qquad = -\dfrac{2r-s}{s+2r}$

97. $\dfrac{3y - 3x}{2x^2 - 4xy + 2y^2} = \dfrac{3(y-x)}{2(x^2 - 2xy + y^2)}$

$\qquad = \dfrac{3(y-x)}{2(x-y)(x-y)}$

$\qquad = \dfrac{-3(x-y)}{2(x-y)(x-y)}$

$\qquad = -\dfrac{3}{2(x-y)}$

99. $\dfrac{2t^2 - 3t}{2t^4 - 13t^3 + 15t^2} = \dfrac{t(2t-3)}{t^2(2t^2 - 13t + 15)}$

$\qquad = \dfrac{t(2t-3)}{t \cdot t(2t-3)(t-5)}$

$\qquad = \dfrac{1}{t(t-5)}$

101. $\dfrac{w^3 - 8}{w^2 + 2w + 4} = \dfrac{(w-2)(w^2 + 2w + 4)}{w^2 + 2w + 4}$

$\qquad = w - 2$

103. $\dfrac{z^2 - 16}{z^3 - 64} = \dfrac{(z-4)(z+4)}{(z-4)(z^2 + 4z + 16)}$

$\qquad = \dfrac{z+4}{z^2 + 4z + 16}$

105. (a) What value of x makes the denominator equal 0? $\begin{array}{l} x + 12 = 0 \\ \quad x = -12 \end{array}$; $x = -12$

(b) $\{x \mid x \text{ is a real number}, x \neq -12\}$

107. (a) What value of a makes the denominator equal 0? $\begin{array}{l} a^2 - a = 0 \\ a(a-1) = 0 \\ a = 0 \quad \text{or} \quad a - 1 = 0 \\ \qquad\qquad a = 1 \end{array}$;

$a = 0, a = 1$

(b) $\{a \mid a \text{ is a real number}, a \neq 0, a \neq -1\}$

109. (a) What value of z makes the denominator equal 0? $\begin{array}{l} z^2 + 4z + 3 = 0 \\ (z+3)(z+1) = 0 \\ z + 3 = 0 \quad \text{or} \quad z + 1 = 0 \\ z = -3 \quad \text{or} \quad z = -1 \end{array}$;

$z = -3, z = -1$

(b) $\{z \mid z \text{ is a real number}, z \neq -3, z \neq -1\}$

Section 7.2 Practice Exercises

1. $\dfrac{3}{5} \cdot \dfrac{1}{2} = \dfrac{3}{10}$

3. $\dfrac{3}{4} \div \dfrac{3}{8} = \dfrac{3}{4} \cdot \dfrac{8}{3} = \dfrac{24}{12} = 2$

5. $6 \cdot \dfrac{5}{12} = \dfrac{6}{1} \cdot \dfrac{5}{12} = \dfrac{30}{12} = \dfrac{5}{2}$

7. $\dfrac{\frac{21}{4}}{\frac{7}{5}} = \dfrac{21}{4} \cdot \dfrac{5}{7} = \dfrac{105}{28} = \dfrac{15}{4}$

9. $\dfrac{2xy}{5x^2} \cdot \dfrac{15}{4y} = \dfrac{2 \cdot x \cdot y}{5 \cdot x \cdot x} \cdot \dfrac{3 \cdot 5}{2 \cdot 2 \cdot y} = \dfrac{3}{2x}$

11. $\dfrac{6x^3}{9x^6y^2} \cdot \dfrac{18x^4y^7}{4y}$

$= \dfrac{2 \cdot 3 \cdot x^3}{3 \cdot 3 \cdot x^6y^2} \cdot \dfrac{2 \cdot 3 \cdot 3 \cdot x^4y^7}{4y}$

$= 3xy^4$

13. $\dfrac{4x-24}{20x} \cdot \dfrac{5x}{8} = \dfrac{4(x-6)}{20x} \cdot \dfrac{5x}{8} = \dfrac{x-6}{8}$

15. $\dfrac{3y+18}{y^2} \cdot \dfrac{4y}{6y+36} = \dfrac{3(y+6)}{y^2} \cdot \dfrac{4y}{6(y+6)} = \dfrac{2}{y}$

17. $\dfrac{10}{2-a} \cdot \dfrac{a-2}{16} = \dfrac{10}{-(a-2)} \cdot \dfrac{a-2}{16} = -\dfrac{5}{8}$

19. $\dfrac{b^2-a^2}{a-b} \cdot \dfrac{a}{a^2-ab} = \dfrac{-(a+b)(a-b)}{a-b} \cdot \dfrac{a}{a(a-b)}$

$= -\dfrac{b+a}{a-b}$

21. $\dfrac{y^2+2y+1}{5y-10} \cdot \dfrac{y^2-3y+2}{y^2-1}$

$= \dfrac{(y+1)^2}{5(y-2)} \cdot \dfrac{(y-1)(y-2)}{(y-1)(y+1)}$

$= \dfrac{y+1}{5}$

23. $\dfrac{10x}{2x^2+3x+1} \cdot \dfrac{x^2+7x+6}{5x}$

$= \dfrac{10x}{(2x+1)(x+1)} \cdot \dfrac{(x+1)(x+6)}{5x}$

$= \dfrac{10_{2}x}{(2x+1)(x+1)} \cdot \dfrac{(x+1)(x+6)}{5x}$

$= \dfrac{2(x+6)}{2x+1}$

25. $\dfrac{4x}{7y} \div \dfrac{2x^2}{21xy} = \dfrac{4x}{7y} \cdot \dfrac{21xy}{2x^2}$

$= \dfrac{2 \cdot 2 \cdot x}{7 \cdot y} \cdot \dfrac{3 \cdot 7 \cdot x \cdot y}{2 \cdot x^2}$

$= 6$

27. $\dfrac{8m^4n^5}{5n^6} \div \dfrac{24mn}{15m^3}$

$= \dfrac{8m^4n^5}{5n^6} \cdot \dfrac{15m^3}{24mn}$

$= \dfrac{2 \cdot 2 \cdot 2 \cdot m^4 \cdot n^5}{5 \cdot n^6} \cdot \dfrac{3 \cdot 5 \cdot m^3}{2 \cdot 2 \cdot 2 \cdot 3 \cdot m \cdot n}$

$= \dfrac{m^6}{n^2}$

29. $\dfrac{4a+12}{6a-18} \div \dfrac{3a+9}{5a-15} = \dfrac{4(a+3)}{6(a-3)} \cdot \dfrac{5(a-3)}{3(a+3)} = \dfrac{10}{9}$

31. $\dfrac{3x-21}{6x^2-42x} \div \dfrac{7}{12x} = \dfrac{3(x-7)}{6x(x-7)} \cdot \dfrac{12x}{7} = \dfrac{6}{7}$

33. $\dfrac{m^2-n^2}{9} \div \dfrac{3n-3m}{27m}$

$= \dfrac{(m+n)(m-n)}{9} \cdot \dfrac{27m}{-3(m-n)}$

$= -m(m+n)$

35. $\dfrac{3p+4q}{p^2+4pq+4q^2} \div \dfrac{4}{p+2q}$

$= \dfrac{3p+4q}{(p+2q)(p+3q)} \cdot \dfrac{p+2q}{4}$

$= \dfrac{3p+4q}{4(p+2q)}$

37. $\dfrac{p^2-2p-3}{p^2-p-6} \div \dfrac{p^2-1}{p^2+2p}$

$= \dfrac{p^2-2p-3}{p^2-p-6} \cdot \dfrac{p^2+2p}{p^2-1}$

$= \dfrac{(p+1)(p-3)}{(p+2)(p-3)} \cdot \dfrac{p(p+2)}{(p-1)(p+1)}$

$= \dfrac{p}{(p-1)}$

191

39. $(w+3) \cdot \dfrac{w}{2w^2 + 5w - 3}$

$= \dfrac{(w+3)}{1} \cdot \dfrac{w}{(2w-1)(w+3)}$

$= \dfrac{w}{2w-1}$

41. $(r-5) \cdot \dfrac{4r}{2r^2 - 7r - 15}$

$= \dfrac{(r-5)}{1} \cdot \dfrac{4r}{(2r+3)(r-5)}$

$= \dfrac{4r}{2r+3}$

43. $\dfrac{\frac{5t-10}{12}}{\frac{4t-8}{8}} = \dfrac{5(t-2)}{12} \cdot \dfrac{8}{4(t-2)} = \dfrac{5}{6}$

45. $\dfrac{2a^2 + 13a - 24}{8a - 12} \div (a+8)$

$= \dfrac{(2a-3)(a+8)}{4(2a-3)} \cdot \dfrac{1}{(a+8)}$

$= \dfrac{1}{4}$

47. $\dfrac{y^2 + 5y - 36}{y^2 - 2y - 8} \cdot \dfrac{y+2}{y-6} = \dfrac{(y+9)(y-4)}{(y-4)(y+2)} \cdot \dfrac{y+2}{y-6}$

$= \dfrac{y+9}{y-6}$

49. $\dfrac{t^2 + 4t - 5}{t^2 + 7t + 10} \cdot \dfrac{t+2}{t-1} = \dfrac{(t+5)(t-1)}{(t+5)(t+2)} \cdot \dfrac{t+2}{t-1}$

$= 1$

51. $(5t-1) \div \dfrac{5t^2 + 9t - 2}{3t + 8}$

$= \dfrac{(5t-1)}{1} \cdot \dfrac{3t+8}{(5t-1)(t+2)}$

$= \dfrac{3t+8}{t+2}$

53. $\dfrac{x^2 + 2x - 3}{x^2 - 3x + 2} \cdot \dfrac{x^2 + 2x - 8}{x^2 + 4x + 3}$

$= \dfrac{(x+3)(x-1)}{(x-2)(x-1)} \cdot \dfrac{(x+4)(x-2)}{(x+3)(x+1)}$

$= \dfrac{x+4}{x+1}$

55. $\dfrac{\frac{w^2 - 6w + 9}{8}}{\frac{9 - w^2}{4w + 12}} = \dfrac{(w-3)(w-3)}{8} \cdot \dfrac{4(w+3)}{-(w+3)(w-3)}$

$= -\dfrac{w-3}{2}$

57. $\dfrac{k^2 + 3k + 2}{k^2 + 5k + 4} \div \dfrac{k^2 + 5k + 6}{k^2 + 10k + 24}$

$= \dfrac{(k+1)(k+2)}{(k+4)(k+1)} \cdot \dfrac{(k+6)(k+4)}{(k+2)(k+3)}$

$= \dfrac{k+6}{k+3}$

59. $\dfrac{ax + a + bx + b}{2x^2 + 4x + 2} \cdot \dfrac{4x + 4}{a^2 + ab}$

$= \dfrac{a(x+1) + b(x+1)}{2(x^2 + 2x + 1)} \cdot \dfrac{4(x+1)}{a(a+b)}$

$= \dfrac{\cancel{(a+b)}\,\cancel{(x+1)}}{2\,\cancel{(x+1)}\,\cancel{(x+1)}} \cdot \dfrac{4\,\cancel{(x+1)}}{a\,\cancel{(a+b)}}$

$= \dfrac{2}{a}$

61. $\dfrac{y^4 - 1}{2y^2 - 3y + 1} \div \dfrac{2y^2 + 2}{8y^2 - 4y}$

$= \dfrac{(y^2 - 1)(y^2 + 1)}{(2y-1)(y-1)} \cdot \dfrac{4y(2y-1)}{2(y^2 + 1)}$

$= \dfrac{\cancel{(y-1)}(y+1)\,\cancel{(y^2+1)}}{\cancel{(2y-1)}\,\cancel{(y-1)}} \cdot \dfrac{4y\,\cancel{(2y-1)}}{2\,\cancel{(y^2+1)}}$

$= 2y(y+1)$

63. $\dfrac{x^2 - xy - 2y^2}{x + 2y} \div \dfrac{x^2 - 4xy + 4y^2}{x^2 - 4y^2}$

$= \dfrac{\cancel{(x-2y)}(x+y)}{\cancel{x+2y}} \cdot \dfrac{\cancel{(x-2y)}\,\cancel{(x+2y)}}{\cancel{(x-2y)}\,\cancel{(x-2y)}}$

$= x + y$

65. $\dfrac{z^2+2z+1}{3z+6}\cdot\dfrac{z^2-4}{z+1}\cdot\dfrac{z}{z^2-z-2}$

$=\dfrac{\cancel{(z+1)}\,\cancel{(z+1)}}{3\cancel{(z+2)}}\cdot\dfrac{\cancel{(z+2)}\,\cancel{(z-2)}}{\cancel{z+1}}\cdot\dfrac{z}{\cancel{(z-2)}\,\cancel{(z+1)}}$

$=\dfrac{z}{3}$

67. $\dfrac{y^3-3y^2+4y-12}{y^4-16}\cdot\dfrac{3b^2+5y-2}{3b^2-10y+3}\div\dfrac{3}{6y-12}$

$=\dfrac{(y^2+4)(y-3)}{(y+2)(y-2)(y^2+4)}\cdot\dfrac{(3y-1)(y+2)}{(3y-1)(y-3)}\cdot\dfrac{6(y-2)}{3}$

$=\dfrac{6}{3}$

$=2$

69. $\dfrac{a^2-5a}{a^2+7a+12}\div\dfrac{a^3-7a^2+10a}{a^2+9a+18}\div\dfrac{a+6}{a+4}$

$=\dfrac{a(a-5)}{(a+3)(a+4)}\cdot\dfrac{(a+6)(a+3)}{a(a-5)(a-2)}\cdot\dfrac{a+4}{a+6}$

$=\dfrac{1}{a-2}$

71. $\dfrac{p^3-q^3}{p-q}\cdot\dfrac{p+q}{2p^2+2pq+2q^2}$

$=\dfrac{(p-q)(p^2+pq+q^2)}{p-q}\cdot\dfrac{p+q}{2(p^2\,pq+q^2)}$

$=\dfrac{p+q}{2}$

Section 7.3 Practice Exercises

1. multiple; denominators

3. Restricted values: $x=1,\,x=1$

$\dfrac{3x+3}{5x^2-5}=\dfrac{3(x+1)}{5(x+1)(x-1)}=\dfrac{3}{5(x-1)}$

5. $\dfrac{a+3}{a+7}\cdot\dfrac{a^2+3a-10}{a^2+a-6}=\dfrac{a+3}{a+7}\cdot\dfrac{(a+5)(a-2)}{(a+3)(a-2)}$

$=\dfrac{a+5}{a+7}$

7. $\dfrac{16y^2}{9y+36}\div\dfrac{8y^3}{3y+12}=\dfrac{16y^2}{9(y+4)}\cdot\dfrac{3(y+4)}{8y^3}=\dfrac{2}{3y}$

9. a, b, c, d

11. x^5 is the greatest power of x that appears in any denominator

13. $3^2\cdot5=45$

15. $2^4\cdot3=48$

17. $7\cdot9=63$

19. $3^2\cdot x^2y^3=9x^2y^3$

21. w^2y

23. $(p+3)(p-1)(p+2)$

25. $9t(t+1)^2$

27. $(y-2)(y+2)(y+3)$

29. $3-x$ or $x-3$

31. Because $(b-1)$ and $(1-b)$ are opposites; they differ by a factor of -1.

33. LCD: $5x^2$; $\dfrac{6}{5x^2},\dfrac{5x}{5x^2}$

35. LCD: $5\cdot6x^3=30x^3$; $\dfrac{24x}{30x^3},\dfrac{5y}{30x^3}$

37. LCD: $12a^2b$; $\dfrac{10}{12a^2b},\dfrac{a^3}{12a^2b}$

39. LCD: $(m+4)(m-1)$;

$\dfrac{6(m-1)}{(m+4)(m-1)}=\dfrac{6m-6}{(m+4)(m-1)}$,

$\dfrac{3}{m-1}=\dfrac{3(m+4)}{(m+4)(m-1)}=\dfrac{3m+12}{(m+4)(m-1)}$

41. LCD: $(2x-5)(x+3)$

$\dfrac{6(x+3)}{(2x-5)(x+3)}=\dfrac{6x+18}{(2x-5)(x+3)}$,

$\dfrac{1(2x-5)}{(x+3)(2x-5)}=\dfrac{2x-5}{(2x-5)(x+3)}$

43. LCD: $(w + 3)(w - 8)(w - 1)$;

$$\frac{6(w+1)}{(w+3)(w-8)(w-1)} = \frac{6w+6}{(w+3)(w-8)(w-1)},$$

$$\frac{w(w+3)}{(w+3)(w-8)(w-1)} = \frac{w^2+3w}{(w+3)(w-8)(w-1)}$$

45. LCD:

$$(p-2)(p+2)(p+2) = (p-2)(p+2)^2$$

$$\frac{6p(p+2)}{(p-2)(p+2)^2} = \frac{6p^2+12p}{(p-2)(p+2)^2},$$

$$\frac{3(p-2)}{(p-2)(p+2)^2} = \frac{3p-6}{(p-2)(p+2)^2}$$

47. LCD: $a - 4$ or $4 - a$;

$$\frac{1}{a-4}, \frac{(-1)a}{(-1)(4-a)} = \frac{1}{a-4}; \frac{-a}{a-4} \text{ or}$$

$$\frac{(-1)(1)}{(-1)(a-4)} = -\frac{1}{4-a}, \frac{a}{4-a}$$

49. LCD: $2(x - 7)$ or $2(7 - x)$;

$$\frac{8}{2(x-7)}, \frac{-1(y)}{(2)(x-7)} = \frac{-y}{2(x-7)} \text{ or}$$

$$\frac{-1(8)}{2(7-x)} = -\frac{8}{2(7-x)}, \frac{y}{2(7-x)}$$

51. LCD: $a + b$;

$$\frac{1}{a+b}, \frac{6}{-1(a+b)} = -\frac{6}{a+b} \text{ or}$$

$$\frac{-1}{-a-b}, \frac{6}{-a-b}$$

53. $24y + 8 = 8(3y + 1)$
$18y + 6 = 6(3y + 1)$
LCD: $24(3y + 1)$

$$\frac{-3}{24y+8} = \frac{-3(3)}{8(3y+1)(3)} = \frac{-9}{24(3y+1)},$$

$$\frac{5}{18y+6} = \frac{5(4)}{6(3y+1)(4)} = \frac{20}{24(3y+1)}$$

55. LCD: $5z(z + 4)$

$$\frac{3(z+4)}{5z(z+4)} = \frac{3z+12}{5z(z+4)}$$

$$\frac{1(5z)}{(z+4)(5z)} = \frac{5z}{5z(z+4)}$$

57. LCD: $(z + 2)(z + 3)(z + 7)$;

$$\frac{z(z+3)}{(z+2)(z+3)(z+7)} = \frac{z^2+3z}{(z+2)(z+3)(z+7)},$$

$$\frac{-3z(z+2)}{(z+2)(z+3)(z+7)} = \frac{-3z^2-6z}{(z+2)(z+3)(z+7)},$$

$$\frac{5(z+7)}{(z+2)(z+3)(z+7)} = \frac{5z+35}{(z+2)(z+3)(z+7)}$$

59. LCD: $(p - 2)(p + 2)(p^2 + 2p + 4)$;

$$\frac{3(p+2)}{(p^2-4)(p^2+2p+4)}$$

$$= \frac{3p+6}{(p^2-4)(p^2+2p+4)},$$

$$\frac{p(p^2+2p+4)}{(p^2-4)(p^2+2p+4)}$$

$$= \frac{p^3+2p^2+4p}{(p^2-4)(p^2+2p+4)}$$

$$\frac{5p(p^2-4)}{(p^2-4)(p^2+2p+4)}$$

$$= \frac{5p^3-20p}{(p^2-4)(p^2+2p+4)}$$

Section 7.4 Practice Exercises

1. (a) $x = 0$; $\dfrac{-5}{10} = -\dfrac{1}{2}$

$x = 1$; $\dfrac{1^2-4(1)-5}{1^2-7(1)+10} = -2$

$x = -1$; $\dfrac{(-1)^2-4(-1)-5}{(-1)^2-7(-1)+10} = \dfrac{0}{18} = 0$

$x = 2;$

$\dfrac{(2)^2 - 4(2) - 5}{(2)^2 - 7(2) - 10} = \dfrac{-9}{0}$ is undefined

$x = 5;\ \dfrac{5^2 - 4(5) - 5}{5^2 - 7(5) + 10} = \dfrac{0}{0}$ is undefined

(b) $(x-5)(x-2);\ x = 5,\ x = 2$

(c) $\dfrac{(x-5)(x+1)}{(x-5)(x-2)} = \dfrac{x+1}{x-2}$

3. $\dfrac{2b^2 - b - 3}{2b^2 - 3b - 9} \div \dfrac{b^2 - 1}{4b + 6}$

$= \dfrac{(2b-3)\cancel{(b+1)}}{\cancel{(2b+3)}(b-3)} \cdot \dfrac{2\cancel{(2b+3)}}{\cancel{(b+1)}(b-1)}$

$= \dfrac{2(2b-3)}{(b-3)(b-1)}$

5. $\dfrac{7}{8} + \dfrac{3}{8} = \dfrac{10}{8} = \dfrac{5}{4}$

7. $\dfrac{9}{16} - \dfrac{3}{16} = \dfrac{6}{16} = \dfrac{3}{8}$

9. $\dfrac{5a}{a+2} - \dfrac{3a-4}{a+2} = \dfrac{5a - (3a-4)}{a+2}$

$= \dfrac{2a+4}{a+2}$

$= \dfrac{2(a+2)}{a+2}$

$= 2$

11. $\dfrac{5c}{c+6} + \dfrac{30}{c+6} = \dfrac{5c+30}{c+6} = \dfrac{5(c+6)}{c+6} = 5$

13. $\dfrac{5}{t-8} - \dfrac{2t+1}{t-8} = \dfrac{5 - 2t - 1}{t-8} = \dfrac{4 - 2t}{t-8} = \dfrac{-2(t-2)}{t-8}$

15. $\dfrac{9x^2}{3x-7} - \dfrac{49}{3x-7} = \dfrac{9x^2 - 49}{3x-7}$

$= \dfrac{\cancel{(3x-7)}(3x+7)}{\cancel{3x-7}}$

$= 3x + 7$

17. $\dfrac{m^2}{m+5} + \dfrac{10m+25}{m+5} = \dfrac{m^2 + 10m + 25}{m+5}$

$= \dfrac{(m+5)^2}{m+5}$

$= m + 5$

19. $\dfrac{2a}{a+2} + \dfrac{4}{a+2} = \dfrac{2a+4}{a+2} = \dfrac{2(a+2)}{a+2} = 2$

21. $\dfrac{x^2}{x+5} - \dfrac{25}{x+5} = \dfrac{x^2 - 25}{x+5}$

$= \dfrac{(x+5)(x-5)}{x+5}$

$= x - 5$

23. $\dfrac{r}{r^2 + 3r + 2} + \dfrac{2}{r^2 + 3r + 2} = \dfrac{r+2}{r^2 + 3r + 2}$

$= \dfrac{\cancel{r+2}}{\cancel{(r+2)}(r+1)}$

$= \dfrac{1}{r+1}$

25. $\dfrac{1}{3y^2 + 22y + 7} - \dfrac{-3y}{3y^2 + 22y + 7}$

$= \dfrac{1 + 3y}{3y^2 + 22y + 7}$

$= \dfrac{\cancel{3y+1}}{\cancel{(3y+1)}(y+7)}$

$= \dfrac{1}{y+7}$

27. $P = \dfrac{2x}{y} + \dfrac{6x}{y} + \dfrac{7x}{y} = \dfrac{15x}{y}$

29. $\dfrac{5}{4} + \dfrac{3}{2a} = \dfrac{5a}{4a} + \dfrac{3(2)}{2a(2)} = \dfrac{5a+6}{4a}$

31. $\dfrac{4}{5xy^3} + \dfrac{2x}{15y^2} = \dfrac{4(3)}{5xy^3(3)} + \dfrac{2x(xy)}{15y^2(xy)}$

$= \dfrac{12 + 2x^2 y}{15xy^3}$

$= \dfrac{2(6 + x^2 y)}{15xy^3}$

33. $\dfrac{2}{s^3t^3} - \dfrac{3}{s^4t} = \dfrac{2s}{s^3t^3 \cdot s} - \dfrac{3t^2}{s^4t \cdot t^2} = \dfrac{2s - 3t^2}{s^4t^3}$

35. $\dfrac{z}{3z-9} - \dfrac{z-2}{z-3} = \dfrac{z}{3(z-3)} - \dfrac{3(z-2)}{3(z-3)}$

$= \dfrac{z - 3z + 6}{3(z-3)}$

$= \dfrac{-2z + 6}{3(z-3)}$

$= \dfrac{-2(z-3)}{3(z-3)}$

$= -\dfrac{2}{3}$

37. $\dfrac{5}{a+1} + \dfrac{4}{3a+3} = \dfrac{5(3)}{3(a+1)} + \dfrac{4}{3(a+1)}$

$= \dfrac{15 + 4}{3(a+1)}$

$= \dfrac{19}{3(a+1)}$

39. $\dfrac{k}{k^2-9} - \dfrac{4}{k-3}$

$= \dfrac{k}{(k+3)(k-3)} - \dfrac{4(k+3)}{(k+3)(k-3)}$

$= \dfrac{k - 4k - 12}{(k+3)(k-3)}$

$= \dfrac{-3(k+4)}{(k+3)(k-3)}$

41. $\dfrac{3a-7}{6a+10} - \dfrac{10}{3a^2+5a}$

$= \dfrac{a(3a-7)}{2a(3a+5)} - \dfrac{10(2)}{2a(3a+5)}$

$= \dfrac{3a^2 - 7a - 20}{2a(3a+5)}$

$= \dfrac{(3a+5)(a-4)}{2a(3a+5)}$

$= \dfrac{a-4}{2a}$

43. $\dfrac{x}{x-4} + \dfrac{3}{x+1} = \dfrac{x(x+1)}{(x-4)(x+1)} + \dfrac{3(x-4)}{(x+1)(x-4)}$

$= \dfrac{x^2 + x + 3x - 12}{(x-4)(x+1)}$

$= \dfrac{x^2 + 4x - 12}{(x-4)(x+1)}$

$= \dfrac{(x+6)(x-2)}{(x-4)(x+1)}$

45. $\dfrac{3x}{x^2+6x+9} + \dfrac{x}{x^2+5x+6}$

$= \dfrac{3x}{(x+3)(x+3)} + \dfrac{x}{(x+3)(x+2)}$

$= \dfrac{3x(x+2)}{(x+3)(x+3)(x+2)} + \dfrac{x(x+3)}{(x+3)(x+2)(x+3)}$

$= \dfrac{3x^2 + 6x + x^2 + 3x}{(x+3)^2(x+2)}$

$= \dfrac{4x^2 + 9x}{(x+3)^2(x+2)}$

$= \dfrac{x(4x+9)}{(x+3)^2(x+2)}$

47. $\dfrac{p}{3} - \dfrac{4p-1}{-3} = \dfrac{p}{3} - \dfrac{-(4p-1)}{3}$

$= \dfrac{p + 4p - 1}{3}$

$= \dfrac{5p-1}{3}$ or $\dfrac{-5p+1}{-3}$

49. $\dfrac{8}{x-3} - \dfrac{1}{3-x} = \dfrac{8}{x-3} - \dfrac{(-1)1}{-1(3-x)}$

$= \dfrac{8}{x-3} - \dfrac{-1}{x-3}$

$= \dfrac{8}{x-3} + \dfrac{1}{x-3}$

$= \dfrac{9}{x-3}$

or

$$\frac{8}{x-3} - \frac{1}{3-x} = \frac{(-1)8}{(-1)(x-3)} - \frac{1}{3-x}$$

$$= \frac{-8}{-x+3} - \frac{1}{3-x}$$

$$= \frac{-8}{3-x} - \frac{1}{3-x}$$

$$= \frac{-9}{3-x}$$

51. $\dfrac{4n}{n-8} - \dfrac{2n-1}{8-n} = \dfrac{4n}{n-8} - \dfrac{2n-1}{-1(n-8)}$

$$= \frac{4n}{n-8} + \frac{2n-1}{n-8}$$

$$= \frac{4n+2n-1}{n-8}$$

$$= \frac{6n-1}{n-8} \text{ or } \frac{-6n+1}{8-n}$$

53. $\dfrac{5}{x} + \dfrac{3}{x+2} = \dfrac{5(x+2)}{x(x+2)} + \dfrac{3x}{x(x+2)}$

$$= \frac{5x+10+3x}{x(x+2)}$$

$$= \frac{8x+10}{x(x+2)}$$

$$= \frac{2(4x+5)}{x(x+2)}$$

55. $\dfrac{y}{4y+2} + \dfrac{3y}{6y+3}$

$$= \frac{y \cdot 3}{2(2y+1)(3)} + \frac{3y \cdot 2}{3(2y+1)(2)}$$

$$= \frac{3y+6y}{3 \cdot 2(2y+1)}$$

$$= \frac{9y}{3 \cdot 2(2y+1)}$$

$$= \frac{3y}{2(2y+1)}$$

57. $\dfrac{4w}{w^2+2w-3} + \dfrac{2}{1-w}$

$$= \frac{4w}{(w+3)(w-1)} + \frac{-2}{w-1}$$

$$= \frac{4w}{(w+3)(w-1)} - \frac{2(w+3)}{(w+3)(w-1)}$$

$$= \frac{4w-2w-6}{(w+3)(w-1)}$$

$$= \frac{12w-6}{(w+3)(w-1)}$$

$$= \frac{2(w-3)}{(w+3)(w-1)}$$

59. $\dfrac{3a-8}{a^2-5a+6} + \dfrac{a+2}{a^2-6a+8} = \dfrac{3a-8}{(a-2)(a-3)} + \dfrac{a+2}{(a-2)(a-4)}$

$$= \frac{(3a-8)(a-4)}{(a-2)(a-3)(a-4)} + \frac{(a+2)(a-3)}{(a-2)(a-3)(a-4)}$$

$$= \frac{3a^2-20a+32+a^2-a-6}{(a-2)(a-3)(a-4)}$$

$$= \frac{4a^2-21a+26}{(a-2)(a-3)(a-4)}$$

$$= \frac{(4a-13)(a-2)}{(a-2)(a-3)(a-4)}$$

$$= \frac{4a-13}{(a-3)(a-4)}$$

61. $\dfrac{4x}{x^2+4x-5} - \dfrac{x}{x^2+10x+25} = \dfrac{4x}{(x-1)(x+5)} - \dfrac{x}{(x+5)(x+5)}$

$$= \dfrac{4x(x+5)}{(x-1)(x+5)(x+5)} - \dfrac{x(x-1)}{(x+5)(x+5)(x-5)}$$

$$= \dfrac{4x^2+20x-(x^2-x)}{(x+5)^2(x-1)}$$

$$= \dfrac{4x^2+20x-x^2+x}{(x+5)^2(x-1)}$$

$$= \dfrac{3x^2+21x}{(x+5)^2(x-1)}$$

$$= \dfrac{3x(x+7)}{(x+5)^2(x-1)}$$

63. $\dfrac{3y}{2y^2-y-1} - \dfrac{4y}{2y^2-7y-4} = \dfrac{3y}{(2y+1)(y-1)} - \dfrac{4y}{(2y+1)(y-4)}$

$$= \dfrac{3y(y-4)}{(2y+1)(y-1)(y-4)} - \dfrac{4y(y-1)}{(2y+1)(y-1)(y-4)}$$

$$= \dfrac{3y^2-12y-4y^2+4y}{(2y+1)(y-1)(y-4)}$$

$$= \dfrac{-y^2-8y}{(2y+1)(y-1)(y-4)}$$

$$= \dfrac{-y(y+8)}{(2y+1)(y-1)(y-4)}$$

65. $\dfrac{3}{2p-1} - \dfrac{4p+4}{4p^2-1} = \dfrac{3}{2p-1} - \dfrac{4p+4}{(2p+1)(2p-1)}$

$$= \dfrac{3(2p+1)}{(2p+1)(2p-1)} - \dfrac{4p+4}{(2p+1)(2p-1)}$$

$$= \dfrac{6p+3-4p-4}{(2p+1)(2p-1)}$$

$$= \dfrac{2p-1}{(2p+1)(2p-1)}$$

$$= \dfrac{1}{2p+1}$$

67. $\dfrac{m}{m+n} - \dfrac{m}{m-n} + \dfrac{1}{m^2-n^2} = \dfrac{m(m-n)}{(m+n)(m-n)} - \dfrac{m(m+n)}{(m-n)(m+n)} + \dfrac{1}{(m+n)(m-n)}$

$$= \dfrac{m^2 - mn - m^2 - mn + 1}{(m+n)(m-n)}$$

$$= \dfrac{-2mn+1}{(m+n)(m-n)}$$

69. $\dfrac{2}{a+b} + \dfrac{2}{a-b} - \dfrac{4a}{a^2-b^2} = \dfrac{2(a-b)}{(a+b)(a-b)} + \dfrac{2(a+b)}{(a+b)(a-b)} - \dfrac{4a}{(a+b)(a-b)}$

$$= \dfrac{2a - 2b + 2a + 2b - 4a}{(a+b)(a-b)}$$

$$= 0$$

71. $P = 2w + 2l$

$$= 2\left(\dfrac{2}{x+3}\right) + 2\left(\dfrac{1}{x+2}\right)$$

$$= \dfrac{4}{x+3} + \dfrac{2}{x+2}$$

$$= \dfrac{4(x+2)}{(x+3)(x+2)} + \dfrac{2(x+3)}{(x+3)(x+2)}$$

$$= \dfrac{4x + 8 + 2x + 6}{(x+3)(x+2)}$$

$$= \dfrac{6x + 14}{(x+3)(x+2)}$$

$$= \dfrac{2(3x+7)}{(x+3)(x+2)}$$

73. $\dfrac{1}{n}$

75. Let n = number; $\dfrac{5}{n+2}$

77. Let n = the number, then $n + \left(7 \cdot \dfrac{1}{n}\right)$; $\quad n + \dfrac{7}{n} = \dfrac{n^2}{n} + \dfrac{7}{n} = \dfrac{n^2+7}{n}$

79. $\dfrac{1}{n} - \dfrac{2}{n}$; $\dfrac{1}{n} - \dfrac{2}{n} = -\dfrac{1}{n}$

81. $\dfrac{-3}{w^3+27}-\dfrac{1}{w^2-9}=\dfrac{-3}{(w+3)(w^2-3w+9)}-\dfrac{1}{(w+3)(w-3)}$

$$=\dfrac{-3(w-3)}{(w-3)(w+3)(w^2-3w+9)}-\dfrac{w^2-3w+9}{(w-3)(w+3)(w^2-3w+9)}$$

$$=\dfrac{-3w+9-w^2+3w-9}{(w-3)(w+3)(w^2-3+9)}$$

$$=\dfrac{-w^2}{(w-3)(w+3)(w^2-3w+9)}$$

83. $\dfrac{2p}{p^2+5p+6}-\dfrac{p+1}{p^2+2p-3}+\dfrac{3}{p^2+p-2}$

$$=\dfrac{2p}{(p+2)(p+3)}-\dfrac{p+1}{(p-1)(p+3)}+\dfrac{3}{(p+2)(p-1)}$$

$$=\dfrac{2p(p-1)}{(p-1)(p+2)(p+3)}-\dfrac{(p+1)(p+2)}{(p-1)(p+2)(p+3)}+\dfrac{3(p+3)}{(p-1)(p+2)(p+3)}$$

$$=\dfrac{2p^2-2p-p^2-3p-2+3p+9}{(p-1)(p+2)(p+3)}$$

$$=\dfrac{p^2-2p+7}{(p-1)(p+2)(p+3)}$$

85. $\dfrac{3m}{m^2+3m-10}+\dfrac{5}{4-2m}-\dfrac{1}{m+5}=\dfrac{3m}{(m+5)(m-2)}+\dfrac{5}{-2(m-2)}-\dfrac{1}{m+5}$

$$=\dfrac{6m}{2(m+5)(m-2)}-\dfrac{5(m+5)}{2(m+5)(m-2)}-\dfrac{2(m-2)}{2(m+5)(m-2)}$$

$$=\dfrac{6m-5m-25-2m+4}{2(m+5)(m-2)}$$

$$=\dfrac{-m-21}{2(m+5)(m-2)}\ \text{ or }\ \dfrac{m+21}{2(m+5)(2-m)}$$

87. $\left(\dfrac{2}{k+1}+3\right)\left(\dfrac{k+1}{4k+7}\right)=\left(\dfrac{2}{k+1}+\dfrac{3(k+1)}{k+1}\right)\left(\dfrac{k+1}{4k+7}\right)=\left(\dfrac{3k+5}{k+1}\right)\left(\dfrac{k+1}{4k+7}\right)=\dfrac{3k+5}{4k+7}$

89. $\left(\dfrac{1}{10a}-\dfrac{b}{10a^2}\right)\div\left(\dfrac{1}{10}-\dfrac{b}{10a}\right)$

$$=\left(\dfrac{a}{10a^2}-\dfrac{b}{10a^2}\right)\div\left(\dfrac{a}{10a}-\dfrac{b}{10a}\right)$$

$$=\left(\dfrac{a-b}{10a^2}\right)\cdot\left(\dfrac{10a}{a-b}\right)$$

$$=\dfrac{1}{a}$$

Problem Recognition Exercises

1. $\dfrac{5}{3x+1} - \dfrac{2x-4}{3x+1} = \dfrac{5-2x+4}{3x+1}$

$\qquad = \dfrac{-2x+9}{3x+1}$

3. $\dfrac{3}{y} \cdot \dfrac{y^2-5y}{6y-9} = \dfrac{\cancel{3}}{\cancel{y}} \cdot \dfrac{\cancel{y}(y-5)}{\cancel{3}(2y-3)}$

$\qquad = \dfrac{y-5}{2y-3}$

5. $\dfrac{x-9}{9x-x^2} = \dfrac{x-9}{x(9-x)}$

$\qquad = \dfrac{\cancel{x-9}}{-x(\cancel{x-9})}$

$\qquad = \dfrac{-1}{x}$

7. $\dfrac{c^2+5c+6}{c^2+c-2} \div \dfrac{c}{c-1} = \dfrac{c^2+5c+6}{c^2+c-2} \cdot \dfrac{c-1}{c}$

$\qquad = \dfrac{(c+2)(c+3)}{(c-1)(c+2)} \cdot \dfrac{c-1}{c}$

$\qquad = \dfrac{c+3}{c}$

9. $\dfrac{6a^2b^3}{72ab^7c} = \dfrac{6aab^3}{6 \cdot 12ab^3b^4c} = \dfrac{a}{12b^4c}$

11. $\dfrac{p^2+10pq+25q^2}{p^2+6pq+5q^2} \div \dfrac{10p+50q}{2p^2-2q^2}$

$\qquad = \dfrac{(p+5q)(p+5q)}{(p+5q)(p+q)} \cdot \dfrac{2(p+q)(p-q)}{10(p+5q)}$

$\qquad = \dfrac{p-q}{5}$

13. $\dfrac{20x^2+10x}{4x^3+4x^2+x} = \dfrac{10x(2x+1)}{x(2x+1)(2x+1)} = \dfrac{10}{2x+1}$

15. $\dfrac{8x^2-18x-5}{4x^2-25} \div \dfrac{4x^2-11x-3}{3x-9}$

$\qquad = \dfrac{8x^2-18x-5}{4x^2-25} \cdot \dfrac{3x-9}{4x^2-11x-3}$

$\qquad = \dfrac{(4x+1)(2x-5)}{(2x-5)(2x+5)} \cdot \dfrac{3(x-3)}{(4x+1)(x-3)}$

$\qquad = \dfrac{3}{2x+5}$

17. $\dfrac{a}{a^2-9} - \dfrac{3}{6a-18}$

$\qquad = \dfrac{a}{(a+3)(a-3)} - \dfrac{3}{6(a-3)}$

$\qquad = \dfrac{2a}{2(a+3)(a-3)} - \dfrac{(a+3)}{2(a+3)(a-3)}$

$\qquad = \dfrac{2a-a-3}{2(a+3)(a-3)}$

$\qquad = \dfrac{a-3}{2(a+3)(a-3)}$

$\qquad = \dfrac{1}{2(a+3)}$

19. $(t^2+5t-24)\left(\dfrac{t+8}{t-3}\right) = \dfrac{(t+8)(t-3)}{1}\left(\dfrac{t+8}{t-3}\right)$

$\qquad = (t+8)^2$

Section 7.5 Practice Exercises

1. complex

3. $a \neq \dfrac{3}{2}, a \neq -5$

$\qquad \dfrac{a+5}{2a^2+7a-15} = \dfrac{a+5}{(2a-3)(a+5)} = \dfrac{1}{2a-3}$

5. $\dfrac{6}{5} - \dfrac{3}{5k-10} = \dfrac{6(k-2)}{5(k-2)} - \dfrac{3}{5(k-2)}$

$\qquad = \dfrac{6k-15}{5(k-2)}$

$\qquad = \dfrac{3(2k-5)}{5(k-2)}$

7. $\dfrac{\frac{7}{18y}}{\frac{2}{9}} = \dfrac{(18y)\left(\frac{7}{18y}\right)}{(18y)\left(\frac{2}{9}\right)} = \dfrac{7}{4y}$

9. $\dfrac{\frac{3x+2y}{2y}}{\frac{6x+4y}{2}} = \dfrac{3x+2y}{2y} \cdot \dfrac{2}{2(3x+2y)} = \dfrac{1}{2y}$

11. $\dfrac{\frac{8a^4b^3}{3c}}{\frac{a^7b^2}{9c}} = \dfrac{8a^4b^3}{3c} \cdot \dfrac{9c}{a^7b^2} = \dfrac{24b}{a^3}$

13. $\dfrac{\frac{4r^3s}{t^5}}{\frac{2s^7}{r^2t^9}} = \dfrac{4r^3s}{t^5} \cdot \dfrac{r^2t^9}{2s^7} = \dfrac{2r^5t^4}{s^6}$

15. $\dfrac{\frac{1}{8}+\frac{4}{3}}{\frac{1}{2}-\frac{5}{12}} = \dfrac{\frac{3}{24}+\frac{32}{24}}{\frac{6}{12}-\frac{5}{12}} = \dfrac{\frac{35}{24}}{\frac{1}{12}} = \dfrac{35}{24} \cdot \dfrac{12}{1} = \dfrac{35}{2}$

17. $\dfrac{\frac{1}{h}+\frac{1}{k}}{\frac{1}{hk}} = \dfrac{\frac{k}{hk}+\frac{h}{hk}}{\frac{1}{hk}} = \dfrac{\frac{k+h}{hk}}{\frac{1}{hk}} = \dfrac{k+h}{hk} \cdot \dfrac{hk}{1} = k+h$

19. $\dfrac{\frac{n+1}{n^2-9}}{\frac{2}{n+3}} = \dfrac{n+1}{(n+3)(n-3)} \cdot \dfrac{n+3}{2} = \dfrac{n+1}{2(n-3)}$

21. $\dfrac{2+\frac{1}{x}}{4+\frac{1}{x}} = \dfrac{\frac{2x}{x}+\frac{1}{x}}{\frac{4x}{x}+\frac{1}{x}}$

$= \dfrac{\frac{2x+1}{x}}{\frac{4x+1}{x}}$

$= \dfrac{2x+1}{x} \cdot \dfrac{x}{4x+1}$

$= \dfrac{2x+1}{4x+1}$

23. $\dfrac{\frac{m}{7}-\frac{7}{m}}{\frac{1}{7}+\frac{1}{m}} = \dfrac{(7m)\left(\frac{m}{7}-\frac{7}{m}\right)}{(7m)\left(\frac{1}{7}+\frac{1}{m}\right)}$

$= \dfrac{m^2-49}{m+7}$

$= \dfrac{(m+7)(m-7)}{m+7}$

$= m-7$

25. $\dfrac{\frac{1}{5}-\frac{1}{y}}{\frac{7}{10}+\frac{1}{y^2}} = \dfrac{10y^2\left(\frac{1}{5}-\frac{1}{y}\right)}{10y^2\left(\frac{7}{10}+\frac{1}{y^2}\right)}$

$= \dfrac{2y^2-10y}{7y^2+10}$

$= \dfrac{2y(y-5)}{7y^2+10}$

27. $\dfrac{\frac{8}{a+4}+2}{\frac{12}{a+4}-2} = \dfrac{(a+4)\left(\frac{8}{a+4}+2\right)}{(a+4)\left(\frac{12}{a+4}-2\right)}$

$= \dfrac{8+2a+8}{12-2a-8}$

$= \dfrac{2a+16}{4-2a}$

$= \dfrac{2(a+8)}{2(2-a)}$

$= \dfrac{a+8}{2-a}$ or $-\dfrac{a+8}{a-2}$

29. $\dfrac{1-\frac{4}{t^2}}{1-\frac{2}{t}-\frac{8}{t^2}} = \dfrac{t^2\left(1-\frac{4}{t^2}\right)}{t^2\left(1-\frac{2}{t}-\frac{8}{t^2}\right)}$

$= \dfrac{t^2-4}{t^2-2t-8}$

$= \dfrac{(t+2)(t-2)}{(t-4)(t+2)}$

$= \dfrac{t-2}{t-4}$

31. $\dfrac{t+4+\frac{3}{t}}{t-4-\frac{5}{t}} = \dfrac{\frac{t^2+4t+3}{t}}{\frac{t^2-4t-5}{t}}$

$= \dfrac{t^2+4t+3}{t^2-4t-5}$

$= \dfrac{(t+1)(t+3)}{(t-5)(t+1)}$

$= \dfrac{t+3}{t-5}$

33. $\dfrac{\frac{1}{k-6}-1}{\frac{2}{k-6}-2} = \dfrac{\frac{1}{k-6}-1}{2\left(\frac{1}{k-6}-1\right)}$

$= \dfrac{1}{2}$

35. $\dfrac{\frac{1}{2}+\frac{2}{3}}{5}$; $\dfrac{\frac{1}{2}+\frac{2}{3}}{5}=\dfrac{\frac{3}{6}+\frac{4}{6}}{5}=\dfrac{\frac{7}{6}}{5}=\dfrac{7}{6}\cdot\dfrac{1}{5}=\dfrac{7}{30}$

37. $\dfrac{3}{\frac{2}{3}+\frac{3}{4}}$; $\dfrac{3}{\frac{8}{12}+\frac{9}{12}}=\dfrac{3}{\frac{17}{12}}=3\cdot\dfrac{12}{17}=\dfrac{36}{17}$

39. **(a)** $R=\dfrac{1}{\frac{1}{2}+\frac{1}{3}}=\dfrac{1}{\frac{5}{6}}=\dfrac{6}{5}\ \Omega$

(b) $R=\dfrac{1}{\frac{1}{10}+\frac{1}{15}}=\dfrac{1}{\frac{5}{30}}=\dfrac{30}{5}=6\ \Omega$

41.
$$\dfrac{x^{-1}-y^{-1}}{x^{-2}-y^{-2}}=\dfrac{\dfrac{1}{x}-\dfrac{1}{y}}{\dfrac{1}{x^2}-\dfrac{1}{y^2}}$$

$$=\dfrac{\dfrac{1(y)}{xy}-\dfrac{1(x)}{xy}}{\dfrac{1(y^2)}{x^2y^2}-\dfrac{1(x^2)}{x^2y^2}}$$

$$=\dfrac{\dfrac{y-x}{xy}}{\dfrac{y^2-x^2}{x^2y^2}}$$

$$=\dfrac{\dfrac{y-x}{xy}}{\dfrac{y^2-x^2}{x^2y^2}}\cdot\dfrac{\dfrac{x^2y^2}{y^2-x^2}}{\dfrac{x^2y^2}{y^2-x^2}}$$

$$=\dfrac{(y-x)\left(x^2y^2\right)}{xy\left(y^2-x^2\right)}$$

$$=\dfrac{\cancel{(y-x)}\,(xy)(xy)}{xy\,\cancel{(y-x)}(y+x)}$$

$$=\dfrac{xy}{y+x}$$

43. $\dfrac{2x^{-1}+8y^{-1}}{4x^{-1}}=\dfrac{\frac{2}{x}+\frac{8}{y}}{\frac{4}{x}}$

$$=\dfrac{\frac{2y+8x}{xy}}{\frac{4}{x}}$$

$$=\dfrac{2y+8x}{xy}\cdot\dfrac{x}{4}$$

$$=\dfrac{y+4x}{2y}$$

45. $\dfrac{(mn)^{-2}}{m^{-2}+n^{-2}}=\dfrac{\frac{1}{m^2n^2}}{\frac{1}{m^2}+\frac{1}{n^2}}$

$$=\dfrac{\frac{1}{m^2n^2}}{\frac{n^2+m^2}{m^2n^2}}$$

$$=\dfrac{1}{m^2n^2}\cdot\dfrac{m^2n^2}{n^2+m^2}$$

$$=\dfrac{1}{n^2+m^2}$$

47. $\dfrac{\frac{1}{z^2-9}+\frac{2}{z+3}}{\frac{3}{z-3}}=\dfrac{(z+3)(z-3)\left(\frac{1}{(z+3)(z-3)}+\frac{2}{z+3}\right)}{(z+3)(z-3)\left(\frac{3}{z-3}\right)}$

$$=\dfrac{1+2z-6}{3z+9}$$

$$=\dfrac{2z-5}{3(z+3)}$$

49. $\dfrac{\frac{2}{x-1}+2}{\frac{2}{x+1}-2}=\dfrac{(x-1)(x+1)\left(\frac{2}{x-1}+2\right)}{(x-1)(x+1)\left(\frac{2}{x+1}-2\right)}$

$$=\dfrac{2(x+1)+2(x+1)(x-1)}{2(x-1)-2(x+1)(x-1)}$$

$$=\dfrac{2x^2+2x}{2x-2x^2}$$

$$=\dfrac{2x(x+1)}{2x(1-x)}$$

$$=\dfrac{x+1}{1-x}\ \text{or}\ -\dfrac{x+1}{x-1}$$

51. $1+\dfrac{1}{1+1}=1+\dfrac{1}{2}=\dfrac{3}{2}$

Section 7.6 Practice Exercises

1. **(a)** linear; quadratic
(b) rational
(c) denominator

3. $\dfrac{2x-6}{x^2+3x+2} \div \dfrac{x^2-5x+6}{x^2-4}$

$= \dfrac{2\cancel{(x-3)}}{\cancel{(x+2)}(x+1)} \cdot \dfrac{\cancel{(x+2)}\,\cancel{(x-2)}}{\cancel{(x-2)}\,\cancel{(x-3)}}$

$= \dfrac{2}{x+1}$

5. $\dfrac{h-\frac{1}{h}}{\frac{1}{5}-\frac{1}{5h}} = \dfrac{5h\left(h-\frac{1}{h}\right)}{5h\left(\frac{1}{5}-\frac{1}{5h}\right)}$

$= \dfrac{5h^2-5}{h-1}$

$= \dfrac{5(h+1)(h-1)}{(h-1)}$

$= 5(h+1)$

7. $1+\dfrac{1}{x}-\dfrac{12}{x^2} = \dfrac{x^2+x+12}{x^2}$

$= \dfrac{(x+4)(x-3)}{x^2}$

9. $\dfrac{5}{2}+\dfrac{1}{2}b = 5-\dfrac{1}{3}b$

$6\left(\dfrac{5}{2}+\dfrac{1}{2}b\right) = 6\left(5-\dfrac{1}{3}b\right)$

$15+3b = 30-2b$

$5b = 15$

$b = 3$

11. $\dfrac{5}{3}-\dfrac{1}{6}k = \dfrac{3k+5}{4}$

$12\left(\dfrac{5}{3}-\dfrac{1}{6}k\right) = 12\left(\dfrac{3k+5}{4}\right)$

$20-2k = 9k+15$

$-11k = -5$

$k = \dfrac{5}{11}$

13. $\dfrac{4y+2}{3}-\dfrac{7}{6} = -\dfrac{y}{6}$

$6\left(\dfrac{4y+2}{3}-\dfrac{7}{6}\right) = 6\left(-\dfrac{y}{6}\right)$

$8y+4-7 = -y$

$9y = 3$

$y = \dfrac{1}{3}$

15. **(a)** $z = 0$

(b) LCD: $5z$

(c) $\dfrac{3}{z}-\dfrac{4}{5} = -\dfrac{1}{5}$

$5z\left(\dfrac{3}{z}-\dfrac{4}{5}\right) = 5z\left(-\dfrac{1}{5}\right)$

$15-4z = -z$

$3z = 15$

$z = 5$

17. $\dfrac{1}{8} = \dfrac{3}{5}+\dfrac{5}{y}$

$40y\left(\dfrac{1}{8}\right) = 40y\left(\dfrac{3}{5}+\dfrac{5}{y}\right)$

$5y = 24y+200$

$-19y = 200$

$y = -\dfrac{200}{19}$

19. $\dfrac{7}{4a} = \dfrac{3}{a-5}$

$7(a-5) = 3(4a)$

$7a-35 = 12a$

$7a-7a-35 = 12a-7a$

$-35 = 5a$

$\dfrac{-35}{5} = \dfrac{5a}{5}$

$-7 = a$

21.
$$\frac{5}{6x} + \frac{7}{x} = 1$$
$$6x\left(\frac{5}{6x} + \frac{7}{x}\right) = 6x(1)$$
$$5 + 42 = 6x$$
$$6x = 47$$
$$x = \frac{47}{6}$$

23.
$$1 - \frac{2}{y} = \frac{3}{y^2}$$
$$y^2\left(1 - \frac{2}{y}\right) = y^2\left(\frac{3}{y^2}\right)$$
$$y^2 - 2y = 3$$
$$y^2 - 2y - 3 = 0$$
$$(y-3)(y+1) = 0$$
$$y - 3 = 0 \quad \text{or} \quad y + 1 = 0$$
$$y = 3 \qquad\qquad y = -1$$

25.
$$\frac{a+1}{a} = 1 + \frac{a-2}{2a}$$
$$2a\left(\frac{a+1}{a}\right) = 2a\left(1 + \frac{a-2}{2a}\right)$$
$$2a + 2 = 2a + a - 2$$
$$-a = -4$$
$$a = 4$$

27.
$$\frac{w}{5} - \frac{w+3}{w} = -\frac{3}{w}$$
$$5w\left(\frac{w}{5} - \frac{w+3}{w}\right) = 5w\left(-\frac{3}{w}\right)$$
$$w^2 - 5w - 15 = -15$$
$$w^2 - 5w = 0$$
$$w(w-5) = 0$$
$w = 0$ or $w = 5$
$w = 0$ is extraneous, $w = 5$ is the solution.

29.
$$\frac{2}{m+3} = \frac{5}{4m+12} - \frac{3}{8}$$
$$\frac{2}{m+3} = \frac{5}{4(m+3)} - \frac{3}{8}$$
$$8(m+3) \cdot \frac{2}{m+3} = 8(m+3)\left(\frac{5}{4(m+3)} - \frac{3}{8}\right)$$
$$16 = 8(m+3)\left(\frac{5}{4(m+3)}\right) - 8(m+3) \cdot \frac{3}{8}$$
$$16 = 10 - 3m - 9$$
$$15 = -3m$$
$$m = -5$$

31.
$$\frac{p}{p-4} - 5 = \frac{4}{p-4}$$
$$(p-4)\left(\frac{p}{p-4} - 5\right) = (p-4) \cdot \frac{4}{p-4}$$
$$(p-4) \cdot \frac{p}{p-4} - 5(p-4) = 4$$
$$p - 5p + 20 = 4$$
$$-4p = -16$$
$$p = 4$$
No solution; $p = 4$ is extraneous.

33.
$$\frac{2t}{t+2} - 2 = \frac{t-8}{t+2}$$
$$\left(t + 2\right)\left(\frac{2t}{t+2} - 2\right) = (t+2) \cdot \frac{t-8}{t+2}$$
$$(t+2)\frac{2t}{t+2} - 2(t+2) = t - 8$$
$$2t - 2t - 4 = t - 8$$
$$-4 = t - 8$$
$$t = 4$$

35.
$$\frac{x^2 - x}{x-2} = \frac{12}{x-2}$$
$$(x-2) \cdot \frac{x^2 - x}{x-2} = (x-2) \cdot \frac{12}{x-2}$$
$$x^2 - x = 12$$
$$x^2 - x - 12 = 0$$
$$(x-4)(x+3) = 0$$
$$x = 4 \text{ or } x = -3$$

37.
$$\frac{x^2 + 3x}{x-1} = \frac{4}{x-1}$$
$$(x-1) \cdot \frac{x^2 + 3x}{x-1} = (x-1) \cdot \frac{4}{x-1}$$
$$x^2 + 3x = 4$$
$$x^2 + 3x - 4 = 0$$
$$(x+4)(x-1) = 0$$
$$x + 4 = 0 \text{ or } x - 1 = 0$$
$x = -4$ is the solution ($x = 1$ is extraneous).

39.
$$\frac{2x}{x+4} - \frac{8}{x-4} = \frac{2x^2 + 32}{x^2 - 16}$$
$$\frac{2x}{x+4} - \frac{8}{x-4} = \frac{2x^2 + 32}{(x+4)(x-4)}$$
Multiply both sides by LCD: $(x + 4)(x - 4)$
$$2x(x-4) - 8(x+4) = 2x^2 + 32$$
$$2x^2 - 8x - 8x - 32 = 2x^2 + 32$$
$$-16x = 64$$
$$x = -4$$
No solution ($x = -4$ is extraneous).

41.
$$\frac{x}{x+6} = \frac{72}{x^2 - 36} + 4$$
$$\frac{x}{x+6} = \frac{72}{(x+6)(x-6)} + 4$$
Multiply both sides by LCD: $(x + 6)(x - 6)$
$$x(x-6) = 72 + 4(x+6)(x-6)$$
$$x^2 - 6x = 72 + 4x^2 - 144$$
$$-3x^2 - 6x + 72 = 0$$
$$-3(x^2 + 2x - 24) = 0$$
$$(x+6)(x-4) = 0$$
$$x + 6 = 0 \text{ or } x - 4 = 0$$
$x = 4$ is the solution ($x = -6$ is extraneous).

43.
$$\frac{5}{3x-3} - \frac{2}{x-2} = \frac{7}{x^2 - 3x + 2}$$
$$\frac{5}{3(x-1)} - \frac{2}{x-2} = \frac{7}{(x-2)(x-1)}$$
Multiply both sides by LCD: $3(x - 1)(x -

2)$
$$5(x-2) - 2(3)(x-1) = 7(3)$$
$$5x - 10 - 6x + 6 = 21$$
$$-x - 4 = 21$$
$$-x = 25$$
$$x = -25$$

45.
$$\frac{w}{w-3} = \frac{17}{w^2 - 7w + 12} + \frac{1}{w-4}$$
$$\frac{w}{w-3} = \frac{17}{(w-3)(w-4)} + \frac{1}{w-4} \quad \text{Multiply}$$
both sides by LCD: $(w - 3)(w - 4)$
$$w(w-4) = 17 + w - 3$$
$$w^2 - 4w = 14 + w$$
$$w^2 - 4w - w = 14 + w - w$$
$$w^2 - 5w - 14 = 14 - 14$$
$$w^2 - 5w - 14 = 0$$
$$(w+2)(w-7) = 0$$
$$w + 2 = 0 \quad \text{or} \quad (w-7) = 0$$
$$w = -2 \quad \text{or} \quad w = 7$$

47. Let $x = $ a number. Then,
$$\frac{1}{x} + 3 = \frac{25}{x}$$
$$x\left(\frac{1}{x} + 3\right) = x \cdot \frac{25}{x}$$
$$1 + 3x = 25$$
$$3x = 24$$
$$x = 8$$
The number is 8.

49. Let $x = $ a number. Then,
$$\frac{x+5}{x-2} = \frac{3}{4}$$
$$4(x-2) \cdot \frac{x+5}{x-2} = 4(x-2) \cdot \frac{3}{4}$$
$$4x + 20 = 3x - 6$$
$$x = -26$$
The number is −26.

51. $K = \dfrac{ma}{F}$

$FK = ma$

$m = \dfrac{FK}{a}$

53. $K = \dfrac{IR}{E}$

$EK = IR$

$E = \dfrac{IR}{K}$

55. $I = \dfrac{E}{R+r}$

$I(R+r) = E$

$IR + Ir = E$

$IR = E - Ir$

$R = \dfrac{E - Ir}{I}$ or $R = \dfrac{E}{I} - r$

57. $h = \dfrac{2A}{B+b}$

$h(B+b) = 2A$

$Bh + bh = 2A$

$Bh = 2A - bh$

$B = \dfrac{2A - bh}{h}$ or $B = \dfrac{2A}{h} - b$

59. $\dfrac{V}{\pi h} = r^2$

$V = r^2 \pi h$

$h = \dfrac{V}{r^2 \pi}$

61. $x = \dfrac{at + b}{t}$

$xt = at + b$

$xt - at = b$

$(x - a)t = b$

$t = \dfrac{b}{x - a}$ or $t = \dfrac{-b}{a - x}$

63. $\dfrac{x - y}{xy} = z$

$x - y = xyz$

$x - xyz = y$

$x(1 - yz) = y$

$x = \dfrac{y}{1 - yz}$ or $x = \dfrac{-y}{yz - 1}$

65. $a + b = \dfrac{2A}{h}$

$h(a + b) = 2A$

$h = \dfrac{2A}{a + b}$

67. $\dfrac{1}{R} = \dfrac{1}{R_1} + \dfrac{1}{R_2}$

$RR_1 R_2 \cdot \dfrac{1}{R} = RR_1 R_2 \left(\dfrac{1}{R_1} + \dfrac{1}{R_2} \right)$

$R_1 R_2 = RR_2 + RR_1$

$R_1 R_2 = R(R_2 + R_1)$

$R = \dfrac{R_1 R_2}{R_1 + R_2}$

Problem Recognition Exercises

1. $\dfrac{y}{2y + 4} - \dfrac{2}{y^2 + 2y} = \dfrac{y}{2(y + 2)} - \dfrac{2}{y(y + 2)}$

$= \dfrac{y^2 - 4}{2y(y + 2)}$

$= \dfrac{(y - 2)(y + 2)}{2y(y + 2)}$

$= \dfrac{y - 2}{2y}$

3.
$$\frac{5t}{2} - \frac{t-2}{3} = 5$$
$$\frac{5t \cdot 3}{2 \cdot 3} - \frac{2(t-2)}{2 \cdot 3} = 5$$
$$\frac{15t - 2t + 4}{2 \cdot 3} = 5$$
$$\frac{13t + 4}{6} = 5$$
$$13t + 4 = 5 \cdot 6 = 30$$
$$13t = 26$$
$$t = 2$$

5.
$$\frac{7}{6p^2} + \frac{2}{9p} + \frac{1}{3p^2}$$
$$= \frac{3 \cdot 7}{3 \cdot 6p^2} + \frac{2 \cdot 2p}{2 \cdot 9p^2} + \frac{6}{3 \cdot 6p^2}$$
$$= \frac{21 + 4p + 6}{18p^2}$$
$$= \frac{27 + 4p}{18p^2}$$

7.
$$4 + \frac{2}{h-3} = 5$$
$$\frac{2}{h-3} = 1$$
$$h - 3 = 2$$
$$h = 5$$

9.
$$\frac{1}{x-6} - \frac{3}{x^2 - 6x} = \frac{4}{x}$$
$$\frac{1}{x-6} - \frac{3}{x(x-6)} = \frac{4}{x}$$
$$\frac{x}{x(x-6)} - \frac{3}{x(x-6)} = \frac{4(x-6)}{x(x-6)}$$
$$\frac{x-3}{x(x-6)} = \frac{4x-24}{x(x-6)}$$
$$\frac{21}{x(x-6)} = \frac{3x}{x(x-6)}$$
$$x = 7$$

11.
$$\frac{7}{2x+2} + \frac{3x}{4x+4} = \frac{7}{2(x+1)} + \frac{3x}{4(x+1)}$$
$$= \frac{7 \cdot 2 + 3x}{4(x+1)}$$
$$= \frac{14 + 3x}{4(x+1)}$$

13.
$$\frac{3}{5x} + \frac{7}{2x} = 1$$

Multiply all terms by common denominator $(5x)(2x)$

$$\frac{3}{5x} + \frac{7}{2x} = 1$$
$$3(2x) + 7(5x) = (2x)(5x)$$
$$6x + 35x = 10x^2$$
$$41x = 10x^2$$
$$\frac{41x}{x} = \frac{10x^2}{x}$$
$$41 = 10x$$
$$\frac{41}{10} = \frac{10x}{10}$$
$$\frac{41}{10} = x$$

15.
$$\frac{5}{2a-1} + 4 = \frac{5}{2a-1} + \frac{4(2a-1)}{2a-1}$$
$$= \frac{5 + 8a - 4}{2a-1}$$
$$= \frac{8a + 1}{2a-1}$$

17.
$$\frac{3}{u} + \frac{12}{u^2 - 3u} = \frac{u+1}{u-3}$$
$$\frac{3}{u} + \frac{12}{u(u-3)} = \frac{u+1}{u-3}$$

Multiply all terms by the common denominator $u(u-3)$

$$3(u-3)+12 = u(u+1)$$

$$3u - 9 + 12 = u^2 + u$$

$$3u - 3u + 3 = u^2 + u - 3u$$

$$3 - 3 = u^2 - 2u - 3$$

$$0 = u^2 - 2u - 3$$

$$0 = (u+1)(u-3)$$

$$u + 1 = 0 \text{ or } u - 3 = 0$$

$$u = -1, \ 3$$

$$u = 3 \text{ does not check.}$$

19. $\dfrac{-2h}{h^2-9} + \dfrac{3}{h-3} = \dfrac{-2h}{(h+3)(h-3)} + \dfrac{3(h+3)}{(h+3)(h-3)}$

$$= \dfrac{-2h + 3h + 9}{(h+3)(h-3)}$$

$$= \dfrac{h+9}{(h+3)(h-3)}$$

Section 7.7 Practice Exercises

1. (a) proportion

(b) proportional

3. Expression;

$$\dfrac{m}{m-1} - \dfrac{2}{m+3}$$

$$= \dfrac{m(m+3)}{(m-1)(m+3)} - \dfrac{2(m-1)}{(m-1)(m+3)}$$

$$= \dfrac{m^2 + 3m - 2m + 2}{(m-1)(m+3)}$$

$$= \dfrac{m^2 + m + 2}{(m-1)(m+3)}$$

5. Expression;

$$\dfrac{3y+6}{20} \div \dfrac{4y+8}{8} = \dfrac{3(y+2)}{20} \cdot \dfrac{8}{4(y+2)} = \dfrac{3}{10}$$

7. Equation; $\dfrac{3}{p+3} = \dfrac{12p+19}{p^2+7p+12} - \dfrac{5}{p+4}$

$$\dfrac{3}{p+3} = \dfrac{12p+19}{(p+3)(p+4)} - \dfrac{5}{p+4}$$

Multiply both sides by the LCD $(p+3)(p+4)$

$$3(p+4) = 12p + 19 - 5(p+3)$$

$$3p + 12 = 12p + 19 - 5p - 15$$

$$3p + 12 = 7p + 4$$

$$-4p = -8$$

$$p = 2$$

9. $\quad \dfrac{8}{5} = \dfrac{152}{p}$

$$5p\left(\dfrac{8}{5}\right) = 5p\left(\dfrac{152}{p}\right)$$

$$8p = 760$$

$$p = 95$$

11. $\quad \dfrac{19}{76} = \dfrac{z}{4}$

$$76\left(\dfrac{19}{76}\right) = 76\left(\dfrac{z}{4}\right)$$

$$19 = 19z$$

$$1 = z$$

13. $\quad \dfrac{5}{3} = \dfrac{a}{8}$

$$24 \cdot \dfrac{5}{3} = 24 \cdot \dfrac{a}{8}$$

$$40 = 3a$$

$$a = \dfrac{40}{3}$$

15. $\quad \dfrac{2}{1.9} = \dfrac{x}{38}$

$$1.9x = 76$$

$$x = \dfrac{76}{1.9} = 40$$

17.
$$\frac{y+1}{2y} = \frac{2}{3}$$
$$6y\left(\frac{y+1}{2y}\right) = 6y\left(\frac{2}{3}\right)$$
$$3(y+1) = 4y$$
$$3y+3 = 4y$$
$$3 = y$$

19.
$$\frac{9}{2z-1} = \frac{3}{z}$$
$$z(2z-1)\left(\frac{9}{2z-1}\right) = z(2z-1)\left(\frac{3}{z}\right)$$
$$9z = 3(2z-1)$$
$$9z = 6z-3$$
$$3z = -3$$
$$z = -1$$

21.
$$\frac{8}{9a-1} = \frac{5}{3a+2}$$
$$8(3a+2) = 5(9a-1)$$
$$24a+16 = 45a-5$$
$$-21a = -21$$
$$a = 1$$

23. (a)
$$\frac{V_i}{V_f} = \frac{T_i}{T_f}$$
$$V_i T_f = T_i V_f$$
$$V_f = \frac{V_i T_f}{T_i}$$

(b)
$$\frac{V_i}{V_f} = \frac{T_i}{T_f}$$
$$V_i T_f = T_i V_f$$
$$T_f = \frac{T_i V_f}{V_i}$$

25. Let x represent the number of miles Toni can drive on 9 gallons of gas.

$$\frac{132 \text{ mi}}{4 \text{ gal}} = \frac{x \text{ mi}}{9 \text{ gal}}$$
$$132(9) = 4x$$
$$1188 = 4x$$
$$297 = x$$

Toni can drive 297 mi on 9 gallons of gas.

27. Let x represent the amount of Grow-It-Right plant food needed for 1 gal (3.8 L) of water.

$$\frac{7.8 \text{ mL plant food}}{2 \text{ L of water}} = \frac{x}{3.8 \text{ L of water}}$$
$$7.8(3.8) = 2x$$
$$29.62 = 2x$$
$$14.82 = x$$

Use 14.82 mL of plant food.

29. Let x represent the carbohydrate.
$$\frac{8 \text{ oz}}{19.2 \text{ g}} = \frac{5 \text{ oz}}{x \text{ g}}$$
$$x = \frac{5(19.2)}{8} = 12$$

The 5 oz pineapple contains 12 g carbohydrate.

31. Let x = length of the ramp,
$$\frac{12 \text{ ft long}}{1 \text{ ft of height}} = \frac{x}{1\frac{2}{3} \text{ ft of height}}$$
$$12 \bullet 1\frac{2}{3} = x$$
$$12 \bullet \frac{5}{3} = x$$
$$20 = x$$

The minimum length is 20 ft.

33.
$$\frac{15}{3} = \frac{20}{x} \qquad\qquad \frac{15}{3} = \frac{25}{y}$$
$$15x = 60 \qquad\qquad 15y = 75$$
$$x = 4 \text{ cm} \qquad\qquad y = 5 \text{ cm}$$

35.
$$\frac{x}{15} = \frac{3}{12}$$
$$12x = 45$$
$$x = 3.75 \text{ cm}$$

$$\frac{y}{18} = \frac{3}{12}$$
$$12y = 54$$
$$y = 4.5 \text{ cm}$$

37.
$$\frac{x}{16.8} = \frac{1}{2.4}$$
$$2.4x = 16.8$$
$$x = 7$$
The height of the pole is 7 m.

39. Let x = the height of the post. Then,
$$\frac{x}{54+18} = \frac{6}{18}$$
$$\frac{x}{72} = \frac{6}{18}$$
$$18x = 432$$
$$x = 24$$
The post is 24 ft.

41. Let x = speed of boat.

	Distance	Rate	Time
Downstream	66	x + 2	$\frac{66}{x+2}$
Upstream	54	x − 2	$\frac{54}{x-2}$

$$\frac{66}{x+2} = \frac{54}{x-2}$$
$$(x+2)(x-2)\frac{66}{x+2} = (x+2)(x-2)\frac{54}{x-2}$$
$$66(x-2) = 54(x+2)$$
$$66x - 132 = 54x + 108$$
$$12x = 240$$
$$x = 20$$
The speed is 20 mph.

43. Let x = speed of plane.

	Distance	Rate	Time
With the wind	700	35 + x	$\frac{700}{35+x}$
Against the wind	500	x-35	$\frac{500}{x-35}$

$$\frac{700}{35+x} = \frac{500}{x-35}$$
$$700(x-35) = 500(35+x)$$
$$7(x-35) = 5(35+x)$$
$$7x - 245 = 175 + 5x$$
$$2x = 420$$
$$x = 210$$
The plane speed is 210 mph.

45. Let x = running speed. $2x$ = biking speed

	Distance	Speed	Time
Biking	20	$2x$	$\frac{20}{2x}$
Running	10	x	$\frac{10}{x}$

$$\frac{20}{2x} + \frac{10}{x} = 2.5$$

$$\frac{10}{x} + \frac{10}{x} = 2.5$$

$$\frac{20}{x} = 2.5$$

$$x = \frac{20}{2.5}$$

$$x = 8, \quad 2x = 16$$

He runs at 8 mph and bikes at 16 mph.

47. Let x = Floyd's speed. $x - 2$ = Rachel's speed

	Distance	Speed	Time
Floyd	12	x	$\frac{12}{x}$
Rachel	12	$x - 2$	$\frac{12}{x-2}$

$$\frac{12}{x} + 3 = \frac{12}{x-2}$$

$$\frac{12}{x} + \frac{3x}{x} = \frac{12}{x-2}$$

$$\frac{12+3x}{x} = \frac{12}{x-2}$$

$$(12+3x)(x-2) = 12x$$

$$12x - 24 + 3x^2 - 6x = 12x$$

$$3x^2 - 6x - 24 = 0$$

$$(3x-12)(x+2) = 0$$

$$x = 4 \text{ or } \cancel{x=-2}$$

$$x - 2 = 2$$

Floyd's speed is 4 mph and Rachelle's speed is 2 mph

49. Let x represent the riding rate. Then $x - 9$ represents the walking rate.

	Distance	Rate	Time
Bike ride	4	x	$\frac{4}{x}$
Walk	4	$x - 9$	$\frac{4}{x-9}$

$$\begin{pmatrix} \text{Time to} \\ \text{ride bike} \end{pmatrix} + (1 \text{ hour}) = \begin{pmatrix} \text{Time to} \\ \text{Walk} \end{pmatrix}$$

$$\frac{4}{x} + 1 = \frac{4}{x-9}$$

$$x(x-9)\left(\frac{4}{x} + 1\right) = x(x-9)\left(\frac{4}{x-9}\right)$$

$$4(x-9) + x(x-9) = 4x$$

$$4x - 36 + x^2 - 9x = 4x$$

$$x^2 - 9x - 36 = 0$$

$$(x-12)(x+3) = 0$$

$$x = 12 \text{ or } \cancel{x = -3}$$

Sergio rode 12 mph and walked 3 mph.

51. In one minute, the cold water can fill $\frac{1}{10}$ of the sink; the hot water can fill $\frac{1}{12}$ of the sink. If $x =$ how long it would take both faucets to fill the sink together, then both faucets can fill $\frac{1}{x}$ of the sink.

$$\frac{1}{10} + \frac{1}{12} = \frac{1}{x}$$

$$60x\left(\frac{1}{10} + \frac{1}{12}\right) = 60x\left(\frac{1}{x}\right)$$

$$60x\left(\frac{1}{10}\right) + 60x\left(\frac{1}{12}\right) = 60x\left(\frac{1}{x}\right)$$

$$6x + 5x = 60$$

$$11x = 60$$

$$x = \frac{60}{11} = 5\frac{5}{11}$$

Both faucets can fill the sink in $5\frac{5}{11}$ or $5.\overline{45}$ minutes.

53. In one minute, one printer can do $\frac{1}{50}$ of the job; the other printer can do $\frac{1}{40}$ of the job. If $x =$ how long it takes both printers to do the job together, then $\frac{1}{x}$ of the job can be completed in 1 minute.

$$\frac{1}{50} + \frac{1}{40} = \frac{1}{x}$$

$$200x\left(\frac{1}{50} + \frac{1}{40}\right) = 200x\left(\frac{1}{x}\right)$$

$$4x + 5x = 200$$

$$9x = 200$$

$$x = \frac{200}{9} = 22\frac{2}{9}$$

Together they can do the job in $22\frac{2}{9}$ or $22.\overline{2}$ minutes.

55. Let $x =$ how long it takes both pipes to fill the reservoir. In one hour, the first pipe

213

can fill $\dfrac{1}{16}$ of the reservoir; the second pipe can empty $\dfrac{1}{24}$ of the reservoir.

Together in 1 hour they can fill $\dfrac{1}{x}$ of the reservoir.

$$\frac{1}{16}-\frac{1}{24}=\frac{1}{x}$$

$$\frac{3}{16\cdot 3}-\frac{2}{24\cdot 2}=\frac{1}{x}$$

$$\frac{3-2}{48}=\frac{1}{x}$$

$$\frac{1}{48}=\frac{1}{x}$$

$$x=48$$

The reservoir would be filled in 48 hours.

57. Let x = how long it will take Al. Then, in 1 day, Al can complete $\dfrac{1}{x}$, Tim can complete $\dfrac{1}{5}$, and together they can complete $\dfrac{1}{2}$ of the job.

$$\frac{1}{x}+\frac{1}{5}=\frac{1}{2}$$

$$10x\left(\frac{1}{x}+\frac{1}{5}\right)=10x\left(\frac{1}{2}\right)$$

$$10+2x=5x$$

$$3x=10$$

$$x=\frac{10}{3}=3\frac{1}{3}$$

It would take Al $3\dfrac{1}{3}$ or $3.\overline{3}$ days.

59. Let x represent the number of smokers. Then $x + 100$ represents the number of nonsmokers.

$$\frac{2}{7}=\frac{x}{x+100}$$

$$7x=2(x+100)$$

$$7x=2x+200$$

$$5x=200$$

$$x=40$$

There are 40 smokers and 140 nonsmokers.

61. Let x represent the number of men. Then $440 - x$ represents the number of women.

$$\frac{6}{5}=\frac{x}{440-x}$$

$$6(440-x)=5x$$

$$2640-6x=5x$$

$$2640=11x$$

$$240=x$$

There are 240 men and 200 women.

Section 7.8 Practice Exercises

1. (a) $\qquad kx$

(b) $\dfrac{k}{x}$

(c) kxw

3.
$$\frac{2y}{3}-\frac{3y-1}{5}=1$$

$$15\left(\frac{2y}{3}-\frac{3y-1}{5}\right)=15(1)$$

$$5(2y)-3(3y-1)=15$$

$$10y-9y+3=15$$

$$y+3=15$$

$$y=12$$

5.
$$\frac{a}{4}+\frac{3}{a}=2$$

$$4a\left(\frac{a}{4}+\frac{3}{a}\right)=4a(2)$$

$$a^2+12=8a$$

$$a^2-8a+12=0$$

$$(a-6)(a-2)=0$$

$$a-6=0 \quad\text{or}\quad a-2=0$$

$$a=6 \qquad\qquad a=2$$

7. $\dfrac{a+\dfrac{a}{b}}{\dfrac{a}{b}-a} = \dfrac{\dfrac{ab}{b}+\dfrac{a}{b}}{\dfrac{a}{b}-\dfrac{ab}{b}}$

$= \dfrac{\dfrac{ab+a}{b}}{\dfrac{a-ab}{b}}$

$= \dfrac{ab+a}{b}\cdot\dfrac{b}{a-ab}$

$= \dfrac{ab+a}{a-ab}$

$= \dfrac{a(b+1)}{a(1-b)} = \dfrac{b+1}{1-b}$

9. Inversely

11. $T = kq$

13. $b = \dfrac{k}{c}$

15. $Q = \dfrac{kx}{y}$

17. $c = kst$

19. $L = kw\sqrt{v}$

21. $x = \dfrac{ky^2}{z}$

23. $y = kx$
$18 = k(4)$
$\dfrac{18}{4} = k$
$\dfrac{9}{2} = k$

25. $p = \dfrac{k}{q}$
$32 = \dfrac{k}{16}$
$512 = k$

27. $y = kwv$
$8.75 = k(50)(0.1)$
$8.75 = k(5)$
$\dfrac{8.75}{5} = k$
$1.75 = k$

29. $x = kp$
$50 = k(10)$
$\dfrac{50}{10} = k$
$5 = k$
$x = 5p = 5(14)$
$= 70$

31. $b = \dfrac{k}{c}$
$4 = \dfrac{k}{3}$
$12 = k$
$b = \dfrac{12}{c} = \dfrac{12}{2}$
$= 6$

33. $Z = kw^2$
$14 = k(4)^2$
$14 = k(16)$
$\dfrac{14}{16} = k$
$\dfrac{7}{8} = k$

The variation model is: $Z = \dfrac{7}{8}w^2$.

$Z = \dfrac{7}{8}(8)^2$
$Z = \dfrac{7}{8}(64)$
$Z = 56$

35. $Q = \dfrac{k}{p^2}$

$4 = \dfrac{k}{3^2}$

$4 = \dfrac{k}{9}$

$36 = k$

The variation model is $Q = \dfrac{36}{p^2}$.

$Q = \dfrac{36}{2^2}$

$Q = \dfrac{36}{4}$

$Q = 9$

37. $L = ka\sqrt{b}$

$72 = k(8)\sqrt{9}$

$72 = k8(3)$

$72 = k(24)$

$3 = k$

The variation model is $L = 3a\sqrt{b}$.

$L = 3\left(\dfrac{1}{2}\right)\sqrt{36}$

$L = 3\left(\dfrac{1}{2}\right)(6)$

$L = 9$

39. $B = k \cdot \dfrac{m}{n}$

$20 = k \cdot \dfrac{10}{3}$

$\dfrac{3}{10} \cdot 20 = k$

$6 = k$

The variation model is $B = (6) \cdot \dfrac{m}{n}$.

$B = (6) \cdot \dfrac{15}{12}$

$B = \dfrac{15}{2}$

41. $h = kw$

$0.75 = 150k$

$0.005 = k$

The variation model is $h = 0.005w$.

a. $h = 0.005w$

$h = 0.00h \cdot 184$

$h = 0.92$

The heart weighs 0.92 lb.

b. Answers will vary.

43. $m = k \cdot w$

$3 = k \cdot (150)$

$\dfrac{3}{150} = k$

$\dfrac{1}{50} = k$

$m = \dfrac{1}{50}w$

a. $m = \dfrac{1}{50}w = \dfrac{1}{50}(180)$

$= 3.6$ g

b. $m = \dfrac{1}{50}w = \dfrac{1}{50}(225)$

$= 4.5$ g

c. $m = \dfrac{1}{50}w = \dfrac{1}{50}(120)$

$= 2.4$ g

45. $c = \dfrac{k}{n}$

$0.48 = \dfrac{k}{5000}$

$0.48 \cdot 5000 = k$

$2400 = k$

$c = \dfrac{2400}{n}$

a. $c = \dfrac{2400}{n} = \dfrac{2400}{6000}$

$\quad = 0.4$

$\quad = \$0.40$

b. $c = \dfrac{2400}{n} = \dfrac{2400}{8000}$

$\quad = 0.3$

$\quad = \$0.30$

c. $c = \dfrac{2400}{n} = \dfrac{2400}{2400}$

$\quad = 1.00$

$\quad = \$1.00$

47. $\qquad A = k \cdot n$

$56{,}800 = k \cdot (80{,}000)$

$\dfrac{56{,}800}{80{,}000} = k$

$\dfrac{71}{100} = k$

The variation model is $A = \dfrac{71}{100} \cdot n.$

$A = \dfrac{71}{100} \cdot (500{,}000)$

$A = 355{,}000$ tons

49. $\qquad d = ks^2$

$109 = k(40)^2$

$109 = k(1600)$

$\dfrac{109}{1600} = k$

The variation model is: $d = \dfrac{109}{1600}s^2.$

$d = \dfrac{109}{1600}(25)^2$

$d = \dfrac{109}{1600}(625)$

$d = 42.6$ feet

51. $\qquad P = kcr^2$

$144 = k(4)(6)^2$

$144 = k(4)(36)$

$144 = k(144)$

$\quad 1 = k$

The variation model is: $P = (1)cr^2.$

$P = (3)(10)^2$

$P = (3)(100)$

$P = 300$ watts

53. $\qquad R = k \cdot \dfrac{l}{d^2}$

$4 = k \cdot \dfrac{40}{0.1^2}$

$4 = k \cdot \dfrac{40}{0.01}$

$4 = k \cdot (4000)$

$\dfrac{4}{4000} = k$

$\dfrac{1}{1000} = k$

The variation model is $R = \left(\dfrac{1}{1000}\right) \cdot \dfrac{l}{d^2}.$

$R = \left(\dfrac{1}{1000}\right) \cdot \dfrac{50}{0.2^2}$

$R = \left(\dfrac{1}{1000}\right) \cdot \dfrac{50}{0.04}$

$R = 1.25$ ohms

55. $\qquad W = kr^3$

$4.32 = k(3)^3$

$4.32 = k(27)$

$\dfrac{4.32}{27} = k$

$0.16 = k$

The variation model is: $W = 0.16r^3.$

$W = 0.16(5)^3$

$W = 0.16(125)$

$W = 20$ pounds

57.
$$i = kpt$$
$$500 = k(2500)(4)$$
$$500 = k10000$$
$$\frac{500}{10000} = k$$
$$0.05 = k$$
$$i = 0.05pt$$
$$i = 0.05(7000)(10)$$
$$= 3500 = \$3500$$

Group Activity

1. Amount down payment $= \$200{,}000 \times 20\%$
$= \$40{,}000$

3. $p = \dfrac{\dfrac{Ar}{12}}{1 - \dfrac{1}{\left(1 + \dfrac{r}{12}\right)^{12t}}}$

$$p = \frac{\dfrac{(160000)(0.075)}{12}}{1 - \dfrac{1}{\left(1 + \dfrac{0.075}{12}\right)^{12 \cdot 30}}}) = \frac{1000}{1 - \dfrac{1}{9.421}}$$

$$= \frac{1000}{1 - 0.10614} = \frac{1000}{0.89386}$$
$$p = 1118.74$$
$$= \$1118.74$$

5. $\$402746.40 - \$160000 = \$242746.40$

Chapter 7 Review Exercises

1. (a) $\dfrac{0-2}{0+9} = -\dfrac{2}{9};\ \dfrac{1-2}{1+9} = -\dfrac{1}{10};\ \dfrac{2-2}{2+9} = 0;$
$\dfrac{-3-2}{-3+9} = -\dfrac{5}{6};\ \dfrac{-9-2}{-9+9} = \dfrac{-11}{0}$

$-\dfrac{2}{9}, -\dfrac{1}{10}, 0, -\dfrac{5}{6}$ is undefined

(b) $t = -9,$

3. (a) $\dfrac{2-1}{1-2} = -1$

(b) $\dfrac{-1-5}{-1+5} = \dfrac{-6}{4} = -\dfrac{3}{2}$

(c) $\dfrac{-x-7}{x+7} = \dfrac{-(-1)-7}{-1+7} = -1$

(d) $\dfrac{(-1)^2 - 4}{4 - (-1)^2} = -1$

a, c, d are the expressions equal to -1.

5. $(3h+1)(h+7) = 0$
$3h+1 = 0$ or $h+7 = 0$
$3h = -1$ $\qquad h = -7\ \ h = -\dfrac{1}{3},\ h$
$h = -\dfrac{1}{3}$
$= -7;$

$$\frac{\cancel{h+7}}{(3h+1)(\cancel{h+7})} = \frac{1}{3h+1}$$

7. $w^2 - 16 = 0$
$(w-4)(w+4) = 0$
$w-4 = 0$ or $w+4 = 0$ $\qquad w = 4,\ w = -4$
$w = 4$ $\qquad w = -4$

$$\frac{2w^2 + 11w + 12}{w^2 - 16} = \frac{(2w+3)\cancel{(w+4)}}{(w-4)\cancel{(w+4)}}$$
$$= \frac{2w+3}{w-4}$$

9. $2k^2 - 10k = 0$
$2k(k-5) = 0$ $\qquad k = 0,\ k = 5$
$2k = 0$ or $k-5 = 0$
$k = 0$ $\qquad k = 5$

$$\frac{15-3k}{2k^2-10k} = \frac{3(5-k)}{2k(k-5)}$$

$$= \frac{-3\cancel{(k-5)}}{2k\cancel{(k-5)}}$$

$$= -\frac{3}{2k}$$

or

$$\frac{15-3k}{2k^2-10k} = \frac{3(5-k)}{2k(k-5)}$$

$$= \frac{3\cancel{(5-k)}}{-2k\cancel{(5-k)}}$$

$$= -\frac{3}{2k}$$

11. $9m+9=0$

$$9m = -9 \quad m = -1$$

$$m = -1$$

$$\frac{3m^2-12m-15}{9m+9} = \frac{3(m^2-4m-5)}{9(m+1)}$$

$$= \frac{\cancel{3}(m-5)\cancel{(m+1)}}{\cancel{9}3\cancel{(m+1)}}$$

$$= \frac{m-5}{3}$$

13. $p^2+14+49=0$

$$(p+7)(p+7)=0 \qquad p=-7$$

$$p+7=0$$

$$n=-7$$

$$\frac{p+7}{p^2+14p+49} = \frac{\cancel{p+7}}{\cancel{(p+7)}(p+7)}$$

$$= \frac{1}{p+7}$$

15. $\dfrac{2u+10}{u} \cdot \dfrac{u^3}{4u+20} = \dfrac{2(u+5)}{u} \cdot \dfrac{u^3}{4(u+5)} = \dfrac{u^2}{2}$

17. $\dfrac{8}{x^2-25} \cdot \dfrac{3x+15}{16} = \dfrac{8}{(x+5)(x-5)} \cdot \dfrac{3(x+5)}{16}$

$$= \frac{3}{2(x-5)}$$

19. $\dfrac{q^2-5q+6}{2q+4} \div \dfrac{2q-6}{q+2}$

$$= \frac{(q-3)(q-2)}{2(q+2)} \cdot \frac{q+2}{2(q-3)}$$

$$= \frac{q-2}{4}$$

21. $(s^2-6s+8)\left(\dfrac{4s}{s-2}\right) = \dfrac{(s-4)(s-2)}{1} \cdot \dfrac{4s}{s-2}$

$$= 4s(s-4)$$

23. $\dfrac{\frac{n^2+n+1}{n^2-4}}{\frac{n^2+n+1}{n+2}} = \dfrac{n^2+n+1}{(n+2)(n-2)} \cdot \dfrac{n+2}{n^2+n+1} = \dfrac{1}{n-2}$

25. $\dfrac{3m-3}{6m^2+18m+12} \cdot \dfrac{2m^2-8}{m^2-3m+2} \div \dfrac{m+3}{m+1}$

$$= \frac{3(m-1)}{6(m+2)(m+1)} \cdot \frac{2(m+2)(m-2)}{(m-2)(m-1)} \cdot \frac{m+1}{m+3}$$

$$= \frac{1}{m+3}$$

27. $\dfrac{4y^2-1}{1+2y} \div \dfrac{y^2-4y-5}{5-y}$

$$= \frac{(2y+1)(2y-1)}{2y+1} \cdot \frac{-1(y-5)}{(y-5)(y+1)}$$

$$= -\frac{2y-1}{y+1}$$

29. $\dfrac{8}{m^2-16} ; \ \dfrac{7}{m^2-m-12}$

$$= \frac{8}{(m+4)(m-4)} ; \ \frac{7}{(m+3)(m-4)}$$

LCD: $(m-4)(m+4)(m+3)$

31. LCD: $3-x$ or $x-3$

219

33. $\dfrac{7}{4x}; \dfrac{11}{6y}$

LCD $= 12xy$;

$$\dfrac{7}{4x} = \dfrac{7(3y)}{4x(3y)} = \dfrac{21y}{12xy}$$

$$\dfrac{11}{6y} = \dfrac{11(2x)}{6y(2x)} = \dfrac{22x}{12xy}$$

35. $\dfrac{5}{ab^3}; \dfrac{3}{ac^2}$

LCD $= ab^3c^2$;

$$\dfrac{5}{ab^3} = \dfrac{5(c^2)}{ab^3(c^2)} = \dfrac{5c^2}{ab^3c^2}$$

$$\dfrac{3}{ac^2} = \dfrac{3(b^3)}{ac^2(b^3)} = \dfrac{3b^3}{ab^3c^2}$$

37. $\dfrac{6}{q}; \dfrac{1}{q+8}$

LCD $= q(q+8)$

$$\dfrac{6}{q} = \dfrac{6(q+8)}{q(q+8)} = \dfrac{6q+48}{q(q+8)}$$

$$\dfrac{1}{q+8} = \dfrac{1(q)}{q(q+8)} = \dfrac{q}{q(q+8)}$$

39. $\dfrac{b-6}{b-2} + \dfrac{b+2}{b-2} = \dfrac{b-6+b+2}{b-2}$

$$= \dfrac{2b-4}{b-2}$$

$$= \dfrac{2(b-2)}{b-2}$$

$$= 2$$

41. $\dfrac{x^2}{x+7} - \dfrac{49}{x+7} = \dfrac{x^2-49}{x+7}$

$$= \dfrac{(x+7)(x-7)}{x+7}$$

$$= x-7$$

43. $\dfrac{3}{4-t^2} + \dfrac{t}{2-t} = \dfrac{3}{(2+t)(2-t)} + \dfrac{t(2+t)}{(2+t)(2-t)}$

$$= \dfrac{3+2t+t^2}{(2+t)(2-t)}$$

$$= \dfrac{t^2+2t+3}{(2+t)(2-t)}$$

45. $\dfrac{5}{2r+12} - \dfrac{1}{r} = \dfrac{5}{2(r+6)} - \dfrac{1}{r}$

$$= \dfrac{5r}{2r(r+6)} - \dfrac{2(r+6)}{2r(r+6)}$$

$$= \dfrac{5r-2r-12}{2r(r+6)}$$

$$= \dfrac{3r-12}{2r(r+6)}$$

$$= \dfrac{3(r-4)}{2r(r+6)}$$

47. $\dfrac{3q}{q^2+7q+10} - \dfrac{2q}{q^2+6q+8}$

$$= \dfrac{3q}{(q+5)(q+2)} - \dfrac{2q}{(q+4)(q+2)}$$

$$= \dfrac{3q(q+4)}{(q+5)(q+4)(q+2)} - \dfrac{2q(q+5)}{(q+5)(q+4)(q+2)}$$

$$= \dfrac{3q^2+12q-2q^2-10q}{(q+5)(q+4)(q+2)}$$

$$= \dfrac{q^2+2q}{(q+5)(q+4)(q+2)}$$

$$= \dfrac{q(q+2)}{(q+5)(q+4)(q+2)}$$

$$= \dfrac{q}{(q+5)(q+4)}$$

49. $\dfrac{x}{3x+9} - \dfrac{3}{x^2+3x} + \dfrac{1}{x}$

$= \dfrac{x}{3(x+3)} - \dfrac{3}{x(x+3)} + \dfrac{1}{x}$

$= \dfrac{x^2}{3x(x+3)} - \dfrac{9}{3x(x+3)} + \dfrac{3(x+3)}{3x(x+3)}$

$= \dfrac{x^2 - 9 + 3x + 9}{3x(x+3)}$

$= \dfrac{x^2 + 3x}{3x(x+3)}$

$= \dfrac{x(x+3)}{3x(x+3)}$

$= \dfrac{1}{3}$

51. $\dfrac{\frac{z+5}{z}}{\frac{z-5}{3}} = \dfrac{z+5}{z} \cdot \dfrac{3}{z-5} = \dfrac{3(z+5)}{z(z-5)}$

53. $\dfrac{\frac{2}{y}+6}{\frac{3y+1}{4}} = \dfrac{4y\left(\frac{2}{y}+6\right)}{4y\left(\frac{3y+1}{4}\right)}$

$= \dfrac{8 + 24y}{3y^2 + y}$

$= \dfrac{8(1 + 3y)}{y(3y + 1)}$

$= \dfrac{8}{y}$

55. $\dfrac{\frac{b}{a}-\frac{a}{b}}{\frac{1}{b}-\frac{1}{a}} = \dfrac{ab\left(\frac{b}{a}-\frac{a}{b}\right)}{ab\left(\frac{1}{b}-\frac{1}{a}\right)}$

$= \dfrac{b^2 - a^2}{a - b}$

$= \dfrac{-(a+b)(a-b)}{a-b}$

$= -(a+b)$

57. $\dfrac{\frac{25}{k+5}+5}{\frac{5}{k+5}-5} = \dfrac{(k+5)\left(\frac{25}{k+5}+5\right)}{(k+5)\left(\frac{5}{k+5}-5\right)}$

$= \dfrac{25 + 5(k+5)}{5 - 5(k+5)}$

$= \dfrac{5k + 50}{-5k - 20}$

$= \dfrac{5(k+10)}{-5(k+4)}$

$= -\dfrac{k+10}{k+4}$

59. $\dfrac{1}{y} + \dfrac{3}{4} = \dfrac{1}{4}$

$4\left(\dfrac{1}{y} + \dfrac{3}{4}\right) = 4 \cdot \dfrac{1}{4}$

$\dfrac{4}{y} + 3 = 1$

$\dfrac{4}{y} = -2$

$y = \dfrac{4}{-2} = -2$

61. $\dfrac{t+1}{3} - \dfrac{t-1}{6} = \dfrac{1}{6}$

$6\left(\dfrac{t+1}{3} - \dfrac{t-1}{6}\right) = 6 \cdot \dfrac{1}{6}$

$2(t+1) - (t-1) = 1$

$2t + 2 - t + 1 = 1$

$t = -2$

63. $\dfrac{w+1}{w-3} - \dfrac{3}{w} = \dfrac{12}{w^2 - 3w}$

$\dfrac{w+1}{w-3} - \dfrac{3}{w} = \dfrac{12}{w(w-3)}$

Multiply both sides by LCD: $w(w-3)$

$$w(w+1) - 3(w-3) = 12$$
$$w^2 + w - 3w + 9 = 12$$
$$w^2 - 2w + 9 - 12 = 12 - 12$$
$$w^2 - 2w + -3 = 0$$
$$(w+1)(w-3) = 0$$
$$w + 1 = 0 \quad \text{or} \quad w - 3 = 0$$
$$w = -1 \quad \text{or} \quad w = 3$$
-1; (the value 3 does not check.)

65. $\dfrac{y+1}{y+3} = \dfrac{y^2 - 11y}{y^2 + y - 6} - \dfrac{y-3}{y-2}$

$$\dfrac{y+1}{y+3} = \dfrac{y^2 - 11y}{(y+3)(y-2)} - \dfrac{y-3}{y-2}$$

Multiply both sides by LCD: $(y+3)(y-2)$

$$(y-2)(y+1) = y^2 - 11y - (y+3)(y-3)$$
$$y^2 - y - 2 = y^2 - 11y - y^2 + 9$$
$$y^2 - y - 2 = -11y + 9$$
$$y^2 + 10y - 11 = 0$$
$$(y-1)(y+11) = 0$$
$$y - 1 = 0 \quad \text{or} \quad y + 11 = 0$$
$$y = 1 \quad \text{or} \quad y = -11$$

67. $\dfrac{V}{h} = \dfrac{\pi r^2}{3}$

$$3h\left(\dfrac{V}{h}\right) = 3h\left(\dfrac{\pi r^2}{3}\right)$$
$$3V = h\pi r^2$$
$$h = \dfrac{3V}{\pi r^2}$$

69. $\dfrac{m+2}{8} = \dfrac{m}{3}$

$$3(m+2) = 8m$$
$$3m + 6 = 8m$$
$$6 = 5m$$
$$m = \dfrac{6}{5}$$

71. Let x represent the number of grams of fat in a 5-oz bag.

$$\dfrac{4 \text{ g}}{2 \text{ oz}} = \dfrac{x \text{ g}}{5 \text{ oz}}$$
$$4(5) = 2x$$
$$20 = 2x$$
$$10 = x$$

It contains 10 g of fat.

73. Let x = time to fill pool if both pumps are working together. Then in 1 minute the first pump can fill $\dfrac{1}{24}$ of the pool, the second pump can fill $\dfrac{1}{56}$ of the pool, and together they can fill $\dfrac{1}{x}$ of the pool.

$$\dfrac{1}{24} + \dfrac{1}{56} = \dfrac{1}{x}$$
$$168x\left(\dfrac{1}{24} + \dfrac{1}{56}\right) = 168x\left(\dfrac{1}{x}\right)$$
$$7x + 3x = 168$$
$$10x = 168$$
$$x = 16.8$$

Together both pumps would fill the pool in 16.8 minutes.

75. (a) $F = kd$

(b) $6 = k \cdot 2$
$$3 = k$$

(c)
$$F = 3d$$
$$F = 3(4.2)$$
$$F = 12.6$$

The force required to stretch 4.2 ft is 12.6 lb.

77. $y = kx\sqrt{z}$

$3 = k \cdot 3\sqrt{4}$

$3 = k \cdot 3(2)$

$\dfrac{1}{2} = k$

The variation model is: $y = \dfrac{1}{2}x\sqrt{z}$

$y = \dfrac{1}{2}(8)\sqrt{9} = \dfrac{1}{2}(8)(3) = 12$

Chapter 7 Test

1. (a) $30(2 - x) = 0$

$2 - x = 0 \quad x = 2$

$2 = x$

(b)

$\dfrac{5(x-2)(x+1)}{30(2-x)} = \dfrac{5(x-2)(x+1)}{-30(x-2)} = -\dfrac{x+1}{6}$

3. (a) $\dfrac{-1+4}{-1-4} = \dfrac{3}{-5} = -\dfrac{3}{5}$

(b) $\dfrac{7-2(-1)}{2(-1)-7} = \dfrac{9}{-9} = -1$

(c) $\dfrac{9(-1)^2 + 16}{-9(-1)^2 - 16} = \dfrac{25}{-25} = -1$

(d) $-\dfrac{-1+5}{-1+5} = -\dfrac{4}{4} = -1$

b, c and d are equal to -1.

5. $\dfrac{2}{y^2 + 4y + 3} + \dfrac{1}{3y + 9}$

$= \dfrac{2}{(y+3)(y+1)} + \dfrac{1}{3(y+3)}$

$= \dfrac{6}{3(y+3)(y+1)} + \dfrac{y+1}{3(y+3)(y+1)}$

$= \dfrac{6+y+1}{3(y+3)(y+1)}$

$= \dfrac{y+7}{3(y+3)(y+1)}$

9.

$\dfrac{1}{x+4} + \dfrac{2}{x^2 + 2x - 8} + \dfrac{x}{x-2}$

$= \dfrac{1(x-2)}{(x+4)(x-2)} + \dfrac{2}{(x+4)(x-2)} + \dfrac{x(x+4)}{(x+4)(x-2)}$

$= \dfrac{x-2+2+x^2+4x}{(x+4)(x-2)}$

$= \dfrac{x^2+5x}{(x+4)(x-2)}$

$= \dfrac{x(x+5)}{(x+4)(x-2)}$

11. $\dfrac{1 - \frac{4}{m}}{m - \frac{16}{m}} = \dfrac{m\left(1 - \frac{4}{m}\right)}{m\left(m - \frac{16}{m}\right)}$

$= \dfrac{m-4}{m^2 - 16}$

$= \dfrac{m-4}{(m+4)(m-4)}$

$= \dfrac{1}{m+4}$

13. $\dfrac{p}{p-1} + \dfrac{1}{p} = \dfrac{p^2+1}{p^2-p}$

$\dfrac{p}{p-1} + \dfrac{1}{p} = \dfrac{p^2+1}{p(p-1)}$

$p(p-1)\left(\dfrac{p}{p-1} + \dfrac{1}{p}\right) = p(p-1)\left(\dfrac{p^2+1}{p(p-1)}\right)$

$p^2 + p - 1 = p^2 + 1$

$p - 1 = 1$

$p = 2$

15. $\dfrac{4x}{x-4} = 3 + \dfrac{16}{x-4}$

$(x-4)\left(\dfrac{4x}{x-4}\right) = (x-4)\left(3 + \dfrac{16}{x-4}\right)$

$4x = 3(x-4) + 16$

$4x = 3x - 12 + 16$

$x = 4$

No solution ($x = 4$ does not check).

17. $\dfrac{C}{2} = \dfrac{A}{r}$

$2A = rC$

$r = \dfrac{2A}{C}$

19. Let x = cups of carrots. Then,

$$\dfrac{\frac{1}{2}\text{ cup}}{6\text{ servings}} = \dfrac{x}{15\text{ servings}}$$

$6x = \dfrac{1}{2}\cdot 15$

$6x = \dfrac{15}{2}$

$x = \dfrac{1}{6}\cdot\dfrac{15}{2} = \dfrac{5}{4} = 1\dfrac{1}{4}$ or 1.25 cups of carrots

21. Let x = time it takes second printer to complete job. Then in one hour both printers can complete $\dfrac{1}{2}$ of the job, the first printer can complete $\dfrac{1}{6}$ of the job, and the second printer can complete $\dfrac{1}{x}$ of the job. Therefore,

$$\dfrac{1}{6} + \dfrac{1}{x} = \dfrac{1}{2}$$

$12x\left(\dfrac{1}{6} + \dfrac{1}{x}\right) = 12x\cdot\dfrac{1}{2}$

$2x + 12 = 6x$

$12 = 4x$

$x = 3$

It takes the second printer 3 hr to do the job working alone.

23. $m = k\cdot w$

$6 = k\cdot(160)$

$\dfrac{6}{160} = k$

$\dfrac{3}{80} = k$

$m = \dfrac{3}{80}w$

$m = \dfrac{3}{80}(220)$

$= 8.25$ ml

Cumulative Review Exercises
Chapters 1–7

1. $\left(\dfrac{1}{2}\right)^{-4} + 2^4 = 2^4 + 2^4 = 16 + 16 = 32$

3. $\dfrac{1}{2} - \dfrac{3}{4}(y-1) = \dfrac{5}{12}$

$12\left(\dfrac{1}{2} - \dfrac{3}{4}(y-1)\right) = 12\cdot\dfrac{5}{12}$

$6 - 9(y-1) = 5$

$6 - 9y + 9 = 5$

$-9y = -10$

$y = \dfrac{10}{9}$

5. Let x = width of the pool. Then, $(2x + 1)$ is the length.

$P = 2w + 2l$

$104 = 2x + 2(2x+1)$

$104 = 2x + 4x + 2$

$102 = 6x$

$x = 17$

The width is 17 m and the length is 35 m.

7. $\left(\dfrac{4x^{-1}y^{-2}}{z^4}\right)^{-2}(2y^{-1}z^3)^3$

$=\left(\dfrac{z^4}{4x^{-1}y^{-2}}\right)^2\left(\dfrac{2z^3}{y}\right)^3$

$=\left(\dfrac{2^8}{16x^{-2}y^{-4}}\right)\left(\dfrac{8z^9}{y^3}\right)$

$=\left(\dfrac{2^8 x^2 y^4}{16}\right)\left(\dfrac{8z^9}{y^3}\right)$

$=\dfrac{x^2 y z^{17}}{2}$

9. $25x^2-30x+9=(5x-3)(5x-3)=(5x-3)^2$

11. $(x-5)(2x+1)=0$

$x-5=0 \quad$ or $\quad 2x-1=0$

$x-5=0$

$x=5 \quad$ or $\quad 2x=1 \qquad x=5,$

$\qquad\qquad\qquad x=\dfrac{1}{2}$

$x=-\dfrac{1}{2}$

13. $\dfrac{2x-6}{x^2-16}\div\dfrac{10x^2-90}{x^2-x-12}$

$=\dfrac{2(x-3)}{(x+4)(x-4)}\cdot\dfrac{(x+3)(x-4)}{10(x+3)(x-3)}$

$=\dfrac{1}{5(x+4)}$

15. $\dfrac{7}{y^2-4}=\dfrac{3}{y-2}+\dfrac{2}{y+2}$

$\dfrac{7}{(y-2)(y+2)}=\dfrac{3(y+2)}{(y-2)(y+2)}+\dfrac{2(y-2)}{(y-2)(y+2)}$

$\dfrac{7}{(y-2)(y+2)}=\dfrac{3y+6}{(y-2)(y+2)}+\dfrac{2y-4}{(y-2)(y+2)}$

$7=3y+6+2y-4$

$5=5y$

$1=y$

17. (a) $-2x+4y=8$

x-intercept: $-2x+4(0)=8$

$-2x=8$

$x=-4$

$(-4,0)$

y-intercept: $-2(0)+4y=8$

$4y=8$

$y=2$

$(0,2)$

(b) $y=5x$

x-intercept: $0=5x$

$0=x$

$(0,0)$

y-intercept: $y=5(0)$

$y=0$

$(0,0)$

19. $y-y_1=m(x-x_1)$

$y-2=5(x-1)$

$y-2=5x-5$

$y=5x-3$

Chapter 8

Chapter 8 opener

1. ANT

3. REM

5. PYT

One application in which square roots are used is with the **PYTHAGOREAN THEOREM.**

Calculator Exercises

1. $\sqrt{5} \approx 2.236$

√(5)
2.236067977

3. $\sqrt{50} \approx 7.071$

√(50)
7.071067812

5. $\sqrt{33} \approx 5.745$

√(33)
5.744562647

7. $\sqrt{80} \approx 8.944$

√(80)
8.94427191

9. $\sqrt[3]{7} \approx 1.913$

³√(7)
1.912931183

11. $\sqrt[3]{65} \approx 4.021$

³√(65)
4.020725759

Section 8.1 Practice Exercises

1. (a) b; a.

(b) principal

(c) rational

(d) $b^n = a$.

(e) index;radicand

(f) cube

(g) is not; is

(h) even;odd

(i) $a^2 + b^2 = c^2$

3. 12 is a square root of 144 because $(12)^2 = 144$. -12 is a square root of 144 because $(-12)^2 = 144$.

5. There are no real-valued square roots of -49.

7. 0 is a square root of 0 because $(0)^2 = 0$.

9. $\frac{1}{5}$ is a square root of $\frac{1}{25}$ because $\left(\frac{1}{5}\right)^2 = \frac{1}{25}$. $-\frac{1}{5}$ is a square root of $\frac{1}{25}$ because $\left(-\frac{1}{5}\right)^2 = \frac{1}{25}$.

11. (a) 13

(b) -13

13. 0

15. 9, 16, 25, 36, 64, 121, and 169

17. $\sqrt{4} = 2$

19. $\sqrt{49} = 7$

21. $\sqrt{0.16} = 0.4$

23. $\sqrt{0.09} = 0.3$

25. $\sqrt{\dfrac{25}{16}} = \dfrac{5}{4}$

27. $\sqrt{\dfrac{1}{144}} = \dfrac{1}{12}$

29. $\sqrt{16+9} = \sqrt{25} = 5$

31. $\sqrt{225-144} = \sqrt{81} = 9$

33. There is no real value of b for which $b^2 = -16$.

35. $-\sqrt{4} = -1\cdot\sqrt{4} = -1\cdot 2 = -2$

37. $\sqrt{-4}$ is not a real number.

39. $\sqrt{-\dfrac{4}{49}}$ is not a real number.

41. $-\sqrt{-\dfrac{1}{36}}$ is not a real number.

43. $-\sqrt{400} = -1\cdot\sqrt{400} = -1\cdot 20 = -20$

45. $\sqrt{-900}$ is not a real number.

47. 0, 1, 27, 125

49. Yes, -3

51. $\sqrt[3]{27} = 3$

53. $\sqrt[3]{64} = 4$

55. $-\sqrt[4]{16} = -2$

57. $\sqrt[4]{-1}$ is not a real number.

59. $\sqrt[4]{-256}$ is not a real number.

61. $\sqrt[5]{-\dfrac{1}{32}} = -\dfrac{1}{2}$

63. $-\sqrt[6]{1} = -1$

65. $\sqrt[6]{0} = 0$

67. $x^2,\ y^4,\ (ab)^6,\ w^8 x^8,\ m^{10}$
The expression is a perfect square if the exponent is even.

69. $\sqrt{(4)^2} = |4| = 4$ $\sqrt{(-4)^2} = |-4| = 4$

71. $\sqrt[3]{(5)^3} = 5$

73. $\sqrt{y^{12}} = y^6$

75. $\sqrt{a^8 b^{30}} = a^4 b^{15}$

77. $\sqrt[3]{q^{24}} = q^8$

79. $\sqrt[3]{8w^6} = 2w^2$

81. $\sqrt{(5x)^2} = 5x$

83. $-\sqrt{25x^2} = -5x$

85. $\sqrt[3]{(5p^2)^3} = 5p^2$

87. $\sqrt[3]{125p^6} = 5p^2$

89. $\sqrt{q} + p^2$

91. $\dfrac{6}{\sqrt[3]{x}}$

93. Let x represent the length of the missing leg.
$$x^2 + 12^2 = 15^2$$
$$x^2 + 144 = 225$$
$$x^2 = 81$$
$$x = \sqrt{81}$$
$$x = 9 \qquad 9 \text{ cm}$$

95. Let x represent the length of the missing leg.
$$x^2 + 12^2 = 13^2$$
$$x^2 + 144 = 169$$
$$x^2 = 25$$
$$x = \sqrt{25}$$
$$x = 5 \qquad 5 \text{ ft}$$

97. Let x represent the length of the hypotenuse.

$$x^2 = (2.4)^2 + (6.5)^2$$

$$x^2 = 5.76 + 42.25$$

$$x^2 = 48.01$$

$$x = \sqrt{48.01}$$

$$x \approx 6.9 \qquad \qquad 6.9 \text{ cm}$$

99. Let x represent the length of the diagonal.

$$x^2 = 12^2 + 12^2$$

$$x^2 = 144 + 144$$

$$x^2 = 288$$

$$x = \sqrt{288}$$

$$x \approx 17.0 \qquad \qquad 17.0 \text{ in.}$$

101. Let x represent the width of the screen.

$$x^2 + 28^2 = 42^2$$

$$x^2 + 784 = 1764$$

$$x^2 = 980$$

$$x = \sqrt{980}$$

$$x \approx 31.3 \qquad \qquad 31.3 \text{ in.}$$

103. Let x represent the distance between Greensboro to Asheville.

$$x^2 + 134^2 = 300^2$$

$$x^2 + 17,956 = 90,000$$

$$x^2 = 72,044$$

$$x = \sqrt{72,044}$$

$$x \approx 268 \qquad \qquad 268 \text{ km}$$

105. For all $x \geq 0$

107. For $\sqrt{a-b}$ to be a real number, the radicand or $(a-b)$ must be greater than or equal to zero. The quantity $(a-b)$ will be greater than or equal to zero if $a \geq b$.

Calculator Exercises

1.

```
√(125)
          11.18033989
5*√(5)
          11.18033989
```

3.

```
³√(54)
          3.77976315
3*³√(2)
          3.77976315
```

Section 8.2 Practice Exercises

1. (a) $\sqrt[n]{a}$

(b) The radical $\sqrt{x^3}$ is not in simplified form because the exponent within the radicand is **not** less than the index.

(c) No. $\sqrt{2}$ is an irrational number, therefore its decimal form is a nonterminating, nonrepeating decimal.

3. $8, 27, y^3, y^9, y^{12},$ and y^{27}

5. $-\sqrt{25} = -5$

7. $-\sqrt[3]{27} = -3$

9. $\sqrt{a^8} = a^4$

11. $\sqrt{4x^2 y^4} = 2xy^2$

13. Let x represent the distance between Portland and Spokane.

$$x^2 = 236^2 + 378^2$$

$$x^2 = 55,696 + 142,884$$

$$x^2 = 198,580$$

$$x = \sqrt{198,580}$$

$$x \approx 446 \qquad \qquad 446 \text{ km}$$

15. $\sqrt{18} = \sqrt{9 \cdot 2} = \sqrt{3^2} \cdot \sqrt{2} = 3\sqrt{2}$

17. $\sqrt{28} = \sqrt{4 \cdot 7} = \sqrt{2^2} \cdot \sqrt{7} = 2\sqrt{7}$

19. $6\sqrt{20} = 6\sqrt{2^2 \cdot 5}$

$$= 6\sqrt{2^2} \cdot \sqrt{5}$$

$$= 6 \cdot 2\sqrt{5}$$

$$= 12\sqrt{5}$$

21. $-2\sqrt{50} = -2\sqrt{5^2 \cdot 2}$
$= -2\sqrt{5^2} \cdot \sqrt{2}$
$= -2 \cdot 5\sqrt{2}$
$= -10\sqrt{2}$

23. $\sqrt{a^5} = \sqrt{a^4 \cdot a} = \sqrt{a^4} \cdot \sqrt{a} = a^2\sqrt{a}$

25. $\sqrt{w^{22}} = w^{11}$

27. $\sqrt{m^4 n^5} = \sqrt{m^4 n^4 \cdot n}$
$= \sqrt{m^4 n^4} \cdot \sqrt{n}$
$= m^2 n^2 \sqrt{n}$

29. $x\sqrt{x^{13} y^{10}} = x\sqrt{x^{12} y^{10} \cdot x}$
$= x\sqrt{x^{12} y^{10}} \cdot \sqrt{x}$
$= x^7 y^5 \sqrt{x}$

31. $3\sqrt{t^{10}} = 3t^5$

33. $\sqrt{8x^3} = \sqrt{2^3 x^3}$
$= \sqrt{2^2 x^2 \cdot 2x}$
$= \sqrt{2^2 x^2} \cdot \sqrt{2x}$
$= 2x\sqrt{2x}$

35. $\sqrt{16z^3} = \sqrt{4^2 z^3}$
$= \sqrt{4^2 z^2 \cdot z}$
$= \sqrt{4^2 z^2} \cdot \sqrt{z}$
$= 4z\sqrt{z}$

37. $-\sqrt{45w^6} = -\sqrt{3^2 \cdot 5w^6}$
$= -\sqrt{3^2 w^6} \cdot \sqrt{5}$
$= -3w^3\sqrt{5}$

39. $\sqrt{z^{25}} = \sqrt{z^{24} \cdot z} = \sqrt{z^{24}} \cdot \sqrt{z} = z^{12}\sqrt{z}$

41. $-\sqrt{15z^{11}} = -\sqrt{z^{10} \cdot 15z}$
$= -\sqrt{z^{10}} \cdot \sqrt{15z}$
$= -z^5\sqrt{15z}$

43. $5\sqrt{104a^2 b^7} = 5\sqrt{2^3 \cdot 13a^2 b^7}$
$= 5\sqrt{2^2 a^2 b^6 \cdot 2 \cdot 13b}$
$= 5\sqrt{2^2 a^2 b^6} \cdot \sqrt{26b}$
$= 10ab^3\sqrt{26b}$

45. $\sqrt{26pq} = \sqrt{26pq}$

This radical is simplified.

47. $m\sqrt{m^{10} n^{16}} = m\sqrt{m^{10} \cdot n^{16}}$
$= m\sqrt{m^{10}} \cdot \sqrt{n^{16}}$
$= m \cdot m^5 \cdot n^8$
$= m^6 n^8$

49. $-\sqrt{48a^3 b^5 c^4} = -\sqrt{4^2 \cdot 3 \cdot a^2 \cdot a \cdot b^4 \cdot b \cdot c^4}$
$= -\sqrt{4^2} \cdot \sqrt{a^2} \cdot \sqrt{b^4} \cdot \sqrt{c^4} \cdot \sqrt{3ab}$
$= -4 \cdot a \cdot b^2 \cdot c^2 \sqrt{3ab}$
$= -4ab^2 c^2 \sqrt{3ab}$

51. $\sqrt{\dfrac{a^9}{a}} = \sqrt{a^8} = a^4$

53. $\sqrt{\dfrac{y^{15}}{y^5}} = \sqrt{y^{10}} = y^5$

55. $\sqrt{\dfrac{5}{20}} = \sqrt{\dfrac{5}{4 \cdot 5}} = \sqrt{\dfrac{1}{4}}$
$= \sqrt{\dfrac{1}{2^2}} = \dfrac{1}{2}$

57. $\sqrt{\dfrac{40}{10}} = \sqrt{4} = 2$

59. $\sqrt{\dfrac{32x^3}{8x}} = \sqrt{\dfrac{4x^3}{x}} = \sqrt{4x^2} = 2x$

61. $\sqrt{\dfrac{50p^7}{2p}} = \sqrt{\dfrac{25p^7}{p}} = \sqrt{25p^6} = 5p^3$

63. $\dfrac{3\sqrt{20}}{2} = \dfrac{3\sqrt{2^2 \cdot 5}}{2}$

$= \dfrac{3\sqrt{2^2} \cdot \sqrt{5}}{2}$

$= \dfrac{3 \cdot 2\sqrt{5}}{2}$

$= 3\sqrt{5}$

65. $\dfrac{5\sqrt{24}}{10} = \dfrac{\sqrt{2^3 \cdot 3}}{2}$

$= \dfrac{\sqrt{2^2} \cdot \sqrt{2} \cdot \sqrt{3}}{2}$

$= \dfrac{2\sqrt{2} \cdot \sqrt{3}}{2}$

$= \sqrt{2} \cdot \sqrt{3} = \sqrt{6}$

67. $\dfrac{10 + \sqrt{4}}{3} = \dfrac{10 + 2}{3}$

$= \dfrac{12}{3}$

$= 4$

69. $\dfrac{20 - \sqrt{36}}{2} = \dfrac{20 - 6}{2}$

$= \dfrac{14}{2} = 7$

71. Let x represent the length of the missing side.

$x^2 = 11^2 + 11^2$

$x^2 = 121 + 121$

$x^2 = 242$

$x = \sqrt{242} = \sqrt{2 \cdot 11^2}$

$x = \sqrt{11^2} \cdot \sqrt{2} = 11\sqrt{2} \qquad 11\sqrt{2}$ ft

73. Let x represent the length of the missing side.

$x^2 + 5^2 = 17^2$

$x^2 + 25 = 289$

$x^2 = 264$

$x = \sqrt{264}$

$x = \sqrt{2^3 \cdot 3 \cdot 11}$

$x = \sqrt{2^2 \cdot 2 \cdot 3 \cdot 11}$

$x = \sqrt{2^2} \cdot \sqrt{66}$

$x = 2\sqrt{66}$

$2\sqrt{66}$ cm

75. $\sqrt[3]{a^8} = \sqrt[3]{a^6 \cdot a^2} = \sqrt[3]{a^6} \cdot \sqrt[3]{a^2} = a^2 \sqrt[3]{a^2}$

77. $7\sqrt[3]{16z^3} = 7\sqrt[3]{2^4 z^3}$

$= 7\sqrt[3]{2^3 z^3 \cdot 2}$

$= 7\sqrt[3]{2^3 z^3} \cdot \sqrt[3]{2}$

$= 14z\sqrt[3]{2}$

79. $\sqrt[3]{16a^5 b^6} = \sqrt[3]{2^3 \cdot 2 \cdot a^3 \cdot a^2 \cdot b^6}$

$= \sqrt[3]{2^3} \cdot \sqrt[3]{a^3} \cdot \sqrt[3]{b^6} \cdot \sqrt[3]{2a^2}$

$= 2ab^2 \sqrt[3]{2a^2}$

81. $\dfrac{\sqrt[3]{z^4}}{\sqrt[3]{z}} = \sqrt[3]{\dfrac{z^4}{z}} = \sqrt[3]{z^3} = z$

83. $\sqrt[3]{-\dfrac{32}{4}} = \sqrt[3]{-8}$

$= \sqrt[3]{-1 \cdot 2^3}$

$= \sqrt[3]{-1} \cdot \sqrt[3]{2^3}$

$= -1 \cdot 2$

$= -2$

85. $-\sqrt[3]{40} = -\sqrt[3]{8 \cdot 5}$

$= -\sqrt[3]{2^3} \cdot \sqrt[3]{5}$

$= -2\sqrt[3]{5}$

87. $\dfrac{\sqrt{3}}{\sqrt{27}} = \sqrt{\dfrac{3}{27}} = \sqrt{\dfrac{1}{9}} = \dfrac{\sqrt{1}}{\sqrt{9}} = \dfrac{1}{3}$

89. $\sqrt{16a^3} = \sqrt{2^4 a^3}$
$= \sqrt{2^4 a^2 \cdot a}$
$= \sqrt{2^4 a^2} \cdot \sqrt{a}$
$= 2^2 a\sqrt{a}$
$= 4a\sqrt{a}$

91. $\sqrt{\dfrac{4x^3}{x}} = \sqrt{4x^2}$
$= \sqrt{2^2} \cdot \sqrt{x^2}$
$= 2x$

93. $\sqrt{8p^2 q} = \sqrt{2^2 \cdot 2 \cdot p^2 \cdot q}$
$= \sqrt{2^2} \cdot \sqrt{p^2} \cdot \sqrt{2q}$
$= 2p\sqrt{2q}$

95. $-\sqrt{32} = -\sqrt{4^2 \cdot 2} = -\sqrt{4^2} \cdot \sqrt{2} = -4\sqrt{2}$

97. $\sqrt{52u^4 v^7} = \sqrt{2^2 \cdot 13u^4 v^7}$
$= \sqrt{2^2 u^4 v^6 \cdot 13v}$
$= \sqrt{2^2 u^4 v^6} \cdot \sqrt{13v}$
$= 2u^2 v^3 \sqrt{13v}$

99. $\sqrt{216} = \sqrt{6^3} = \sqrt{6^2 \cdot 6} = \sqrt{6^2} \cdot \sqrt{6} = 6\sqrt{6}$

101. $\sqrt[3]{216} = \sqrt[3]{2^3 \cdot 3^3} = 2 \cdot 3 = 6$

103. $\sqrt[3]{16a^3} = \sqrt[3]{2^3 \cdot 2 \cdot a^3}$
$= \sqrt[3]{2^3} \cdot \sqrt[3]{2} \cdot \sqrt[3]{a^3}$
$= 2a\sqrt[3]{2}$

105. $\sqrt[3]{\dfrac{x^5}{x^2}} = \sqrt[3]{x^3} = x$

107. $\dfrac{-6\sqrt{20}}{12} = \dfrac{-\sqrt{20}}{2} = \dfrac{-\sqrt{2^2 \cdot 5}}{2}$
$= \dfrac{-\sqrt{2^2}\sqrt{5}}{2} = \dfrac{-2\sqrt{5}}{2} = -\sqrt{5}$

109. $\dfrac{-4 - \sqrt{25}}{18} = \dfrac{-4 - 5}{18} = \dfrac{-9}{18} = -\dfrac{1}{2}$

111. $\sqrt{(-2-5)^2 + (-4+3)^2} = \sqrt{(-7)^2 + (-1)^2}$
$= \sqrt{49+1}$
$= \sqrt{50}$
$= \sqrt{5^2 \cdot 2}$
$= 5\sqrt{2}$

113. $\sqrt{x^2 + 10x + 25} = \sqrt{(x+5)^2} = x + 5$

Section 8.3 Practice Exercises

1. index; radicand

3. $\sqrt[3]{8y^3} = 2y$

5. $\sqrt{36x^3} = \sqrt{36x^2 \cdot x} = \sqrt{36x^2} \cdot \sqrt{x} = 6x\sqrt{x}$

7. $\dfrac{\sqrt{2x^3}}{\sqrt{x}} = \sqrt{\dfrac{12x^3}{3x}} = \sqrt{4x^2} = \sqrt{2^2 \cdot x^2} = 2x$

9. $\sqrt{-25}$ Not a real number.

11. For example: $2\sqrt{3},\ 6\sqrt[3]{3}$

13. c

15. $3\sqrt{2} + 5\sqrt{2} = (3+5)\sqrt{2} = 8\sqrt{2}$

17. $5\sqrt{7} - 3\sqrt{7} + 2\sqrt{7} = (5-3+2)\sqrt{7} = 4\sqrt{7}$

19. $\sqrt[3]{10} + \sqrt[3]{10} = 1\sqrt[3]{10} + 1\sqrt[3]{10}$
$= (1+1)\sqrt[3]{10}$
$= 2\sqrt[3]{10}$

21. $15\sqrt{y} - 4\sqrt{y} = (15-4)\sqrt{y} = 11\sqrt{y}$

23. $5\sqrt{c} - 6\sqrt{c} + \sqrt{c} = 5\sqrt{c} - 6\sqrt{c} + 1\sqrt{c}$
$= (5-6+1)\sqrt{c}$
$= 0$

25. $8y\sqrt{15} - 3y\sqrt{15} = (8y - 3y)\sqrt{15} = 5y\sqrt{15}$

27. $x\sqrt{y} - y\sqrt{x} = x\sqrt{y} - y\sqrt{x}$
Cannot be simplified—the radicals are not *like* radicals.

29. $2\sqrt{12} + \sqrt{48} = 2\sqrt{2^2 \cdot 3} + \sqrt{2^4 \cdot 3}$
$\qquad = 2 \cdot 2\sqrt{3} + 2^2 \cdot \sqrt{3}$
$\qquad = 4\sqrt{3} + 4\sqrt{3}$
$\qquad = (4+4)\sqrt{3}$
$\qquad = 8\sqrt{3}$

31. $4\sqrt{45} - 6\sqrt{20} = 4\sqrt{3^2 \cdot 5} - 6\sqrt{2^2 \cdot 5}$
$\qquad = 4 \cdot 3\sqrt{5} - 6 \cdot 2\sqrt{5}$
$\qquad = 12\sqrt{5} - 12\sqrt{5}$
$\qquad = 0$

33. $\dfrac{1}{2}\sqrt{8} + \dfrac{1}{3}\sqrt{18} = \dfrac{1}{2}\sqrt{2^3} + \dfrac{1}{3}\sqrt{2 \cdot 3^2}$
$\qquad = \dfrac{1}{2}\sqrt{2^2 \cdot 2} + \dfrac{1}{3}\sqrt{2 \cdot 3^2}$
$\qquad = \dfrac{1}{2} \cdot 2\sqrt{2} + \dfrac{1}{3} \cdot 3\sqrt{2}$
$\qquad = 1\sqrt{2} + 1\sqrt{2}$
$\qquad = (1+1)\sqrt{2}$
$\qquad = 2\sqrt{2}$

35. $6p\sqrt{20p^2} + p^2\sqrt{80}$
$\qquad = 6p\sqrt{2^2 p^2 \cdot 5} + p^2\sqrt{2^4 \cdot 5}$
$\qquad = 6p \cdot 2p\sqrt{5} + p^2 \cdot 2^2\sqrt{5}$
$\qquad = 12p^2\sqrt{5} + 4p^2\sqrt{5}$
$\qquad = (12p^2 + 4p^2)\sqrt{5}$
$\qquad = 16p^2\sqrt{5}$

37. $-2\sqrt{2k} + 6\sqrt{8k} = -2\sqrt{2k} + 6\sqrt{2^2 \cdot 2k}$
$\qquad = -2\sqrt{2k} + 6 \cdot 2\sqrt{2k}$
$\qquad = -2\sqrt{2k} + 12\sqrt{2k}$
$\qquad = (-2+12)\sqrt{2k}$
$\qquad = 10\sqrt{2k}$

39. $11\sqrt{a^4 b} - a^2\sqrt{b} - 9a\sqrt{a^2 b}$
$\qquad = 11 \cdot a^2\sqrt{b} - a^2\sqrt{b} - 9a \cdot a\sqrt{b}$
$\qquad = 11a^2\sqrt{b} - 1a^2\sqrt{b} - 9a^2\sqrt{b}$
$\qquad = (11a^2 - 1a^2 - 9a^2)\sqrt{b}$
$\qquad = a^2\sqrt{b}$

41. $4\sqrt{5} - \sqrt{5} = 4\sqrt{5} - 1\sqrt{5} = (4-1)\sqrt{5} = 3\sqrt{5}$

43. $\dfrac{5}{6}z\sqrt{6} + \dfrac{7}{9}z\sqrt{6} = \dfrac{15}{18}z\sqrt{6} + \dfrac{14}{18}z\sqrt{6}$
$\qquad = \left(\dfrac{15}{18}z + \dfrac{14}{18}z\right)\sqrt{6}$
$\qquad = \dfrac{29}{18}z\sqrt{6}$

45. $1.1\sqrt{10} - 5.6\sqrt{10} + 2.8\sqrt{10}$
$\qquad = (1.1 - 5.6 + 2.8)\sqrt{10}$
$\qquad = -1.7\sqrt{10}$

47. $4\sqrt{x^3} - 2x\sqrt{x} = 4\sqrt{x^2 \cdot x} - 2x\sqrt{x}$
$\qquad = 4 \cdot x\sqrt{x} - 2x\sqrt{x}$
$\qquad = 4x\sqrt{x} - 2x\sqrt{x}$
$\qquad = (4x - 2x)\sqrt{x}$
$\qquad = 2x\sqrt{x}$

49. $4\sqrt{7} + \sqrt{63} - 2\sqrt{28}$
$\qquad = 4\sqrt{7} + \sqrt{3^2 \cdot 7} - 2\sqrt{2^2 \cdot 7}$
$\qquad = 4\sqrt{7} + 3\sqrt{7} - 2 \cdot 2\sqrt{7}$
$\qquad = 4\sqrt{7} + 3\sqrt{7} - 4\sqrt{7}$
$\qquad = (4 + 3 - 4)\sqrt{7}$
$\qquad = 3\sqrt{7}$

51. $\sqrt{16w} + \sqrt{24w} + \sqrt{40w}$
$\qquad = \sqrt{4^2 \cdot w} + \sqrt{2^2 \cdot 6w} + \sqrt{2^2 \cdot 10w}$
$\qquad = 4\sqrt{w} + 2\sqrt{6w} + 2\sqrt{10w}$

53. $\sqrt{x^6 y} + 5x^2 \sqrt{x^2 y} = \sqrt{x^6 \cdot y} + 5x^2 \sqrt{x^2 \cdot y}$
$$= x^3 \sqrt{y} + 5x^2 \cdot x\sqrt{y}$$
$$= x^3 \sqrt{y} + 5x^3 \sqrt{y}$$
$$= (x^3 + 5x^3)\sqrt{y}$$
$$= 6x^3 \sqrt{y}$$

55. $4\sqrt{6} + 2\sqrt{3} - 8\sqrt{6} = (4-8)\sqrt{6} + 2\sqrt{3}$
$$= 2\sqrt{3} - 4\sqrt{6}$$

57. $x\sqrt{8} - 2\sqrt{18x^2} + \sqrt{2x}$
$$= x\sqrt{2^2 \cdot 2} - 2\sqrt{3^2 \cdot 2 \cdot x^2} + \sqrt{2x}$$
$$= 2x\sqrt{2} - 6x\sqrt{2} + \sqrt{2x}$$
$$= (2x - 6x)\sqrt{2} + \sqrt{2x}$$
$$= -4x\sqrt{2} + \sqrt{2x}$$

59. The perimeter is the sum of the lengths of the sides.
$$\sqrt{18} + \sqrt{8} + \sqrt{32} = \sqrt{3^2 \cdot 2} + \sqrt{2^2 \cdot 2} + \sqrt{2^4 \cdot 2}$$
$$= 3\sqrt{2} + 2\sqrt{2} + 2^2 \sqrt{2}$$
$$= (3 + 2 + 4)\sqrt{2}$$
$$= 9\sqrt{2} \text{ m}$$

61. The perimeter is the sum of the lengths of the four sides of the rectangle. Remember that opposite sides are the same.
$$2\sqrt{3} + 2\sqrt{3} + 3\sqrt{12} + 3\sqrt{12}$$
$$= 2\sqrt{3} + 2\sqrt{3} + 3\sqrt{2^2 \cdot 3} + 3\sqrt{2^2 \cdot 3}$$
$$= 2\sqrt{3} + 2\sqrt{3} + 3 \cdot 2\sqrt{3} + 3 \cdot 2\sqrt{3}$$
$$= 2\sqrt{3} + 2\sqrt{3} + 6\sqrt{3} + 6\sqrt{3}$$
$$= (2 + 2 + 6 + 6)\sqrt{3}$$
$$= 16\sqrt{3} \text{ in.}$$

63. Radicands are not the same.

65. One term has a radical. One does not.

67. The indices are different.

69. $m = \dfrac{2\sqrt{3} - \sqrt{3}}{4 - 1} = \dfrac{\sqrt{3}}{3}$

71. $x = 23t\sqrt{3}$

   (a) $x = 23(2)\sqrt{3}$
      $x \approx 80$ m

   (b) $x = 23(4)\sqrt{3}$
      $x \approx 159$ m

Section 8.4 Practice Exercises

1. (a) $\sqrt[n]{ab}$

   (b) a

   (c) conjugates

3. $\sqrt{100} - \sqrt{4} + \sqrt{9} = 10 - 2 + 3 = 11$

5. $10\sqrt{zw^4} - w^2\sqrt{49z} = 10w^2\sqrt{z} - w^2 \cdot 7\sqrt{z}$
$$= 10w^2\sqrt{z} - 7w^2\sqrt{z}$$
$$= 3w^2\sqrt{z}$$

1. The radicand has no factor raised to a power greater than or equal to the index.

3. The radicand does not contain a fraction

7. $\sqrt{5} \cdot \sqrt{3} = \sqrt{5 \cdot 3} = \sqrt{15}$

9. $\sqrt{47} \cdot \sqrt{47} = \sqrt{47 \cdot 47} = 47$

11. $\sqrt{b} \cdot \sqrt{b} = \sqrt{b \cdot b} = \sqrt{b^2} = b$

13. $2\sqrt{15} \cdot 3\sqrt{p} = 6\sqrt{15p}$

15. $\sqrt{10} \cdot \sqrt{5} = \sqrt{10 \cdot 5} = \sqrt{50} = \sqrt{25 \cdot 2} = 5\sqrt{2}$

17. $(-\sqrt{7})(-2\sqrt{14}) = 2\sqrt{7 \cdot 14} = 2\sqrt{98}$
$$= 2\sqrt{49 \cdot 2} = 2 \cdot 7\sqrt{2}$$
$$= 14\sqrt{2}$$

19. $3x\sqrt{2} \cdot \sqrt{14} = 3x\sqrt{2} \cdot \sqrt{2 \cdot 7} = 3x\sqrt{2} \cdot \sqrt{2} \cdot \sqrt{7}$
$$= 3x \cdot 2\sqrt{7} = 6x\sqrt{7}$$

21. $\left(\dfrac{1}{6}x\sqrt{xy}\right)\left(24x\sqrt{x}\right) = 4x^2\sqrt{xy}\cdot\sqrt{x}$

$\qquad\qquad\qquad = 4x^2\sqrt{x}\cdot\sqrt{y}\cdot\sqrt{x}$

$\qquad\qquad\qquad = 4x^2x\cdot\sqrt{y}$

$\qquad\qquad\qquad = 4x^3\sqrt{y}$

23. $6w\sqrt{5}\cdot w\sqrt{8} = 6w^2\sqrt{5}\cdot\sqrt{8}$

$\qquad\qquad\quad = 6w^2\sqrt{5}\cdot\sqrt{2^2\cdot 2}$

$\qquad\qquad\quad = 6w^2\sqrt{5}\cdot\sqrt{2^2}\cdot\sqrt{2}$

$\qquad\qquad\quad = 6w^2\sqrt{5}\cdot 2\cdot\sqrt{2}$

$\qquad\qquad\quad = 12w^2\sqrt{5}\cdot\sqrt{2} = 12w^2\sqrt{10}$

25. $-2\sqrt{3}\cdot 4\sqrt{5} = -8\sqrt{15}$

27. Perimeter: $P = 2l + 2w$

$\qquad\qquad P = 2\left(\sqrt{20}\right) + 2\left(\sqrt{5}\right)$

$\qquad\qquad P = 2\sqrt{4\cdot 5} + 2\sqrt{5}$

$\qquad\qquad P = 2\cdot 2\sqrt{5} + 2\sqrt{5}$

$\qquad\qquad P = 4\sqrt{5} + 2\sqrt{5}$

$\qquad\qquad P = 6\sqrt{5}\ \text{ft}$

Area: $A = lw$

$\qquad\quad A = \sqrt{20}\cdot\sqrt{5}$

$\qquad\quad A = \sqrt{20\cdot 5}$

$\qquad\quad A = \sqrt{100}$

$\qquad\quad A = 10\ \text{ft}^2$

29. Area: $A = \dfrac{1}{2}bh$

$\qquad\qquad A = \dfrac{1}{2}\cdot\sqrt{3}\cdot\sqrt{12}$

$\qquad\qquad A = \dfrac{1}{2}\cdot\sqrt{3\cdot 12}$

$\qquad\qquad A = \dfrac{1}{2}\cdot\sqrt{36}$

$\qquad\qquad A = \dfrac{1}{2}\cdot 6$

$\qquad\qquad A = 3\ \text{cm}^2$

31. $\sqrt{3w}\cdot\sqrt{3w} = \sqrt{3w\cdot 3w} = \sqrt{9w^2} = 3w$

35. $\sqrt{2}\left(\sqrt{6}-\sqrt{3}\right) = \sqrt{2}\cdot\sqrt{6}-\sqrt{2}\cdot\sqrt{3}$

$\qquad\qquad\qquad = \sqrt{12}-\sqrt{6}$

$\qquad\qquad\qquad = \sqrt{4\cdot 3}-\sqrt{6}$

$\qquad\qquad\qquad = 2\sqrt{3}-\sqrt{6}$

37. $4\sqrt{x}\left(\sqrt{x}+5\right) = 4\sqrt{x}\cdot\sqrt{x}+4\sqrt{x}\cdot 5$

$\qquad\qquad\qquad = 4x+20\sqrt{x}$

39. $\left(\sqrt{3}+2\sqrt{10}\right)\left(4\sqrt{3}-\sqrt{10}\right) = \sqrt{3}\cdot 4\sqrt{3}-\sqrt{3}\cdot\sqrt{10}+2\sqrt{10}\cdot 4\sqrt{3}-2\sqrt{10}\cdot\sqrt{10}$

$\qquad\qquad\qquad\qquad\qquad = 4\sqrt{9}-\sqrt{30}+8\sqrt{30}-2\sqrt{100}$

$\qquad\qquad\qquad\qquad\qquad = 4\cdot 3-1\sqrt{30}+8\sqrt{30}-2\cdot 10$

$\qquad\qquad\qquad\qquad\qquad = 12+7\sqrt{30}-20$

$\qquad\qquad\qquad\qquad\qquad = -8+7\sqrt{30}$

41. $\left(\sqrt{a}-3b\right)\left(9\sqrt{a}-b\right) = \sqrt{a}\cdot 9\sqrt{a}-\sqrt{a}\cdot b-3b\cdot 9\sqrt{a}+3b\cdot b$

$\qquad\qquad\qquad\qquad = 9\left(\sqrt{a}\right)^2-28b\sqrt{a}+3b^2$

$\qquad\qquad\qquad\qquad = 9a-28b\sqrt{a}+3b^2$

43. $\left(p+2\sqrt{p}\right)\left(8p+3\sqrt{p}-4\right)=p\cdot 8p+p\cdot 3\sqrt{p}-p\cdot 4+2\sqrt{p}\cdot 8p+2\sqrt{p}\cdot 3\sqrt{p}-2\sqrt{p}\cdot 4$

$$=8p^2+3p\sqrt{p}-4p+16p\sqrt{p}+6\left(\sqrt{p}\right)^2-8\sqrt{p}$$

$$=8p^2+3p\sqrt{p}-4p+16p\sqrt{p}+6p-8\sqrt{p}$$

$$=8p^2+19p\sqrt{p}+2p-8\sqrt{p}$$

45. $\left(\sqrt{10}\right)^2=10$

47. $\left(\sqrt[3]{4}\right)^3=4$

49. $\left(\sqrt[4]{t}\right)^4=t$

51. $\left(4\sqrt{c}\right)^2=4^2\cdot\left(\sqrt{c}\right)^2=16c$

53. $\left(\sqrt{13}+4\right)^2=\left(\sqrt{13}\right)^2+2\left(\sqrt{13}\right)(4)+4^2$

$$=13+8\sqrt{13}+16$$

$$=29+8\sqrt{13}$$

55. $\left(\sqrt{a}-2\right)^2=\left(\sqrt{a}\right)^2-2\left(\sqrt{a}\right)(2)+2^2$

$$=a-4\sqrt{a}+4$$

57. $\left(2\sqrt{a}-3\right)^2=\left(2\sqrt{a}\right)^2-2\left(2\sqrt{a}\right)(3)+3^2$

$$=4a-12\sqrt{a}+9$$

59. $\left(\sqrt{10}-\sqrt{11}\right)^2$

$$=\left(\sqrt{10}\right)^2-2\left(\sqrt{10}\right)\left(\sqrt{11}\right)+\left(\sqrt{11}\right)^2$$

$$=10-2\sqrt{110}+11$$

$$=21-2\sqrt{110}$$

61. $\left(\sqrt{5}+2\right)\left(\sqrt{5}-2\right)=\left(\sqrt{5}\right)^2-2^2=5-4=1$

63. $\left(\sqrt{x}+\sqrt{y}\right)\left(\sqrt{x}-\sqrt{y}\right)=\left(\sqrt{x}\right)^2-\left(\sqrt{y}\right)^2$

$$=x-y$$

65. $\left(\sqrt{10}-\sqrt{11}\right)\left(\sqrt{10}+\sqrt{11}\right)=\left(\sqrt{10}\right)^2-\left(\sqrt{11}\right)^2$

$$=10-11$$

$$=-1$$

67. $\left(\sqrt{6}+\sqrt{2}\right)\left(\sqrt{6}-\sqrt{2}\right)=\left(\sqrt{6}\right)^2-\left(\sqrt{2}\right)^2$

$$=6-2$$

$$=4$$

69. $\left(8\sqrt{x}-2\sqrt{y}\right)\left(8\sqrt{x}+2\sqrt{y}\right)$

$$=8^2\left(\sqrt{x}\right)^2-2^2\left(\sqrt{y}\right)^2$$

$$=64x-4y$$

71. $\left(5\sqrt{3}-\sqrt{2}\right)\left(5\sqrt{3}+\sqrt{2}\right)=\left(5\sqrt{3}\right)^2-\left(\sqrt{2}\right)^2$

$$=25\cdot 3-2$$

$$=75-2$$

$$=73$$

73. (a) $3(x+2)=3\cdot x+3\cdot 2=3x+6$

(b) $\sqrt{3}\left(\sqrt{x}+\sqrt{2}\right)=\sqrt{3}\cdot\sqrt{x}+\sqrt{3}\cdot\sqrt{2}$

$$=\sqrt{3x}+\sqrt{6}$$

75. (a) $(2a+3)^2=(2a)^2+2(2a)(3)+3^2$

$$=4a^2+12a+9$$

(b) $\left(2\sqrt{a}+3\right)^2$

$$=\left(2\sqrt{a}\right)^2+2\left(2\sqrt{a}\right)(3)+3^2$$

$$=2^2\cdot\left(\sqrt{a}\right)^2+\left(2\cdot 2\cdot 3\sqrt{a}\right)+9$$

$$=4a+12\sqrt{a}+9$$

77. (a) $(b-5)(b+5)=b^2-5^2=b^2-25$

(b) $\left(\sqrt{b}-5\right)\left(\sqrt{b}+5\right)=\left(\sqrt{b}\right)^2-5^2=b-25$

79. (a) $(x-2y)^2=(x)^2-2(x)(2y)+(-2y)^2$

$$=x^2-4xy+4y^2$$

(b) $\left(\sqrt{x} - 2\sqrt{y}\right)^2$

$= \left(\sqrt{x}\right)^2 - 2\left(\sqrt{x}\right)\left(2\sqrt{y}\right) + \left(-2\sqrt{y}\right)^2$

$= x - 4\sqrt{x}\sqrt{y} + 4y$

$= x - 4\sqrt{xy} + 4y$

81. (a) $(p-q)(p+q)$

$= (p)^2 + (p)(q) - (p)(q) - (q)^2$

$= p^2 - q^2$

(b) $\left(\sqrt{p} - \sqrt{q}\right)\left(\sqrt{p} + \sqrt{q}\right)$

$= \left(\sqrt{p}\right)^2 + \left(\sqrt{p}\right)\left(\sqrt{q}\right)$

$\qquad - \left(\sqrt{p}\right)\left(\sqrt{q}\right) - \left(\sqrt{q}\right)^2$

$= p - q$

83. (a) $(y-3)^2$

$= (y)^2 + 2(y)(-3) + (-3)^2$

$= y^2 - 6y + 9$

(b) $\left(\sqrt{x-2} - 3\right)^2$

$= \left(\sqrt{x-2}\right)^2 - 2\left(\sqrt{x-2}\right)(3) + (3)^2$

$= x - 2 - 6\sqrt{x-2} + 9$

$= x - 6\sqrt{x-2} + 7$

Section 8.5 Practice Exercises

1. (a) index

(b) denominator

(c) rationalizing

(d) $\dfrac{\sqrt[n]{a}}{\sqrt[n]{b}}$

(e) denominator

3. $\left(2\sqrt{y} + 3\right)\left(3\sqrt{y} + 7\right)$

$= 2\sqrt{y} \cdot 3\sqrt{y} + 2\sqrt{y} \cdot 7 + 3 \cdot 3\sqrt{y} + 3 \cdot 7$

$= 6\left(\sqrt{y}\right)^2 + 14\sqrt{y} + 9\sqrt{y} + 21$

$= 6y + 23\sqrt{y} + 21$

5. $4\sqrt{3} + \sqrt{5} \cdot \sqrt{15} = 4\sqrt{3} + \sqrt{5 \cdot 15}$

$= 4\sqrt{3} + \sqrt{75}$

$= 4\sqrt{3} + \sqrt{25 \cdot 3}$

$= 4\sqrt{3} + 5\sqrt{3}$

$= 9\sqrt{3}$

7. $\left(5 - \sqrt{a}\right)^2 = 5^2 - 2(5)\sqrt{a} + \left(\sqrt{a}\right)^2$

$= 25 - 10\sqrt{a} + a$

9. $\left(\sqrt{2} + \sqrt{7}\right)\left(\sqrt{2} - \sqrt{7}\right) = \left(\sqrt{2}\right)^2 - \left(\sqrt{7}\right)^2$

$= 2 - 7$

$= -5$

11. $\sqrt{\dfrac{3}{16}} = \dfrac{\sqrt{3}}{\sqrt{4 \cdot 4}} = \dfrac{\sqrt{3}}{\sqrt{4^2}} = \dfrac{\sqrt{3}}{4}$

13. $\sqrt{\dfrac{a^4}{b^4}} = \dfrac{\sqrt{a^2 \cdot a^2}}{\sqrt{b^2 \cdot b^2}} = \dfrac{a^2}{b^2}$

15. $\sqrt{\dfrac{c^3}{4}} = \dfrac{\sqrt{c^2 \cdot c}}{\sqrt{2 \cdot 2}} = \dfrac{c\sqrt{c}}{2}$

17. $\sqrt[3]{\dfrac{x^2}{27}} = \sqrt[3]{\dfrac{x^2}{3^3}} = \dfrac{\sqrt[3]{x^2}}{3}$

19. $\sqrt[3]{\dfrac{y^5}{27y^3}} = \sqrt[3]{\dfrac{y^2}{27}} = \sqrt[3]{\dfrac{y^2}{3^3}} = \dfrac{\sqrt[3]{y^2}}{3}$

21. $\sqrt{\dfrac{200}{81}} = \sqrt{\dfrac{2 \cdot 100}{9^2}} = \sqrt{\dfrac{2 \cdot 10^2}{9^2}} = \dfrac{10\sqrt{2}}{9}$

23. $\dfrac{\sqrt{8}}{\sqrt{50}} = \sqrt{\dfrac{8}{50}} = \sqrt{\dfrac{4}{25}} = \sqrt{\dfrac{2^2}{5^2}} = \dfrac{2}{5}$

25. $\dfrac{\sqrt{p}}{\sqrt{4p^3}} = \sqrt{\dfrac{p}{4p^3}} = \sqrt{\dfrac{1}{4p^2}} = \sqrt{\dfrac{1}{2^2 p^2}} = \dfrac{1}{2p}$

27. $\dfrac{\sqrt[3]{z^5}}{\sqrt[3]{z^2}} = \sqrt[3]{\dfrac{z^5}{z^2}} = \sqrt[3]{z^3} = z$

29. $\dfrac{\sqrt[3]{24x^5}}{\sqrt[3]{3x^4}} = \sqrt[3]{\dfrac{24x^5}{3x^4}} = \sqrt[3]{8x} = \sqrt[3]{2^3 x} = 2\sqrt[3]{x}$

31. $\dfrac{1}{\sqrt{6}} = \dfrac{1}{\sqrt{6}} \cdot \dfrac{\sqrt{6}}{\sqrt{6}} = \dfrac{\sqrt{6}}{\sqrt{6 \cdot 6}} = \dfrac{\sqrt{6}}{\sqrt{6^2}} = \dfrac{\sqrt{6}}{6}$

33. $\dfrac{15}{\sqrt{5}} = \dfrac{15}{\sqrt{5}} \cdot \dfrac{\sqrt{5}}{\sqrt{5}}$

$\quad = \dfrac{15\sqrt{5}}{\sqrt{5 \cdot 5}}$

$\quad = \dfrac{15\sqrt{5}}{\sqrt{5^2}}$

$\quad = \dfrac{15\sqrt{5}}{5}$

$\quad = 3\sqrt{5}$

35. $\dfrac{6}{\sqrt{x+1}} = \dfrac{6}{\sqrt{x+1}} \cdot \dfrac{\sqrt{x+1}}{\sqrt{x+1}}$

$\quad = \dfrac{6\sqrt{x+1}}{\sqrt{(x+1)^2}}$

$\quad = \dfrac{6\sqrt{x+1}}{x+1}$

37. $\sqrt{\dfrac{6}{x}} = \dfrac{\sqrt{6}}{\sqrt{x}} = \dfrac{\sqrt{6}}{\sqrt{x}} \cdot \dfrac{\sqrt{x}}{\sqrt{x}} = \dfrac{\sqrt{6x}}{\sqrt{x^2}} = \dfrac{\sqrt{6x}}{x}$

39. $\sqrt{\dfrac{3}{7}} = \dfrac{\sqrt{3}}{\sqrt{7}} \cdot \dfrac{\sqrt{7}}{\sqrt{7}} = \dfrac{\sqrt{21}}{\sqrt{7^2}} = \dfrac{\sqrt{21}}{7}$

41. $\dfrac{10}{\sqrt{6y}} = \dfrac{10}{\sqrt{6y}} \cdot \dfrac{\sqrt{6y}}{\sqrt{6y}}$

$\quad = \dfrac{10\sqrt{6y}}{\sqrt{6y \cdot 6y}}$

$\quad = \dfrac{10\sqrt{6y}}{\sqrt{6^2 y^2}}$

$\quad = \dfrac{10\sqrt{6y}}{6y}$

$\quad = \dfrac{5\sqrt{6y}}{3y}$

43. $\dfrac{9}{2\sqrt{6}} \cdot \dfrac{\sqrt{6}}{\sqrt{6}} = \dfrac{9\sqrt{6}}{2\sqrt{6^2}} = \dfrac{9\sqrt{6}}{2 \cdot 6} = \dfrac{9\sqrt{6}}{12} = \dfrac{3\sqrt{6}}{4}$

45. $\sqrt{\dfrac{p}{27}} = \dfrac{\sqrt{p}}{\sqrt{27}}$

$\quad = \dfrac{\sqrt{p}}{\sqrt{3^2 \cdot 3}}$

$\quad = \dfrac{\sqrt{p}}{3\sqrt{3}} \cdot \dfrac{\sqrt{3}}{\sqrt{3}}$

$\quad = \dfrac{\sqrt{3p}}{3\sqrt{3^2}}$

$\quad = \dfrac{\sqrt{3p}}{9}$

47. $\dfrac{5}{\sqrt{20}} = \dfrac{5}{\sqrt{2^2 \cdot 5}}$

$\quad = \dfrac{5}{2\sqrt{5}} \cdot \dfrac{\sqrt{5}}{\sqrt{5}}$

$\quad = \dfrac{5\sqrt{5}}{2\sqrt{5^2}}$

$\quad = \dfrac{5\sqrt{5}}{2 \cdot 5}$

$\quad = \dfrac{\sqrt{5}}{2}$

49. $\sqrt{\dfrac{x^2}{y^3}} = \dfrac{\sqrt{x^2}}{\sqrt{y^3}}$

$= \dfrac{x}{\sqrt{y^2 \cdot y}}$

$= \dfrac{x}{y\sqrt{y}} \cdot \dfrac{\sqrt{y}}{\sqrt{y}}$

$= \dfrac{x\sqrt{y}}{y\sqrt{y^2}}$

$= \dfrac{x\sqrt{y}}{y^2}$

51. $\left(\sqrt{2}+3\right)\left(\sqrt{2}-3\right) = \left(\sqrt{2}\right)^2 - 3^2 = 2 - 9 = -7$

53. The conjugate is $\sqrt{5} + \sqrt{3}$.

$\left(\sqrt{5}-\sqrt{3}\right)\left(\sqrt{5}+\sqrt{3}\right) = \left(\sqrt{5}\right)^2 - \left(\sqrt{3}\right)^2$

$= 5 - 3$

$= 2$

$\sqrt{5} + \sqrt{3}; 2$

55. The conjugate is $\sqrt{x} - 10$.

$\left(\sqrt{x}+10\right)\left(\sqrt{x}-10\right) = \left(\sqrt{x}\right)^2 - 10^2 = x - 100$

$\sqrt{x} - 10; x - 100$

57. $\dfrac{4}{\sqrt{2}+3} = \dfrac{4}{\sqrt{2}+3} \cdot \dfrac{\sqrt{2}-3}{\sqrt{2}-3}$

$= \dfrac{4\left(\sqrt{2}-3\right)}{\left(\sqrt{2}\right)^2 - 3^2}$

$= \dfrac{4\sqrt{2}-12}{2-9}$

$= \dfrac{4\sqrt{2}-12}{-7}$ or $\dfrac{12-4\sqrt{2}}{7}$

59. $\dfrac{1}{\sqrt{5}-\sqrt{2}} = \dfrac{1}{\sqrt{5}-\sqrt{2}} \cdot \dfrac{\sqrt{5}+\sqrt{2}}{\sqrt{5}+\sqrt{2}}$

$= \dfrac{\sqrt{5}+\sqrt{2}}{\left(\sqrt{5}\right)^2 - \left(\sqrt{2}\right)^2}$

$= \dfrac{\sqrt{5}+\sqrt{2}}{5-2}$

$= \dfrac{\sqrt{5}+\sqrt{2}}{3}$

61. $\dfrac{\sqrt{8}}{\sqrt{3}+1} = \dfrac{\sqrt{8}}{\sqrt{3}+1} \cdot \dfrac{\sqrt{3}-1}{\sqrt{3}-1}$

$= \dfrac{\sqrt{8}\left(\sqrt{3}-1\right)}{\left(\sqrt{3}\right)^2 - 1^2}$

$= \dfrac{\sqrt{24}-\sqrt{8}}{3-1}$

$= \dfrac{\sqrt{4\cdot 6}-\sqrt{4\cdot 2}}{2}$

$= \dfrac{2\sqrt{6}-2\sqrt{2}}{2}$

$= \dfrac{2\left(\sqrt{6}-\sqrt{2}\right)}{2}$

$= \sqrt{6} - \sqrt{2}$

63. $\dfrac{1}{\sqrt{x}-\sqrt{3}} = \dfrac{1}{\sqrt{x}-\sqrt{3}} \cdot \dfrac{\sqrt{x}+\sqrt{3}}{\sqrt{x}+\sqrt{3}}$

$= \dfrac{\sqrt{x}+\sqrt{3}}{\left(\sqrt{x}\right)^2 - \left(\sqrt{3}\right)^2}$

$= \dfrac{\sqrt{x}+\sqrt{3}}{x-3}$

65. $\dfrac{2-\sqrt{3}}{2+\sqrt{3}} \cdot \dfrac{2-\sqrt{3}}{2-\sqrt{3}} = \dfrac{4-4\sqrt{3}+3}{4-3}$

$= \dfrac{7-4\sqrt{3}}{1}$

$= 7 - 4\sqrt{3}$

67. $\dfrac{\sqrt{5}+4}{2-\sqrt{5}} = \dfrac{\sqrt{5}+4}{2-\sqrt{5}} \cdot \dfrac{2+\sqrt{5}}{2+\sqrt{5}}$

$\qquad = \dfrac{\left(\sqrt{5}+4\right)\left(2+\sqrt{5}\right)}{2^2 - \left(\sqrt{5}\right)^2}$

$\qquad = \dfrac{\sqrt{5}\cdot 2 + \sqrt{5}\cdot\sqrt{5} + 4\cdot 2 + 4\cdot\sqrt{5}}{4-5}$

$\qquad = \dfrac{2\sqrt{5} + \sqrt{5^2} + 8 + 4\sqrt{5}}{-1}$

$\qquad = \dfrac{2\sqrt{5} + 5 + 8 + 4\sqrt{5}}{-1}$

$\qquad = \dfrac{13 + 6\sqrt{5}}{-1}$

$\qquad = -13 - 6\sqrt{5}$

69. $\dfrac{10-\sqrt{50}}{5} = \dfrac{10-\sqrt{5^2\cdot 2}}{5}$

$\qquad = \dfrac{10 - 5\sqrt{2}}{5}$

$\qquad = \dfrac{5\left(2-\sqrt{2}\right)}{5}$

$\qquad = 2 - \sqrt{2}$

71. $\dfrac{21+\sqrt{98}}{14} = \dfrac{21+\sqrt{7^2\cdot 2}}{14}$

$\qquad = \dfrac{21 + 7\sqrt{2}}{14}$

$\qquad = \dfrac{7\left(3+\sqrt{2}\right)}{7\cdot 2}$

$\qquad = \dfrac{3+\sqrt{2}}{2}$

73. $\dfrac{2-\sqrt{28}}{2} = \dfrac{2-\sqrt{2^2\cdot 7}}{2}$

$\qquad = \dfrac{2 - 2\sqrt{7}}{2}$

$\qquad = \dfrac{2\left(1-\sqrt{7}\right)}{2}$

$\qquad = 1 - \sqrt{7}$

75. $\dfrac{14+\sqrt{72}}{6} = \dfrac{14+\sqrt{6^2\cdot 2}}{6}$

$\qquad = \dfrac{14 + 6\sqrt{2}}{6}$

$\qquad = \dfrac{2\left(7+3\sqrt{2}\right)}{2\cdot 3}$

$\qquad = \dfrac{7+3\sqrt{2}}{3}$

77. (a) Condition 1 fails;

$\sqrt{8x^9} = \sqrt{2^3\cdot x^9} = \sqrt{2\cdot 2^2\cdot x\cdot x^8}$

$\qquad = 2x^4\sqrt{2x}$

(b) Condition 2 fails;

$\dfrac{5}{\sqrt{5x}} = \dfrac{5}{\sqrt{5x}}\cdot\dfrac{\sqrt{5x}}{\sqrt{5x}} = \dfrac{5\sqrt{5x}}{5x} = \dfrac{\sqrt{5x}}{x}$

(c) Condition 3 fails; $\sqrt{\dfrac{1}{3}} = \dfrac{1}{\sqrt{3}}\cdot\dfrac{\sqrt{3}}{\sqrt{3}} = \dfrac{\sqrt{3}}{3}$

79. (a) Condition 2 fails;

$\dfrac{3}{\sqrt{x}+1} = \dfrac{3}{\sqrt{x}+1}\cdot\dfrac{\sqrt{x}-1}{\sqrt{x}-1} = \dfrac{3\sqrt{x}-3}{x-1}$

(b) Condition 1 and 3 fail;

$\sqrt{\dfrac{9w^2}{t}} = \sqrt{\dfrac{3^2 w^2}{t}\cdot\dfrac{t}{t}} = 3w\dfrac{\sqrt{t}}{t}$

(c) Condition 1 fails;

$\sqrt{24a^5 b^9} = \sqrt{6\cdot 2^2\,a^5 b^9}$

$\qquad = \sqrt{6\cdot 2^2\cdot a\cdot a^4\cdot b\cdot b^8}$

$\qquad = 2a^2 b^4\sqrt{6\cdot a\cdot b}$

$\qquad = 2a^2 b^4\sqrt{6ab}$

81. $\sqrt{45} = \sqrt{3^2\cdot 5} = 3\sqrt{5}$

83. $-\sqrt{\dfrac{18w^2}{25}} = -\dfrac{\sqrt{3^2 w^2\cdot 2}}{\sqrt{5^2}} = -\dfrac{3w\sqrt{2}}{5}$

85. $\sqrt{-36}$ is not a real number.

87. $\sqrt{\dfrac{s^2}{t}} = \dfrac{\sqrt{s^2}}{\sqrt{t}} = \dfrac{s}{\sqrt{t}} \cdot \dfrac{\sqrt{t}}{\sqrt{t}} = \dfrac{s\sqrt{t}}{\sqrt{t^2}} = \dfrac{s\sqrt{t}}{t}$

89. $\dfrac{\sqrt{2m^5}}{\sqrt{8m}} = \sqrt{\dfrac{2m^5}{8m}} = \sqrt{\dfrac{m^4}{4}} = \dfrac{\sqrt{m^4}}{\sqrt{4}} = \dfrac{m^2}{2}$

91. $\sqrt{\dfrac{81}{t^3}} = \dfrac{\sqrt{3^4}}{\sqrt{t^2 \cdot t}}$

$= \dfrac{3^2}{t\sqrt{t}}$

$= \dfrac{9}{t\sqrt{t}} \cdot \dfrac{\sqrt{t}}{\sqrt{t}}$

$= \dfrac{9\sqrt{t}}{t\sqrt{t^2}}$

$= \dfrac{9\sqrt{t}}{t \cdot t}$

$= \dfrac{9\sqrt{t}}{t^2}$

93. $\dfrac{3}{\sqrt{11}+\sqrt{5}} = \dfrac{3}{\sqrt{11}+\sqrt{5}} \dfrac{\sqrt{11}-\sqrt{5}}{\sqrt{11}-\sqrt{5}}$

$= \dfrac{3\sqrt{11}-3\sqrt{5}}{11-5} = \dfrac{3\sqrt{11}-3\sqrt{5}}{6}$

$= \dfrac{\sqrt{11}-\sqrt{5}}{2}$

95. $\dfrac{\sqrt{a}+\sqrt{b}}{\sqrt{a}-\sqrt{b}} = \dfrac{\sqrt{a}+\sqrt{b}}{\sqrt{a}-\sqrt{b}} \cdot \dfrac{\sqrt{a}+\sqrt{b}}{\sqrt{a}+\sqrt{b}}$

$= \dfrac{a+2\sqrt{a}\sqrt{b}+b}{a-b} = \dfrac{a+2\sqrt{ab}+b}{a-b}$

97. $m = \dfrac{6-3}{\sqrt{2}-5\sqrt{2}}$

$= \dfrac{3}{-4\sqrt{2}}$

$= -\dfrac{3}{4\sqrt{2}} \cdot \dfrac{\sqrt{2}}{\sqrt{2}}$

$= -\dfrac{3\sqrt{2}}{4\sqrt{2^2}}$

$= -\dfrac{3\sqrt{2}}{4 \cdot 2}$

$= -\dfrac{3\sqrt{2}}{8}$

99. $m = \dfrac{0-(-1)}{4\sqrt{3}-\sqrt{3}}$

$= \dfrac{1}{3\sqrt{3}}$

$= \dfrac{1}{3\sqrt{3}} \cdot \dfrac{\sqrt{3}}{\sqrt{3}}$

$= \dfrac{\sqrt{3}}{3\sqrt{3^2}}$

$= \dfrac{\sqrt{3}}{3 \cdot 3}$

$= \dfrac{\sqrt{3}}{9}$

Problem Recognition Exercises

1 **(a)** $\left(\sqrt{3}\right)\left(\sqrt{6}\right) = \sqrt{3 \cdot 6} = \sqrt{3 \cdot 3 \cdot 2} = 3\sqrt{2}$

(b) $\sqrt{3} + \sqrt{6}$ Cannot be simplified further

(c) $\dfrac{\sqrt{6}}{\sqrt{3}} = \sqrt{\dfrac{6}{3}} = \sqrt{2}$

3. **(a)** $\left(3\sqrt{z}\right)^2 = 3^2 \cdot \left(\sqrt{z}\right)^2 = 9z$

(b) $\left(3+\sqrt{z}\right)^2 = 3^2 + 2(3)\left(\sqrt{z}\right) + \left(\sqrt{z}\right)^2$

$= 9 + 6\sqrt{z} + z$

(c) $\left(3+\sqrt{z}\right)\left(3-\sqrt{z}\right) = 3^2 - \left(\sqrt{z}\right)^2 = 9 - z$

5. (a) $\dfrac{12}{\sqrt{2x}} = \dfrac{12}{\sqrt{2x}} \cdot \dfrac{\sqrt{2x}}{\sqrt{2x}}$

$\qquad = \dfrac{12\sqrt{2x}}{2x} = \dfrac{6\sqrt{2x}}{x}$

(b) $\sqrt{\dfrac{12}{2x}} = \sqrt{\dfrac{12}{2x}} \cdot \dfrac{\sqrt{2x}}{\sqrt{2x}}$

$\qquad = \dfrac{\sqrt{24x}}{2x} = \dfrac{\sqrt{2^2 \cdot 6x}}{2x} = \dfrac{2\sqrt{6x}}{2x} = \dfrac{\sqrt{6x}}{x}$

(c) $\dfrac{12}{\sqrt{2}+x} = \dfrac{12}{\sqrt{2}+x} \cdot \dfrac{\sqrt{2}-x}{\sqrt{2}-x}$

$\qquad = \dfrac{12\sqrt{2}-12 \cdot x}{2-x^2} = \dfrac{12\sqrt{2}-12x}{2-x^2}$

7. (a) $\left(2\sqrt{5}+1\right)+\left(\sqrt{5}-2\right)$

$\qquad = (2+1)\sqrt{5}+(1-2)$

$\qquad = 3\sqrt{5}-1$

(b) $\left(2\sqrt{5}+1\right)\left(\sqrt{5}-2\right)$

$\qquad = 2\sqrt{5} \cdot \sqrt{5}-2\sqrt{5} \cdot 2+\sqrt{5}-2$

$\qquad = 2 \cdot 5-3\sqrt{5}-2$

$\qquad = 8-3\sqrt{5}$

(c) $2\sqrt{5}\left(\sqrt{5}-2\right) = 2\sqrt{5} \cdot \sqrt{5}-2\sqrt{5} \cdot 2$

$\qquad\qquad = 2 \cdot 5-4\sqrt{5}$

$\qquad\qquad = 10-4\sqrt{5}$

9. (a) $\sqrt{16a^{15}} = \sqrt{4^2 a^{14} a^1} = 4a^7 \sqrt{a}$

(b) $\sqrt[3]{16a^{15}} = \sqrt[3]{2^3 \cdot 2^1 a^{15}} = 2a^5 \sqrt[3]{2}$

Section 8.6 Practice Exercises

1. (a) radical

(b) extraneous

(c) The first step to solving the equation $\sqrt{x+2}-3=7$ is to isolate the radical by adding 3 to both sides of the equation.

(d) third.

3. $\dfrac{1}{\sqrt{2}+\sqrt{10}} = \dfrac{1}{\sqrt{2}+\sqrt{10}} \cdot \dfrac{\sqrt{2}-\sqrt{10}}{\sqrt{2}-\sqrt{10}}$

$\qquad = \dfrac{\sqrt{2}-\sqrt{10}}{\left(\sqrt{2}\right)^2-\left(\sqrt{10}\right)^2}$

$\qquad = \dfrac{\sqrt{2}-\sqrt{10}}{2-10}$

$\qquad = \dfrac{\sqrt{2}-\sqrt{10}}{-8}$ or $\dfrac{\sqrt{10}-\sqrt{2}}{8}$

5. $\dfrac{2\sqrt{2}}{\sqrt{3}} = \dfrac{2\sqrt{2}}{\sqrt{3}} \cdot \dfrac{\sqrt{3}}{\sqrt{3}} = \dfrac{2\sqrt{6}}{\sqrt{3^2}} = \dfrac{2\sqrt{6}}{3}$

7. $(x+4)^2 = x^2 + 2(x)(4) + 4^2 = x^2 + 8x + 16$

9. $\left(\sqrt{x}+4\right)^2 = x + 2\sqrt{x} \cdot 4 + 4^2$

$\qquad\qquad = x + 8\sqrt{x} + 16$

11. $\left(\sqrt{2x-3}\right)^2 = 2x-3$

In this problem the binomial is the radicand. Raising a radicand to a power equivalent to the index gives the value of the radicand.

13. $(t+1)^2 = t^2 + 2 \cdot t + 1$

$\qquad\qquad = t^2 + 2t + 1$

15. $\sqrt{t} = 6$

$\qquad \left(\sqrt{t}\right)^2 = 6^2$

$\qquad\quad t = 36$

17. $\sqrt{x+1} = 4$

$\qquad \left(\sqrt{x+1}\right)^2 = 4^2$

$\qquad\quad x+1 = 16$

$\qquad\qquad x = 15$

Check: $\sqrt{15+1} = \sqrt{16} = 4 \checkmark$

19. $\sqrt{y-4} = -5$

The principal square root of a value is never negative. No solution.

241

21.
$$\sqrt{5-t} = 0$$
$$\left(\sqrt{5-t}\right)^2 = 0^2$$
$$5-t = 0$$
$$-t = -5$$
$$t = 5$$

Check: $\sqrt{5-5} = \sqrt{0} = 0$ ✓

23.
$$\sqrt{2n+10} = 3$$
$$\left(\sqrt{2n+10}\right)^2 = (3)^2$$
$$2n+10 = 9$$
$$2n = -1$$
$$n = -\frac{1}{2}$$

Check: $\sqrt{2\left(-\frac{1}{2}\right)+10} \overset{?}{=} 3$

$$\sqrt{-1+10} \overset{?}{=} 3$$
$$\sqrt{9} \overset{?}{=} 3$$
$$3 = 3 ✓$$

25.
$$\sqrt{6w}-8 = -2$$
$$\sqrt{6w} = 6$$
$$\left(\sqrt{6w}\right)^2 = (6)^2$$
$$6w = 36$$
$$w = 6$$

Check: $\sqrt{6\cdot6}-8 \overset{?}{=} -2$

$$\sqrt{36}-8 \overset{?}{=} -2$$
$$6-8 \overset{?}{=} -2$$
$$-2 = -2 ✓$$

27.
$$\sqrt{5a-4}-2 = 4$$
$$\sqrt{5a-4} = 6$$
$$\left(\sqrt{5a-4}\right)^2 = 6^2$$
$$5a-4 = 36$$
$$5a = 40$$
$$a = 8$$

Check: $\sqrt{5(8)-4}-2 \overset{?}{=} 4$

$$\sqrt{40-4}-2 \overset{?}{=} 4$$
$$\sqrt{36}-2 \overset{?}{=} 4$$
$$6-2 \overset{?}{=} 4$$
$$4 = 4 ✓$$

29.
$$\sqrt{2x-3}+7 = 3$$
$$\sqrt{2x-3} = -4$$
$$\left(\sqrt{2x-3}\right)^2 = (-4)^2$$
$$2x-3 = 16$$
$$2x = 19$$
$$x = \frac{19}{2}$$

Check: $\sqrt{2\left(\frac{19}{2}\right)-3}+7 \overset{?}{=} 3$

$$\sqrt{19-3}+7 \overset{?}{=} 3$$
$$\sqrt{16}+7 \overset{?}{=} 3$$
$$4+7 \overset{?}{=} 3$$
$$11 \neq 3$$

The solution does not check. There is no solution.

31.
$$5\sqrt{c} = \sqrt{10c+15}$$
$$\left(5\sqrt{c}\right)^2 = \left(\sqrt{10c+15}\right)^2$$
$$25c = 10c+15$$
$$15c = 15$$
$$c = 1$$

Check: $5\sqrt{1} \overset{?}{=} \sqrt{10(1)+15}$

$5(1) \overset{?}{=} \sqrt{10+15}$

$5 \overset{?}{=} \sqrt{25}$

$5 = 5 \checkmark$

33. $\sqrt{x^2 - x} = \sqrt{12}$

$\left(\sqrt{x^2 - x}\right)^2 = \left(\sqrt{12}\right)^2$

$x^2 - x = 12$

$x^2 - x - 12 = 0$

$(x-4)(x+3) = 0$

$x - 4 = 0 \quad$ or $\quad x + 3 = 0$

$x = 4 \qquad\qquad x = -3$

Check: $\sqrt{(4)^2 - 4} \overset{?}{=} \sqrt{12}$

$\sqrt{16 - 4} \overset{?}{=} \sqrt{12}$

$\sqrt{12} = \sqrt{12} \checkmark$

$\sqrt{(-3)^2 - (-3)} \overset{?}{=} \sqrt{12}$

$\sqrt{9 + 3} \overset{?}{=} \sqrt{12}$

$\sqrt{12} = \sqrt{12} \checkmark$

35. $\sqrt{9y^2 - 8y + 1} = 3y + 1$

$\left(\sqrt{9y^2 - 8y + 1}\right)^2 = (3y+1)^2$

$9y^2 - 8y + 1 = 9y^2 + 6y + 1$

$-8y = 6y$

$-14y = 0$

$y = 0$

Check: $\sqrt{9(0)^2 - 8(0) + 1} \overset{?}{=} 3(0) + 1$

$\sqrt{1} \overset{?}{=} 1$

$1 = 1 \checkmark$

37. $\sqrt{x^2 + 4x + 16} = x$

$\left(\sqrt{x^2 + 4x + 16}\right)^2 = x^2$

$x^2 + 4x + 16 = x^2$

$4x + 16 = 0$

$4(x + 4) = 0$

$x = -4$

Check: $\sqrt{(-4)^2 + 4(-4) + 16} \overset{?}{=} -4$

$\sqrt{16 - 16 + 16} \overset{?}{=} -4$

$\sqrt{16} \overset{?}{=} -4$

$4 \neq -4$

The solution does not check. There is no solution

39. $\sqrt{2k^2 - 3k - 4} = k$

$\left(\sqrt{2k^2 - 3k - 4}\right)^2 = (k)^2$

$2k^2 - 3k - 4 = k^2$

$k^2 - 3k - 4 = 0$

$(k-4)(k+1) = 0$

$k - 4 = 0 \quad$ or $\quad k + 1 = 0$

$k = 4 \qquad\qquad k = -1$

Check: $\sqrt{2(4)^2 - 3(4) - 4} \overset{?}{=} 4$

$\sqrt{2(16) - 3(4) - 4} \overset{?}{=} 4$

$\sqrt{32 - 12 - 4} \overset{?}{=} 4$

$\sqrt{16} \overset{?}{=} 4$

$4 = 4 \checkmark$

$\sqrt{2(-1)^2 - 3(-1) - 4} \overset{?}{=} -1$

$\sqrt{2(1) - 3(-1) - 4} \overset{?}{=} -1$

$\sqrt{2 + 3 - 4} \overset{?}{=} -1$

$\sqrt{1} \overset{?}{=} -1$

$1 \neq -1$

The solution $k = -1$ does not check.

41.
$$\sqrt{y+1} = y+1$$
$$\left(\sqrt{y+1}\right)^2 = (y+1)^2$$
$$y+1 = y^2 + 2y + 1$$
$$0 = y^2 + y$$
$$0 = y(y+1)$$
$$y = 0 \quad \text{or} \quad y+1 = 0$$
$$y = -1$$

Check: $\sqrt{0+1} \stackrel{?}{=} 0+1$
$$\sqrt{1} \stackrel{?}{=} 1$$
$$1 = 1 \checkmark$$

$$\sqrt{-1+1} \stackrel{?}{=} -1+1$$
$$\sqrt{0} \stackrel{?}{=} 0$$
$$0 = 0 \checkmark$$

43. $\sqrt{2m+1} + 7 = m$
$$\sqrt{2m+1} = m - 7$$
$$\left(\sqrt{2m+1}\right)^2 = (m-7)^2$$
$$2m+1 = m^2 - 14m + 49$$
$$0 = m^2 - 16m + 48$$
$$0 = (m-4)(m-12)$$
$$m-4 = 0 \quad \text{or} \quad m-12 = 0$$
$$m = 4 \qquad\qquad m = 12$$

Check: $\sqrt{2(4)+1} + 7 \stackrel{?}{=} 4$
$$\sqrt{8+1} + 7 \stackrel{?}{=} 4$$
$$\sqrt{9} + 7 \stackrel{?}{=} 4$$
$$3 + 7 \stackrel{?}{=} 4$$
$$10 \neq 4$$
$m = 4$ does not check.

$$\sqrt{2(12)+1} + 7 \stackrel{?}{=} 12$$
$$\sqrt{24+1} + 7 \stackrel{?}{=} 12$$
$$\sqrt{25} + 7 \stackrel{?}{=} 12$$
$$5 + 7 \stackrel{?}{=} 12$$
$$12 = 12 \checkmark$$

45. $\sqrt[3]{p-5} - \sqrt[3]{2p+1} = 0$
$$\sqrt[3]{p-5} = \sqrt[3]{2p+1}$$
$$\left(\sqrt[3]{p-5}\right)^3 = \left(\sqrt[3]{2p+1}\right)^3$$
$$p-5 = 2p+1$$
$$-p-5 = 1$$
$$-p = 6$$
$$p = -6$$

Check: $\sqrt[3]{-6-5} \stackrel{?}{=} \sqrt[3]{2(-6)+1}$
$$\sqrt[3]{-11} \stackrel{?}{=} \sqrt[3]{-12+1}$$
$$\sqrt[3]{-11} = \sqrt[3]{-11} \checkmark$$

47. $\sqrt[3]{a-3} = \sqrt[3]{5a+1}$
$$\left(\sqrt[3]{a-3}\right)^3 = \left(\sqrt[3]{5a+1}\right)^3$$
$$a-3 = 5a+1$$
$$-4a = 4$$
$$a = -1$$

49. $\sqrt{x+10} = 1$
$$\left(\sqrt{x+10}\right)^2 = (1)^2$$
$$x+10 = 1$$
$$x = -9$$

Check: $\sqrt{-9+10} \stackrel{?}{=} 1$
$$\sqrt{1} \stackrel{?}{=} 1$$
$$1 = 1 \checkmark$$

51.
$$\sqrt{2x} = x - 4$$
$$\left(\sqrt{2x}\right)^2 = (x-4)^2$$
$$2x = x^2 - 8x + 16$$
$$x^2 - 10x + 16 = 0$$
$$(x-8)(x-2) = 0$$
$$x - 8 = 0 \quad \text{or} \quad x - 2 = 0$$
$$x = 8 \qquad\qquad x = 2$$

Check: $\sqrt{2(8)} \overset{?}{=} 8 - 4$
$$\sqrt{16} \overset{?}{=} 4$$
$$4 = 4 \ \checkmark$$

$$\sqrt{2(2)} \overset{?}{=} 2 - 4$$
$$\sqrt{4} \overset{?}{=} -2$$
$$2 \neq -2$$
The solution $x = 2$ does not check.

53.
$$\sqrt[3]{x+1} = 2$$
$$\left(\sqrt[3]{x+1}\right)^3 = (2)^3$$
$$x + 1 = 8$$
$$x = 7$$

Check: $\sqrt[3]{7+1} \overset{?}{=} 2$
$$\sqrt[3]{8} \overset{?}{=} 2$$
$$2 = 2 \ \checkmark$$

55. $v = 8\sqrt{x}$

(a) $v = 8\sqrt{100}$
$$v = 8(10)$$
$$v = 80$$
The velocity of an object that has fallen 100 ft is 80 ft/s.

(b) $136 = 8\sqrt{x}$
$$17 = \sqrt{x}$$
$$(17)^2 = \left(\sqrt{x}\right)^2$$
$$289 = x$$
An object with a velocity of 136 ft/s will fall 289 ft.

57. $y = 8\sqrt{t}; \ 0 \leq t \leq 40$

(a) $y = 8\sqrt{4}$
$$y = 8(2)$$
$$y = 16$$
The height of the plant after 4 weeks is 16 in.

(b) $40 = 8\sqrt{t}$
$$5 = \sqrt{t}$$
$$(5)^2 = \left(\sqrt{t}\right)^2$$
$$25 = t$$
It will take about 25 weeks for the plant to reach a height of 40 inches.

59.
$$\sqrt{5x-9} = \sqrt{5x} - 3$$
$$\left(\sqrt{5x-9}\right)^2 = (\sqrt{5x}-3)^2$$
$$5x - 9 = 5x + 9 - 6\sqrt{5x}$$
$$18 = -6\sqrt{5x}$$
$$-3 = \sqrt{5x}$$
$$9 = 5x \ \text{ or } x = \frac{9}{5}$$

Check: $\sqrt{5\left(\frac{9}{5}\right)-9} \overset{?}{=} \sqrt{5\left(\frac{9}{5}\right)} - 3$
$$\sqrt{9-9} \overset{?}{=} \sqrt{9} - 3$$
$$0 \overset{?}{=} 3 - 3$$
$$0 = 0 \ \checkmark$$

61.

$$\sqrt{2m+6} = 1 + \sqrt{7-2m}$$

$$\left(\sqrt{2m+6}\right)^2 = \left(1 + \sqrt{7-2m}\right)^2$$

$$2m+6 = 1 + 2\sqrt{7-2m} + 7 - 2m$$

$$4m = 2 + 2\sqrt{7-2m}$$

$$4m - 2 = 2\sqrt{7-2m}$$

$$\left(2m-1\right)^2 = \left(\sqrt{7-2m}\right)^2$$

$$4m^2 - 4m + 1 = 7 - 2m$$

$$4m^2 - 2m - 6 = 0$$

$$0 = (4m-6)(m+1)$$

$$m = -1 \ \text{ or } m = \frac{3}{2}$$

Check:

$$\sqrt{2(-1)+6} \overset{?}{=} 1 + \sqrt{7-2(-1)}$$

$$\sqrt{-2+6} \overset{?}{=} 1 + \sqrt{7-2(-1)}$$

$$\sqrt{4} \overset{?}{=} 1 + \sqrt{7+2}$$

$$2 \overset{?}{=} 1 + 3$$

$$2 \neq 4$$

The solution $m = -1$ does not check

$$\sqrt{2\left(\frac{3}{2}\right)+6} \overset{?}{=} 1 + \sqrt{7-2\left(\frac{3}{2}\right)}$$

$$\sqrt{9} \overset{?}{=} 1 + \sqrt{7-3}$$

$$3 \overset{?}{=} 1 + 2$$

$$3 = 3 ✓$$

Section 8.7 Practice Exercises

1. (a) $\sqrt[n]{a}$

(b) $\sqrt[n]{a^m}$ or $\left(\sqrt[n]{a}\right)^m$

3. $\left(\sqrt[4]{81}\right)^3 = 3^3 = 27$

5. $\sqrt[3]{(a+1)^3} = a+1$

7. $81^{1/2} = \sqrt{81} = 9$

9. $125^{1/3} = \sqrt[3]{125} = 5$

11. $81^{1/4} = \sqrt[4]{81} = 3$

13. $(-8)^{1/3} = \sqrt[3]{-8} = -2$

15. $-8^{1/3} = -\sqrt[3]{8} = -2$

17. $36^{-1/2} = \dfrac{1}{36^{1/2}} = \dfrac{1}{\sqrt{36}} = \dfrac{1}{6}$

19. $x^{1/3} = \sqrt[3]{x}$

21. $(4a)^{1/2} = \sqrt{4a} = 2\sqrt{a}$

23. $(yz)^{1/5} = \sqrt[5]{yz}$

25. $(u^2)^{1/3} = \sqrt[3]{u^2}$

27. $5q^{1/2} = 5\sqrt{q}$

29. $\left(\dfrac{x}{9}\right)^{1/2} = \sqrt{\dfrac{x}{9}} = \dfrac{\sqrt{x}}{\sqrt{9}} = \dfrac{\sqrt{x}}{3}$

31. $a^{m/n} = \sqrt[n]{a^m}$ or $\left(\sqrt[n]{a}\right)^m$ provided the root exists.

33. $16^{3/4} = \left(\sqrt[4]{16}\right)^3 = 2^3 = 8$

35. $27^{-2/3} = \dfrac{1}{27^{2/3}} = \dfrac{1}{\left(\sqrt[3]{27}\right)^2} = \dfrac{1}{3^2} = \dfrac{1}{9}$

37. $(-8)^{5/3} = \left(\sqrt[3]{-8}\right)^5 = (-2)^5 = -32$

39. $\left(\dfrac{1}{4}\right)^{-1/2} = 4^{1/2} = \sqrt{4} = 2$

41. $y^{9/2} = \left(\sqrt{y}\right)^9$

43. $(c^5 d)^{1/3} = \sqrt[3]{c^5 d}$

45. $(qr)^{-1/5} = \dfrac{1}{(qr)^{1/5}} = \dfrac{1}{\sqrt[5]{qr}}$

47. $6y^{2/3} = 6\left(\sqrt[3]{y}\right)^2$

49. $\sqrt[3]{x} = x^{1/3}$

51. $5\sqrt{x} = 5x^{1/2}$

53. $\sqrt[3]{y^2} = y^{2/3}$

55. $\sqrt[4]{m^3 n} = (m^3 n)^{1/4}$

57. $x^{1/4} x^{3/4} = x^{\frac{1}{4}+\frac{3}{4}} = x^{4/4} = x$

59. $(y^{1/5})^{10} = y^{\frac{1}{5}(10)} = y^{10/5} = y^2$

61. $6^{-1/5} 6^{6/5} = 6^{-\frac{1}{5}+\frac{6}{5}} = 6^{5/5} = 6$

63. $(a^{1/3} a^{1/4})^{12} = a^{12/3} a^{12/4} = a^4 a^3 = a^7$

65. $\dfrac{y^{5/3}}{y^{1/3}} = y^{\frac{5}{3}-\frac{1}{3}} = y^{4/3}$

67. $\dfrac{2^{4/3}}{2^{1/3}} = 2^{\frac{4}{3}-\frac{1}{3}} = 2^{3/3} = 2$

69. $\left(x^{-2} y^{1/3}\right)^{1/2} = x^{(-2)(1/2)} y^{(1/3)(1/2)}$

$\qquad = x^{-2/2} y^{1/6} = x^{-1} y^{1/6}$

$\qquad = \dfrac{y^{1/6}}{x}$

71. $\left(\dfrac{w^{-2}}{z^{-4}}\right)^{-3/2} = \dfrac{w^{(-2)(-3/2)}}{z^{(-4)(-3/2)}} = \dfrac{w^{6/2}}{z^{12/2}} = \dfrac{w^3}{z^6}$

73. $(5a^2 c^{-1/2} d^{1/2})^2 = 5^2 a^4 c^{-2/2} d^{2/2}$

$\qquad = 25a^4 c^{-1} d$

$\qquad = \dfrac{25a^4 d}{c}$

75. $\left(\dfrac{x^{-2/3}}{y^{-3/4}}\right)^{12} = \dfrac{x^{-24/3}}{y^{-36/4}} = \dfrac{x^{-8}}{y^{-9}} = \dfrac{y^9}{x^8}$

77. $\left(\dfrac{16w^{-2}z}{2wz^{-8}}\right)^{1/3} = (8w^{-3}z^9)^{1/3}$

$\qquad = 8^{1/3} w^{-3/3} z^{9/3}$

$\qquad = \sqrt[3]{8} w^{-1} z^3$

$\qquad = \dfrac{2z^3}{w}$

79. $(25x^2 y^4 z^3)^{1/2} = 25^{1/2} x^{2/2} y^{4/2} z^{3/2}$

$\qquad = \sqrt{25} x^1 y^2 z^{3/2}$

$\qquad = 5xy^2 z^{3/2}$

81. $s = A^{1/2}$

(a) $s = 100^{1/2}$

$s = 10$

The length of the sides of a square with an area of 100 in.^{2} is 10 in.

(b) $s = 72^{1/2}$

$s \approx 8.49$

The length of the sides of a square with an area of 72 in.^{2} is about 8.49 in.

83. $r = \left(\dfrac{A}{P}\right)^{1/t} - 1$

(a) $r = \left(\dfrac{16{,}802}{10{,}000}\right)^{1/5} - 1$

$r \approx 0.109$

The rate needed to grow \$10,000 to \$16,802 in 5 years is 10.9%.

(b) $r = \left(\dfrac{18{,}000}{10{,}000}\right)^{1/7} - 1$

$r \approx 0.088$

The rate needed to grow \$10,000 to \$18,000 in 7 years is 8.8%.

(c) The account in part a

85. They are *not* the same. For example:

$(64 + 36)^{1/2} = 100^{1/2} = 10$

but

$64^{1/2} + 36^{1/2} = 8 + 6 = 14$

As you can see, the resulting answers are not equal.

87. $\left(\dfrac{1}{8}\right)^{-2/3} + \left(\dfrac{1}{4}\right)^{-1/2} = 8^{2/3} + 4^{1/2}$

$$= \left(\sqrt[3]{8}\right)^2 + \sqrt{4}$$
$$= 2^2 + 2$$
$$= 4 + 2$$
$$= 6$$

89. $\left(\dfrac{1}{16}\right)^{1/4} - \left(\dfrac{1}{49}\right)^{1/2} = \sqrt[4]{\dfrac{1}{16}} - \sqrt{\dfrac{1}{49}}$

$$= \dfrac{1}{2} - \dfrac{1}{7}$$
$$= \dfrac{7}{14} - \dfrac{2}{14}$$
$$= \dfrac{5}{14}$$

91. $\left(\dfrac{a^2 b^{1/2} c^{-2}}{a^{-3/4} b^0 c^{1/8}}\right)^8 = \dfrac{a^{16} b^{8/2} c^{-16}}{a^{-24/4} b^0 c^{8/8}}$

$$= \dfrac{a^{16} b^4 c^{-16}}{a^{-6}(1) c^1}$$
$$= a^{16-(-6)} b^4 c^{-16-1}$$
$$= a^{22} b^4 c^{-17}$$
$$= \dfrac{a^{22} b^4}{c^{17}}$$

Group Activity

1. The mean or average can be found by adding up all of the values and then dividing by the total number of values.

$$\dfrac{6+10+13+22+30+4+12+23}{8} = \dfrac{120}{8} = 15$$

The data points have a mean (or average) value of 15.

3. $\sqrt{\dfrac{81+25+4+49+225+121+9+64}{8-1}}$

$$= \sqrt{\dfrac{578}{7}} \approx 9$$

The standard deviation is approximately 9.

5. Answers will vary.

Chapter 8 Review Exercises

Section 8.1

1. Principal square root: 14
Negative square root: −14

3. Principal square root: 0.8
Negative square root: −0.8

5. There is no real number b such that $b^2 = -64$.

7. $-\sqrt{144} = -12$

9. $\sqrt{-144}$ is not a real number.

11. $\sqrt{y^2} = y$

13. $\sqrt[3]{8p^3} = \sqrt[3]{2^3 p^3} = 2p$

15. $-\sqrt[4]{625} = -5$

17. $\sqrt[3]{\dfrac{64}{t^6}} = \dfrac{\sqrt[3]{64}}{\sqrt[3]{t^6}} = \dfrac{\sqrt[3]{4^3}}{\sqrt[3]{t^6}} = \dfrac{4}{t^2}$

19. $r = \sqrt{\dfrac{A}{\pi}}$

(a) $r = \sqrt{\dfrac{160}{\pi}}$
$r \approx 7.1$
The radius of the garden is about 7.1 m.

(b) $r = \sqrt{\dfrac{1600}{\pi}}$
$r \approx 22.6$
The radius of the fountain is about 22.6 ft.

21. $b^2 + \sqrt{5}$

23. The quotient of 2 and the principal square root of p

25. Let x represent the distance up the house the ladder is placed.
$$5^2 + x^2 = 13^2$$
$$25 + x^2 = 169$$
$$x^2 = 144$$
$$x = 12$$
The ladder is placed 12 ft up the side of the house.

Section 8.2

27. $\sqrt{x^{17}} = \sqrt{x^{16} \cdot x} = x^8 \sqrt{x}$

29. $\sqrt{28} = \sqrt{4 \cdot 7} = 2\sqrt{7}$

31. $\sqrt[3]{27y^{10}} = \sqrt[3]{27y^9 \cdot y} = 3y^3 \sqrt[3]{y}$

33. $\sqrt{\dfrac{c^5}{c^3}} = \sqrt{c^2} = c$

35. $\sqrt{\dfrac{200y^5}{2y}} = \sqrt{100y^4} = 10y^2$

37. $\sqrt[3]{\dfrac{48x^4}{6x}} = \sqrt[3]{8x^3} = \sqrt[3]{2^3 x^3} = 2x$

39. $\dfrac{5\sqrt{12}}{2} = \dfrac{5\sqrt{2^2 \cdot 3}}{2} = \dfrac{5 \cdot 2\sqrt{3}}{2} = 5\sqrt{3}$

41. $\dfrac{12 - \sqrt{49}}{5} = \dfrac{12 - \sqrt{7^2}}{5} = \dfrac{12 - 7}{5} = \dfrac{5}{5} = 1$

Section 8.3

43. $8\sqrt{6} - \sqrt{6} = 7\sqrt{6}$

45. $x\sqrt{20} - 2\sqrt{45x^2} = x\sqrt{4 \cdot 5} - 2\sqrt{9x^2 \cdot 5}$
$$= 2x\sqrt{5} - 2 \cdot 3x\sqrt{5}$$
$$= 2x\sqrt{5} - 6x\sqrt{5}$$
$$= -4x\sqrt{5}$$

47. $3\sqrt{75} - 4\sqrt{28} + \sqrt{7}$
$$= 3\sqrt{5^2 \cdot 3} - 4\sqrt{2^2 \cdot 7} + \sqrt{7}$$
$$= 3 \cdot 5\sqrt{3} - 4 \cdot 2\sqrt{7} + \sqrt{7}$$
$$= 15\sqrt{3} - 8\sqrt{7} + \sqrt{7}$$
$$= 15\sqrt{3} - 7\sqrt{7}$$

49. $7\sqrt{3x^9} - 3x^4\sqrt{75x} = 7\sqrt{3x^8 \cdot x} - 3x^4\sqrt{25 \cdot 3x}$
$$= 7\sqrt{3x^8 \cdot x} - 3x^4\sqrt{5^2 \cdot 3x}$$
$$= 7x^4\sqrt{3x} - 3 \cdot 5x^4\sqrt{3x}$$
$$= 7x^4\sqrt{3x} - 15x^4\sqrt{3x}$$
$$= (7 - 15)x^4\sqrt{3x}$$
$$= -8x^4\sqrt{3x}$$

51. $\sqrt{2} + \sqrt{98} + \sqrt{32} = \sqrt{2} + \sqrt{49 \cdot 2} + \sqrt{16 \cdot 2}$
$$= \sqrt{2} + 7\sqrt{2} + 4\sqrt{2}$$
$$= 12\sqrt{2} \text{ ft}$$

Section 8.4

53. $\sqrt{5} \cdot \sqrt{125} = \sqrt{5 \cdot 125} = \sqrt{625} = 25$

55. $\left(5\sqrt{6}\right)\left(7\sqrt{2x}\right) = (5 \cdot 7)\left(\sqrt{6} \cdot \sqrt{2x}\right)$
$$= 35\sqrt{6 \cdot 2x}$$
$$= 35\sqrt{12x}$$
$$= 35\sqrt{4 \cdot 3x}$$
$$= 35 \cdot 2\sqrt{3x}$$
$$= 70\sqrt{3x}$$

57. $8\sqrt{m}\left(\sqrt{m} + 3\right) = 8\sqrt{m} \cdot \sqrt{m} + 8\sqrt{m} \cdot 3$
$$= 8\sqrt{m^2} + 24\sqrt{m}$$
$$= 8m + 24\sqrt{m}$$

59. $\left(5\sqrt{2}+\sqrt{13}\right)\left(-\sqrt{2}-3\sqrt{13}\right)=-5\sqrt{2}\cdot\sqrt{2}-5\sqrt{2}\cdot3\sqrt{13}-\sqrt{13}\cdot\sqrt{2}-\sqrt{13}\cdot3\sqrt{13}$

$$=-5\sqrt{2\cdot2}-(5\cdot3)\sqrt{2\cdot13}-\sqrt{13\cdot2}-3\sqrt{13\cdot13}$$
$$=-5\sqrt{4}-15\sqrt{26}-\sqrt{26}-3\sqrt{169}$$
$$=-5\cdot2-15\sqrt{26}-\sqrt{26}-3\cdot13$$
$$=-10-16\sqrt{26}-39$$
$$=-49-16\sqrt{26}$$

61. $\left(8\sqrt{w}-\sqrt{z}\right)\left(8\sqrt{w}+\sqrt{z}\right)$

$$=\left(8\sqrt{w}\right)^2-\left(\sqrt{z}\right)^2$$
$$=64w-z$$

63. $V=\sqrt{10}\cdot\sqrt{5}\cdot\sqrt{6}$

$$V=\sqrt{300}$$
$$V=\sqrt{100\cdot3}$$
$$V=10\sqrt{3}\ \text{m}^3$$

Section 8.5

65. $\dfrac{\sqrt{a^{11}}}{\sqrt{a}}=\sqrt{\dfrac{a^{11}}{a}}=\sqrt{a^{10}}=a^5$

67. $\dfrac{\sqrt{96y^3}}{\sqrt{6y^2}}=\sqrt{\dfrac{96y^3}{6y^2}}=\sqrt{16y}=\sqrt{4^2\,y}=4\sqrt{y}$

69. Multiply both the numerator and denominator by the quantity in b

$$\frac{w}{\sqrt{w}-4}=\frac{w}{\sqrt{w}-4}\cdot\frac{\sqrt{w}+4}{\sqrt{w}+4}$$
$$=\frac{w\sqrt{w}+4w}{w-16}$$

71. $\sqrt{\dfrac{18}{y}}=\dfrac{\sqrt{18}}{\sqrt{y}}$

$$=\frac{\sqrt{18}}{\sqrt{y}}\cdot\frac{\sqrt{y}}{\sqrt{y}}$$
$$=\frac{\sqrt{18y}}{\sqrt{y^2}}$$
$$=\frac{\sqrt{9\cdot2y}}{y}$$
$$=\frac{3\sqrt{2y}}{y}$$

73. $\dfrac{10}{\sqrt{7}-\sqrt{2}}=\dfrac{10}{\sqrt{7}-\sqrt{2}}\cdot\dfrac{\sqrt{7}+\sqrt{2}}{\sqrt{7}+\sqrt{2}}$

$$=\frac{10\left(\sqrt{7}+\sqrt{2}\right)}{\left(\sqrt{7}\right)^2-\left(\sqrt{2}\right)^2}$$
$$=\frac{10\left(\sqrt{7}+\sqrt{2}\right)}{7-2}$$
$$=\frac{10\left(\sqrt{7}+\sqrt{2}\right)}{5}$$
$$=2\left(\sqrt{7}+\sqrt{2}\right)$$
$$=2\sqrt{7}+2\sqrt{2}$$

75. $\dfrac{\sqrt{7}+3}{\sqrt{7}-3}=\dfrac{\sqrt{7}+3}{\sqrt{7}-3}\cdot\dfrac{\sqrt{7}+3}{\sqrt{7}+3}$

$=\dfrac{\left(\sqrt{7}+3\right)^2}{\left(\sqrt{7}\right)^2-3^2}$

$=\dfrac{\left(\sqrt{7}\right)^2+2\left(\sqrt{7}\right)(3)+3^2}{7-9}$

$=\dfrac{7+6\sqrt{7}+9}{-2}$

$=\dfrac{16+6\sqrt{7}}{-2}$

$=\dfrac{2\left(8+3\sqrt{7}\right)}{-2}$

$=-8-3\sqrt{7}$

Section 8.6

77. $\sqrt{p+6}=12$

$\left(\sqrt{p+6}\right)^2=12^2$

$p+6=144$

$p=138$

Check: $\sqrt{138+6}\overset{?}{=}12$

$\sqrt{144}\overset{?}{=}12$

$12=12\checkmark$

79. $\sqrt{3x-17}-10=0$

$\sqrt{3x-17}=10$

$\left(\sqrt{3x-17}\right)^2=10^2$

$3x-17=100$

$3x=117$

$x=39$

Check: $\sqrt{3(39)-17}-10\overset{?}{=}0$

$\sqrt{117-17}-10\overset{?}{=}0$

$\sqrt{100}-10\overset{?}{=}0$

$10-10\overset{?}{=}0$

$0=0\checkmark$

81. $\sqrt{2z+2}=\sqrt{3z-5}$

$\left(\sqrt{2z+2}\right)^2=\left(\sqrt{3z-5}\right)^2$

$2z+2=3z-5$

$-z+2=-5$

$-z=-7$

$z=7$

Check: $\sqrt{2(7)+2}\overset{?}{=}\sqrt{3(7)-5}$

$\sqrt{14+2}\overset{?}{=}\sqrt{21-5}$

$\sqrt{16}\overset{?}{=}\sqrt{16}$

$4=4\checkmark$

83. $\sqrt{2m+5}=m+1$

$\left(\sqrt{2m+5}\right)^2=(m+1)^2$

$2m+5=m^2+2m+1$

$0=m^2-4$

$0=(m-2)(m+2)$

$m-2=0$ or $m+2=0$

$m=2\qquad\qquad m=-2$

Check: $\sqrt{2(2)+5}\overset{?}{=}2+1$

$\sqrt{4+5}\overset{?}{=}3$

$\sqrt{9}\overset{?}{=}3$

$3=3\checkmark$

$\sqrt{2(-2)+5}\overset{?}{=}-2+1$

$\sqrt{-4+5}\overset{?}{=}-1$

$\sqrt{1}\neq-1$

The principal square root of 1 is not -1,

therefore $m = -2$ is not a solution of the equation.

85.
$$\sqrt[3]{2y+13} = -5$$
$$\left(\sqrt[3]{2y+13}\right)^3 = (-5)^3$$
$$2y+13 = -125$$
$$2y = -138$$
$$y = -69$$

Check: $\sqrt[3]{2(-69)+13} \overset{?}{=} -5$

$$\sqrt[3]{-138+13} \overset{?}{=} -5$$

$$\sqrt[3]{-125} \overset{?}{=} -5$$

$$-5 = -5 \checkmark$$

Section 8.7

87. $(-27)^{1/3} = \sqrt[3]{-27} = -3$

89. $-16^{1/4} = -\sqrt[4]{16} = -2$

91. $4^{-3/2} = \dfrac{1}{4^{3/2}} = \dfrac{1}{\left(\sqrt{4}\right)^3} = \dfrac{1}{2^3} = \dfrac{1}{8}$

93. $z^{1/5} = \sqrt[5]{z}$

95. $(w^3)^{1/4} = \sqrt[4]{w^3}$

97. $\sqrt[5]{a^2} = a^{2/5}$

99. $\sqrt[5]{a^2b^4} = (a^2b^4)^{1/5}$

101. $y^{2/3}y^{4/3} = y^{\frac{2}{3}+\frac{4}{3}} = y^{6/3} = y^2$

103. $\dfrac{6^{4/5}}{6^{1/5}} = 6^{\frac{4}{5}-\frac{1}{5}} = 6^{3/5}$

105. $(64a^3b^6)^{1/3} = 64^{1/3}a^{3/3}b^{6/3} = 4ab^2$

107. $r = \left(\dfrac{V}{\pi h}\right)^{1/2}$

$r = \left(\dfrac{150.8}{12\pi}\right)^{1/2}$

$r \approx 2.0$

The radius is about 2.0 cm.

Chapter 8 Test

1. 1. The radicand has no factor raised to a power greater than or equal to the index.

   3. The radicand does not contain a fraction.

3. $\sqrt[3]{48y^4} = \sqrt[3]{8y^3 \cdot 6y} = 2y\sqrt[3]{6y}$

5. $\sqrt{\dfrac{5a^6}{81}} = \dfrac{\sqrt{5a^6}}{\sqrt{81}} = \dfrac{a^3\sqrt{5}}{9}$

7. $\dfrac{2}{\sqrt{5}+6} = \dfrac{2}{\sqrt{5}+6} \cdot \dfrac{\sqrt{5}-6}{\sqrt{5}-6}$

$= \dfrac{2\left(\sqrt{5}-6\right)}{\left(\sqrt{5}\right)^2 - 6^2}$

$= \dfrac{2\sqrt{5}-12}{5-36}$

$= \dfrac{2\sqrt{5}-12}{-31}$ or $\dfrac{12-2\sqrt{5}}{31}$

9. $x = 56t\sqrt{3}$

(a) $x = 56(1)\sqrt{3}$

$x \approx 97$

Its horizontal distance is about 97 ft.

(b) $x = 56(3.5)\sqrt{3}$

$x \approx 339$

The horizontal distance is about 339 ft.

11. $\sqrt{3}\left(4\sqrt{2}-5\sqrt{3}\right)=\sqrt{3}\cdot4\sqrt{2}-\sqrt{3}\cdot5\sqrt{3}$

$\qquad\qquad\qquad\quad=4\sqrt{2\cdot3}-5\sqrt{3\cdot3}$

$\qquad\qquad\qquad\quad=4\sqrt{6}-5\sqrt{3^2}$

$\qquad\qquad\qquad\quad=4\sqrt{6}-5\cdot3$

$\qquad\qquad\qquad\quad=4\sqrt{6}-15$

13. $\sqrt{360}+\sqrt{250}-\sqrt{40}$

$=\sqrt{36\cdot10}+\sqrt{25\cdot10}-\sqrt{4\cdot10}$

$=6\sqrt{10}+5\sqrt{10}-2\sqrt{10}$

$=9\sqrt{10}$

19. $\dfrac{6}{\sqrt{7}-\sqrt{3}}\cdot\dfrac{\sqrt{7}+\sqrt{3}}{\sqrt{7}+\sqrt{3}}=\dfrac{6\left(\sqrt{7}+\sqrt{3}\right)}{7-3}$

$\qquad\qquad\qquad\qquad\quad=\dfrac{6\left(\sqrt{7}+\sqrt{3}\right)}{4}$

$\qquad\qquad\qquad\qquad\quad=\dfrac{3\left(\sqrt{7}+\sqrt{3}\right)}{2}$

$\qquad\qquad\qquad\qquad\quad=\dfrac{3\sqrt{7}+3\sqrt{3}}{2}$

21. $\sqrt{2x+7}+6=2$

$\qquad\sqrt{2x+7}=-4$

The principal square root of a value is never negative, therefore this equation has no solution.

23. $\sqrt[3]{x+6}=\sqrt[3]{2x-8}$

$\left(\sqrt[3]{x+6}\right)^3=\left(\sqrt[3]{2x-8}\right)^3$

$\qquad x+6=2x-8$

$\qquad -x+6=-8$

$\qquad\quad -x=-14$

$\qquad\qquad x=14$

Check: $\sqrt[3]{14+6}\overset{?}{=}\sqrt[3]{2(14)-8}$

$\qquad\qquad\sqrt[3]{20}=\sqrt[3]{20}\ \checkmark$

25. $10{,}000^{3/4}=\left(\sqrt[4]{10{,}000}\right)^3=10^3=1000$

15. $\left(3\sqrt{5}-1\right)^2=\left(3\sqrt{5}\right)^2-2\left(3\sqrt{5}\right)+(1)^2$

$\qquad\qquad\qquad\ =9\cdot5-6\sqrt{5}+1$

$\qquad\qquad\qquad\ =45-6\sqrt{5}+1$

$\qquad\qquad\qquad\ =46-6\sqrt{5}$

17. $\left(4-3\sqrt{x}\right)\left(4+3\sqrt{x}\right)=(4)^2-\left(3\sqrt{x}\right)^2$

$\qquad\qquad\qquad\qquad\qquad=16-9x$

27. $x^{3/5}=\sqrt[5]{x^3}\ \text{or}\ \left(\sqrt[5]{x}\right)^3$

29. $\sqrt[4]{ab^3}=(ab^3)^{1/4}$

31. $\dfrac{5^{4/5}}{5^{1/5}}=5^{\frac{4}{5}-\frac{1}{5}}=5^{3/5}$

Cumulative Review Exercises, Chapters 1–8

1. $\dfrac{\left|-3-12\div6+2\right|}{\sqrt{5^2-4^2}}=\dfrac{\left|-3-2+2\right|}{\sqrt{25-16}}$

$\qquad\qquad\qquad\quad=\dfrac{\left|-3\right|}{\sqrt{9}}$

$\qquad\qquad\qquad\quad=\dfrac{3}{3}$

$\qquad\qquad\qquad\quad=1$

3. $\left(\dfrac{1}{3}\right)^0-\left(\dfrac{1}{4}\right)^{-2}=1-4^2=1-16=-15$

5. $\dfrac{14x^3y-7x^2y^2+28xy^2}{7x^2y^2}$

$=\dfrac{14x^3y}{7x^2y^2}-\dfrac{7x^2y^2}{7x^2y^2}+\dfrac{28xy^2}{7x^2y^2}$

$=\dfrac{2x}{y}-1+\dfrac{4}{x}$

7.
$$10x^2 = x + 2$$
$$10x^2 - x - 2 = 0$$
$$10x^2 + 4x - 5x - 2 = 0$$
$$2x(5x + 2) - 1(5x + 2) = 0$$
$$(5x + 2)(2x - 1) = 0$$

$5x + 2 = 0 \quad$ or $\quad 2x - 1 = 0$

$\qquad 5x = -2 \qquad\qquad 2x = 1$

$\qquad x = -\dfrac{2}{5} \qquad\qquad x = \dfrac{1}{2}$

9.
$$\frac{1}{5} + \frac{z}{z-5} = \frac{5}{z-5}$$
$$5(z-5)\left(\frac{1}{5}\right) + 5(z-5)\left(\frac{z}{z-5}\right) = 5(z-5)\left(\frac{5}{z-5}\right)$$
$$(z-5) + 5z = 5(5)$$
$$z - 5 + 5z = 25$$
$$6z - 5 = 25$$
$$6z = 30$$
$$z = 5$$

$z = 5$ is not in the domain, so there is no solution.

11. $3y = 6$

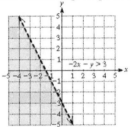

13. First, use the slope formula to determine the slope of the line passing through the points.
$$m = \frac{4 - (-1)}{-3 - 2} = \frac{5}{-5} = -1$$
Second, use the value found for the slope, one of the given points, and the point-

slope formula to determine the equation of the line.
$$y - (-1) = -1(x - 2)$$
$$y + 1 = -x + 2$$
$$y = -x + 1$$

15. $-2x - y > 3$
First, graph the associated equation with a dashed line. Using $(0, 0)$ as a test point results in a false statement. Using $(-4, 0)$ as a test point results in a true statement. Shade the region on the side of the dashed line containing the point $(-4, 0)$.

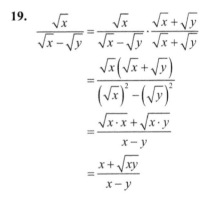

17. $\sqrt{99} = \sqrt{9 \cdot 11} = 3\sqrt{11}$

19.
$$\frac{\sqrt{x}}{\sqrt{x} - \sqrt{y}} = \frac{\sqrt{x}}{\sqrt{x} - \sqrt{y}} \cdot \frac{\sqrt{x} + \sqrt{y}}{\sqrt{x} + \sqrt{y}}$$
$$= \frac{\sqrt{x}\left(\sqrt{x} + \sqrt{y}\right)}{\left(\sqrt{x}\right)^2 - \left(\sqrt{y}\right)^2}$$
$$= \frac{\sqrt{x \cdot x} + \sqrt{x \cdot y}}{x - y}$$
$$= \frac{x + \sqrt{xy}}{x - y}$$

Chapter 9

Chapter 9 opener

$3 - O;\ 2 - W;\ 1 - T$

A quadratic equation has at most $\dfrac{T}{1}\dfrac{W}{2}\dfrac{0}{3}$ solution(s).

Section 9.1 Practice Exercises

1. (a) $0, 0$

(b) 0

(c) $\sqrt{k}$, $-\sqrt{k}$

(d) $4, -3, 3$

3. (a) Linear

(b) Quadratic

(c) Linear

5. $(t+5)(2t-1)=0$
$$t+5=0 \quad \text{or} \quad 2t-1=0$$
$$t=-5 \quad \text{or} \quad t=\frac{1}{2}$$

7. $y^2-2y-35=0$
$$(y-7)(y+5)=0$$
$$y-7=0 \quad \text{or} \quad y+5=0$$
$$y=7 \quad \text{or} \quad y=-5$$

9. $\qquad 6p^2=-13p-2$
$$6p^2+13p+2=0$$
$$(6p+1)(p+2)=0$$
$$6p+1=0 \quad \text{or} \quad p+2=0$$
$$p=-\frac{1}{6} \quad \text{or} \quad p=-2$$

11. $\qquad 2x^2+10x=-7(x+3)$
$$2x^2+10x=-7x-21$$
$$2x^2+17x+21=0$$
$$(2x+3)(x+7)=0$$
$$2x+3=0 \quad \text{or} \quad x+7=0$$
$$x=-\frac{3}{2} \quad \text{or} \quad x=-7$$

13. $\qquad c^2=144$
$$c^2-144=0$$
$$(c-12)(c+12)=0$$
$$c-12=0 \quad \text{or} \quad c+12=0$$
$$c=12 \quad \text{or} \quad c=-12$$

15. $\qquad (x-3)^2=25$
$$(x-3)^2-25=0$$
$$(x-3-5)(x-3+5)=0$$
$$(x-8)(x+2)=0$$
$$x-8=0 \quad \text{or} \quad x+2=0$$
$$x=8 \quad \text{or} \quad x=-2$$

17. $\qquad 4a^2+7a=2$
$$4a^2+7a-2=0$$
$$(4a-1)(a+2)=0$$
$$4a-1=0 \quad \text{or} \quad a+2=0$$
$$a=\frac{1}{4} \quad \text{or} \quad a=-2$$

19. $(x+2)(x+6)=5$
$$x^2+8x+12=5$$
$$x^2+8x+7=0$$
$$(x+7)(x+1)=0$$
$$x+7=0 \quad \text{or} \quad x+1=0$$
$$x=-7 \quad \text{or} \quad x=-1$$

21. $x^2=49$
$$x=\pm\sqrt{49}$$
$$x=\pm 7$$

23. $k^2 - 100 = 0$

$$k^2 = 100$$

$$k = \pm\sqrt{100}$$

$$k = \pm 10$$

25. $p^2 = -24$

$$p = \pm\sqrt{-24}$$

There are no real-valued solutions.

27. $3w^2 - 9 = 0$

$$3w^2 = 9$$

$$w^2 = 3$$

$$w = \pm\sqrt{3}$$

29. $(a-5)^2 = 16$

$$a - 5 = \pm\sqrt{16}$$

$$a - 5 = \pm 4$$

$$a = 5 \pm 4$$

$$a = 5 + 4 \quad \text{or} \quad a = 5 - 4$$

$$a = 9 \qquad\qquad a = 1$$

31. $(y-5)^2 = 36$

$$y - 5 = \pm\sqrt{36}$$

$$y - 5 = \pm 6$$

$$y = 5 \pm 6$$

$$y = 5 + 6 \quad \text{or} \quad y = 5 - 6$$

$$y = 11 \qquad\qquad y = -1$$

33. $(x-11)^2 = 5$

$$x - 11 = \pm\sqrt{5}$$

$$x = 11 \pm \sqrt{5}$$

35. $(a+1)^2 = 18$

$$a + 1 = \pm\sqrt{18}$$

$$a + 1 = \pm\sqrt{9 \cdot 2}$$

$$a + 1 = \pm 3\sqrt{2}$$

$$a = -1 \pm 3\sqrt{2}$$

37. $\left(t - \dfrac{1}{4}\right)^2 = \dfrac{7}{16}$

$$t - \frac{1}{4} = \pm\sqrt{\frac{7}{16}}$$

$$t - \frac{1}{4} = \pm\frac{\sqrt{7}}{4}$$

$$t = \frac{1}{4} \pm \frac{\sqrt{7}}{4}$$

39. $\left(x - \dfrac{1}{2}\right)^2 + 5 = 20$

$$\left(x - \frac{1}{2}\right)^2 = 15$$

$$x - \frac{1}{2} = \pm\sqrt{15}$$

$$x = \frac{1}{2} \pm \sqrt{15}$$

41. $(p-3)^2 = -16$

$$p - 3 = \pm\sqrt{-16}$$

$$p = 3 \pm \sqrt{-16}$$

There are no real-valued solutions.

43. $12t^2 - 20 = 55$

$$12t^2 = 75$$

$$t^2 = \frac{75}{12}$$

$$t^2 = \frac{25}{4}$$

$$t = \pm\sqrt{\frac{25}{4}}$$

$$t = \pm\frac{5}{2}$$

45. $(x+3)^2 = 5$

For $x = -3 + \sqrt{5}$:

$$\left(-3 + \sqrt{5} + 3\right)^2 \overset{?}{=} 5$$

$$\left(\sqrt{5}\right)^2 \overset{?}{=} 5$$

$$5 = 5$$

The solution checks.

47. False; -8 is also a solution.

49. $d = 16t^2$

    (a) $d = 16(2)^2$
$$d = 16(4)$$
$$d = 64 \qquad\qquad 64 \text{ ft}$$

    (b) $200 = 16t^2$
$$12.5 = t^2$$
$$\pm 3.5 \approx t$$
Since the answer to this problem is a measurement, only the positive value is appropriate.   3.5 sec

    (c) $1250 = 16t^2$
$$78.125 = t^2$$
$$\pm 8.8 \approx t$$
Since the answer to this problem is a measurement, only the positive value is appropriate.   8.8 sec

51. Use the Pythagorean Theorem.
$$x^2 + x^2 = 10^2$$
$$2x^2 = 100$$
$$x^2 = 50$$
$$x \approx \pm 7.1$$
Since the answer to this problem is a measurement, only the positive answer is appropriate.   7.1 m

53. Use the area of the circle.
$$A = \pi r^2$$
$$200 = \pi r^2$$
$$r^2 = \frac{200}{\pi}$$
$$r = \pm\sqrt{\frac{200}{\pi}}$$
$$r \approx \pm 7.9788$$
$$r \approx \pm 8.0$$
Since the answer to this problem is a measurement, only the positive answer is appropriate.   8.0 ft.

Section 9.2 Practice Exercises

1. (a) completing

    (b) 100

    (c) 5;1

    (d) 8

3. $(x-5)^2 = 21$
$$x - 5 = \pm\sqrt{21}$$
$$x = 5 \pm \sqrt{21}$$

5. $n = \frac{1}{2}$ of 4, squared. $\left[\frac{1}{2}(4)\right]^2 = 4$.
$$y^2 + 4y$$
$$y^2 + 4y + 4 = (y+2)^2$$

7. $n = \frac{1}{2}$ of -12, squared. $\left[\frac{1}{2}(-12)\right]^2 = 36$.
$$p^2 - 12p$$
$$p^2 - 12p + 36 = (p-6)^2$$

9. $n = \frac{1}{2}$ of -9, squared. $\left[\frac{1}{2}(-9)\right]^2 = \frac{81}{4}$.
$$x^2 - 9x$$
$$x^2 - 9x + \frac{81}{4} = \left(x - \frac{9}{2}\right)^2$$

11. $n = \frac{1}{2}$ of $\frac{5}{3}$, squared. $\left[\frac{1}{2}\left(\frac{5}{3}\right)\right]^2 = \frac{25}{36}$.
$$d^2 + \frac{5}{3}d$$
$$d^2 + \frac{5}{3}d + \frac{25}{36} = \left(d + \frac{5}{6}\right)^2$$

13. $n = \frac{1}{2}$ of $-\frac{1}{5}$, squared. $\left[\frac{1}{2}\left(-\frac{1}{5}\right)\right]^2 = \frac{1}{100}$.
$$m^2 - \frac{1}{5}m$$
$$m^2 - \frac{1}{5}m + \frac{1}{100} = \left(m - \frac{1}{10}\right)^2$$

15. $n = $ ½ of 1, squared. $\left[\dfrac{1}{2}(1)\right]^2 = \dfrac{1}{4}$.

$u^2 + u$

$u^2 + u + \dfrac{1}{4} = \left(u + \dfrac{1}{2}\right)^2$

17.
$$x^2 + 4x = 12$$
$$x^2 + 4x + 4 = 12 + 4$$
$$(x + 2)^2 = 16$$
$$x + 2 = \pm\sqrt{16}$$
$$x = 2 \pm 4$$
$$x = 2 + 4 = 6$$
$$\text{or } x = 2 - 4 = -2$$

19.
$$y^2 + 6y = -5$$
$$y^2 + 6y + 9 = -5 + 9$$
$$(y + 3)^2 = 4$$
$$y + 3 = \pm\sqrt{4}$$
$$y = -3 \pm 2$$
$$y = -3 + 2 = -1$$
$$\text{or } y = -3 - 2 = -5$$

21.
$$x^2 = 2x + 1$$
$$x^2 - 2x = 1$$
$$x^2 - 2x + 1 = 1 + 1$$
$$(x - 1)^2 = 2$$
$$x - 1 = \pm\sqrt{2}$$
$$x = 1 \pm \sqrt{2}$$

23.
$$3x^2 - 6x - 15 = 0$$
$$\dfrac{3x^2}{3} - \dfrac{6x}{3} - \dfrac{15}{3} = 0$$
$$x^2 - 2x = 5$$
$$x^2 - 2x + 1 = 5 + 1$$
$$(x - 1)^2 = 6$$
$$x - 1 = \pm\sqrt{6}$$
$$x = 1 \pm \sqrt{6}$$

25.
$$4p^2 + 16p = -4$$
$$\dfrac{4p^2}{4} + \dfrac{16x}{4} = -\dfrac{4}{4}$$
$$p^2 + 4p = -1$$
$$p^2 + 4p + 4 = -1 + 4$$
$$(p + 2)^2 = 3$$
$$p + 2 = \pm\sqrt{3}$$
$$p = -2 \pm \sqrt{3}$$

27.
$$w^2 + w - 3 = 0$$
$$w^2 + w = 3$$
$$w^2 + w + \dfrac{1}{4} = 3 + \dfrac{1}{4}$$
$$\left(w + \dfrac{1}{2}\right)^2 = \dfrac{13}{4}$$
$$w + \dfrac{1}{2} = \pm\sqrt{\dfrac{13}{4}}$$
$$w = -\dfrac{1}{2} \pm \dfrac{\sqrt{13}}{2}$$

29.
$$x(x + 2) = 40$$
$$x^2 + 2x = 40$$
$$x^2 + 2x + 1 = 40 + 1$$
$$(x + 1)^2 = 41$$
$$x + 1 = \pm\sqrt{41}$$
$$x = -1 \pm \sqrt{41}$$

31.
$$a^2 - 4a - 1 = 0$$
$$a^2 - 4a = 1$$
$$a^2 - 4a + 4 = 1 + 4$$
$$(a - 2)^2 = 5$$
$$a - 2 = \pm\sqrt{5}$$
$$a = 2 \pm \sqrt{5}$$

33. $2r^2 + 12r + 16 = 0$

$r^2 + 6r + 8 = 0$

$r^2 + 6r + 9 = -8 + 9$

$(r + 3)^2 = 1$

$r + 3 = \pm\sqrt{1}$

$r = -3 \pm 1$

$r = -3 + 1 = -2$

or $r = -3 - 1 = -4$

35. $h(h - 11) = -24$

$h^2 - 11h = -24$

$h^2 - 11h + \dfrac{121}{4} = -24 + \dfrac{121}{4}$

$\left(h - \dfrac{11}{2}\right)^2 = \dfrac{25}{4}$

$h - \dfrac{11}{2} = \pm\sqrt{\dfrac{25}{4}}$

$h = \dfrac{11}{2} \pm \dfrac{5}{2}$

$h = \dfrac{11}{2} + \dfrac{5}{2} = 8$

or $h = \dfrac{11}{2} - \dfrac{5}{2} = 3$

37. $y^2 = 121$

$y = \pm\sqrt{121}$

$y = \pm 11$

39. $(p + 2)^2 = 2$

$p + 2 = \pm\sqrt{2}$

$p = -2 \pm\sqrt{2}$

41. $(k + 13)(k - 5) = 0$

$k + 13 = 0$ or $k - 5 = 0$

$k = -13$ $k = 5$

43. $(x - 13)^2 = 0$

$x - 13 = 0$ or $x - 13 = 0$

$x = 13$ $x = 13$

45. $z^2 - 8z - 20 = 0$

$(z - 10)(z + 2) = 0$

$z - 10 = 0$ or $z + 2 = 0$

$z = 10$ $z = -2$

47. $(x - 3)^2 = 16$

$x - 3 = \pm\sqrt{16}$

$x = 3 \pm 4$

$x = 3 + 4 = 7$

or $x = 3 - 4 = -1$

49. $a^2 - 8a + 1 = 0$

$a^2 - 8a = -1$

$a^2 - 8a + 16 = -1 + 16$

$(a - 4)^2 = 15$

$a - 4 = \pm\sqrt{15}$

$a = 4 \pm\sqrt{15}$

51. $2y^2 + 4y = 10$

$y^2 + 2y = 5$

$y^2 + 2y + 1 = 5 + 1$

$(y + 1)^2 = 6$

$y + 1 = \pm\sqrt{6}$

$y = -1 \pm\sqrt{6}$

53. $x^2 - 9x - 22 = 0$

$(x - 11)(x + 2) = 0$

$x - 11 = 0$ or $x + 2 = 0$

$x = 11$ $x = -2$

55. $5h(h - 7) = 0$

$5h = 0$ or $h - 7 = 0$

$h = 0$ $h = 7$

57. $8t^2 + 2t - 3 = 0$

$(4t + 3)(2t - 1) = 0$

$4t + 3 = 0$ or $2t - 1 = 0$

$t = -\dfrac{3}{4}$ $t = \dfrac{1}{2}$

59. $t^2 = 14$

$t = \pm\sqrt{14}$

259

61. $c^2 + 9 = 0$

$c^2 = -9$

$c = \pm\sqrt{-9}$

There are no real-valued solutions.

63. $4x^2 - 8x = -4$

$x^2 - 2x = -1$

$x^2 - 2x + 1 = 0$

$(x-1)^2 = 0$

$x - 1 = 0$

$x = 1$

65. Let h represent the height of the suitcase. The width of the suitcase is represented by $h + 4$.

$30h(h+4) = 4200$

$30h^2 + 120h = 4200$

$h^2 + 4h = 140$

$h^2 + 4h - 140 = 0$

$(h+14)(h-10) = 0$

$h + 14 = 0$ or $h - 10 = 0$

$h = -14$ $h = 10$

Since the answer to this problem is a measurement, only the positive answer is appropriate. The suitcase is 10 in. by 14 in. by 30 in.

$10 + 14 + 30 = 54 > 45$

The bag must be checked because the combined linear measurement of length, width, and height is greater than 45 in.

Calculator Exercises

1.
```
( -5+J(17))/4
       -.2192235936
( -5-J(17))/4
       -2.280776406
```

Section 9.3 Practice Exercises

1. (a) $\dfrac{-b \pm \sqrt{b^2 - 4ac}}{2a}$.

(b) $ax^2 + bx + c = 0$

(c) 5; -24; -36.

(d) 5; 73

3. $p^2 = 1$

$p = \pm\sqrt{1}$

$p = \pm 1$

5. $(y+3)^2 = 7$

$y + 3 = \pm\sqrt{7}$

$y = -3 \pm \sqrt{7}$

7. $3a^2 - 12a - 12 = 0$

$a^2 - 4a = 4$

$a^2 - 4a + 4 = 4 + 4$

$(a-2)^2 = 8$

$a - 2 = \pm\sqrt{8}$

$a = 2 \pm \sqrt{2 \cdot 4}$

$a = 2 \pm 2\sqrt{2}$

9. $2x^2 - x = 5$

$2x^2 - x - 5 = 0$

$a = 2,\ b = -1,\ c = -5$

11. $-3x(x-4) = -2x$

$-3x^2 + 12x = -2x$

$-3x^2 + 14x = 0$

$-3x^2 + 14x + 0 = 0$

$a = -3,\ b = 14,\ c = 0$

13. $x^2 - 9 = 0$

$x^2 + 0x - 9 = 0$

$a = 1,\ b = 0,\ c = -9$

15. $6k^2 - k - 2 = 0$

$$k = \frac{-(-1) \pm \sqrt{(-1)^2 - 4(6)(-2)}}{2(6)}$$

$$= \frac{1 \pm \sqrt{1 + 48}}{12}$$

$$= \frac{1 \pm \sqrt{49}}{12}$$

$$= \frac{1 \pm 7}{12}$$

$$k = \frac{1 + 7}{12} = \frac{8}{12} = \frac{2}{3}$$

$$k = \frac{1 - 7}{12} = -\frac{6}{12} = -\frac{1}{2}$$

$$n = \frac{-5 - 7}{6} = \frac{-12}{6} = -2$$

17. $t^2 + 16t + 64 = 0$

$$t = \frac{-16 \pm \sqrt{16^2 - 4(1)(64)}}{2(1)}$$

$$= \frac{-16 \pm \sqrt{256 - 256}}{2}$$

$$= \frac{-16 \pm \sqrt{0}}{2}$$

$$= -8$$

19. $5t^2 - t = 3$

$5t^2 - t - 3 = 0$

$$t = \frac{-(-1) \pm \sqrt{(-1)^2 - 4(5)(-3)}}{2(5)}$$

$$= \frac{1 \pm \sqrt{1 + 60}}{10}$$

$$= \frac{1 \pm \sqrt{61}}{10}$$

21. $x(x - 2) = 1$

$x^2 - 2x - 1 = 0$

$$x = \frac{-(-2) \pm \sqrt{(-2)^2 - 4(1)(-1)}}{2(1)}$$

$$= \frac{2 \pm \sqrt{4 + 4}}{2}$$

$$= \frac{2 \pm \sqrt{8}}{2}$$

$$= \frac{2 \pm \sqrt{4 \cdot 2}}{2}$$

$$= \frac{2 \pm 2\sqrt{2}}{2}$$

$$= 1 \pm \sqrt{2}$$

23. $\qquad 2p^2 = -10p - 11$

$2p^2 + 10p + 11 = 0$

$$p = \frac{-10 \pm \sqrt{(10)^2 - 4(2)(11)}}{2(2)}$$

$$= \frac{-10 \pm \sqrt{100 - 88}}{4}$$

$$= \frac{-10 \pm \sqrt{12}}{4}$$

$$= \frac{-10 \pm \sqrt{4 \cdot 3}}{4}$$

$$= \frac{-10 \pm 2\sqrt{3}}{4}$$

$$= \frac{-5 \pm \sqrt{3}}{2}$$

25. $-4y^2 - y + 1 = 0$

$$y = \frac{-(-1) \pm \sqrt{(-1)^2 - 4(-4)(1)}}{2(-4)}$$

$$y = \frac{1 \pm \sqrt{1 + 16}}{-8}$$

$$y = \frac{1 \pm \sqrt{17}}{-8} \text{ or } \frac{-1 \pm \sqrt{17}}{8}$$

27.
$$2x(x+1) = 3 - x$$
$$2x^2 + 2x = 3 - x$$
$$2x^2 + 3x - 3 = 0$$
$$x = \frac{-3 \pm \sqrt{3^2 - 4(2)(-3)}}{2(2)}$$
$$x = \frac{-3 \pm \sqrt{9 + 24}}{4}$$
$$x = \frac{-3 \pm \sqrt{33}}{4}$$

29.
$$0.2y^2 = -1.5y - 1$$
$$0.2y^2 + 1.5y + 1 = 0$$
$$2y^2 + 15y + 10 = 0$$
$$y = \frac{-15 \pm \sqrt{(15)^2 - 4(2)(10)}}{2(2)}$$
$$y = \frac{-15 \pm \sqrt{225 - 80}}{4}$$
$$y = \frac{-15 \pm \sqrt{145}}{4}$$

31.
$$\frac{2}{3}x^2 + \frac{4}{9}x = \frac{1}{3}$$
$$6x^2 + 4x = 3$$
$$6x^2 + 4x - 3 = 0$$
$$x = \frac{-4 \pm \sqrt{4^2 - 4(6)(-3)}}{2(6)}$$
$$x = \frac{-4 \pm \sqrt{16 + 72}}{12}$$
$$x = \frac{-4 \pm \sqrt{88}}{12}$$
$$x = \frac{-4 \pm 2\sqrt{22}}{12}$$
$$x = \frac{2\left(-2 \pm \sqrt{22}\right)}{12}$$
$$x = \frac{-2 \pm \sqrt{22}}{6}$$

33.
$$16x^2 - 9 = 0$$
$$(4x - 3)(4x + 3) = 0$$
$$4x - 3 = 0 \quad \text{or} \quad 4x + 3 = 0$$
$$4x = 3 \qquad\qquad 4x = -3$$
$$x = \frac{3}{4} \qquad\qquad x = -\frac{3}{4}$$

35. $(x - 5)^2 = -21$
$$x - 5 = \pm\sqrt{-21}$$
There are no real-valued solutions.

37. $\dfrac{1}{9}x^2 + \dfrac{8}{3}x + 11 = 0$
$$x^2 + 24x + 99 = 0$$
$$x = \frac{-24 \pm \sqrt{24^2 - 4(1)(99)}}{2(1)}$$
$$x = \frac{-24 \pm \sqrt{576 - 396}}{2}$$
$$x = \frac{-24 \pm \sqrt{180}}{2}$$
$$x = \frac{-24 \pm 6\sqrt{5}}{2}$$
$$x = -12 \pm 3\sqrt{5}$$

39. $2x^2 - 6x - 3 = 0$
$$x = \frac{-(-6) \pm \sqrt{(-6)^2 - 4(2)(-3)}}{2(2)}$$
$$x = \frac{6 \pm \sqrt{36 + 24}}{4}$$
$$x = \frac{6 \pm \sqrt{60}}{4}$$
$$x = \frac{6 \pm 2\sqrt{15}}{4}$$
$$x = \frac{2\left(3 \pm \sqrt{15}\right)}{4}$$
$$x = \frac{3 \pm \sqrt{15}}{2}$$

41.
$$9x^2 = 11x$$
$$9x^2 - 11x = 0$$
$$x(9x - 11) = 0$$
$$x = 0 \quad \text{or} \quad 9x - 11 = 0$$
$$9x = 11$$
$$x = \frac{11}{9}$$

43. $(2y - 3)^2 = 5$
$$2y - 3 = \pm\sqrt{5}$$
$$2y = 3 \pm \sqrt{5}$$
$$y = \frac{3 \pm \sqrt{5}}{2}$$

45.
$$0.4x^2 = 0.2x + 1$$
$$4x^2 = 2x + 10$$
$$4x^2 - 2x - 10 = 0$$
$$x = \frac{-(-2) \pm \sqrt{(-2)^2 - 4(4)(-10)}}{2(4)}$$
$$= \frac{2 \pm \sqrt{4 + 160}}{8}$$
$$= \frac{2 \pm \sqrt{164}}{8}$$
$$= \frac{2 \pm \sqrt{4 \cdot 41}}{8}$$
$$= \frac{2 \pm 2\sqrt{41}}{8}$$
$$= \frac{1 \pm \sqrt{41}}{4}$$

47. $9z^2 - z = 0$
$$z(9z - 1) = 0$$
$$z = 0 \quad \text{or} \quad 9z - 1 = 0$$
$$z = \frac{1}{9}$$

49. $r^2 - 52 = 0$
$$r^2 = 52$$
$$r = \pm\sqrt{52}$$
$$r = \pm\sqrt{4 \cdot 13}$$
$$r = \pm 2\sqrt{13}$$

51.
$$-2.5t(t - 4) = 1.5$$
$$-2.5t^2 + 10t = 1.5$$
$$-2.5t^2 + 10t - 1.5 = 0$$
$$-25t^2 + 100t - 15 = 0$$
$$t = \frac{-100 \pm \sqrt{100^2 - 4(-25)(-15)}}{2(-25)}$$
$$t = \frac{-100 \pm \sqrt{10,000 - 1500}}{-50}$$
$$t = \frac{-100 \pm \sqrt{8500}}{-50}$$
$$t = \frac{-100 \pm 10\sqrt{85}}{-50}$$
$$t = \frac{10\left(-10 \pm \sqrt{85}\right)}{-50}$$
$$t = \frac{-10 \pm \sqrt{85}}{-5} \quad \text{or} \quad \frac{10 \pm \sqrt{85}}{5}$$

53. $(m - 3)(m + 2) = 9$
$$m^2 - m - 6 = 9$$
$$m^2 - m - 15 = 0$$
$$m = \frac{-(-1) \pm \sqrt{(-1)^2 - 4(1)(-15)}}{2(1)}$$
$$m = \frac{1 \pm \sqrt{1 + 60}}{2}$$
$$m = \frac{1 \pm \sqrt{61}}{2}$$

263

55. $x^2 + x + 3 = 0$

$$x = \frac{-1 \pm \sqrt{1^2 - 4(1)(3)}}{2(1)}$$

$$= \frac{-1 \pm \sqrt{1 - 12}}{2}$$

$$= \frac{-1 \pm \sqrt{-11}}{2}$$

There are no real-valued solutions.

57. Let x represent the width of the rectangle. The length of the rectangle is represented by $2x - 1$.

$$x(2x - 1) = 100$$

$$2x^2 - x = 100$$

$$2x^2 - x - 100 = 0$$

$$x = \frac{-(-1) \pm \sqrt{(-1)^2 - 4(2)(-100)}}{2(2)}$$

$$x = \frac{1 \pm \sqrt{1 + 800}}{4}$$

$$x = \frac{1 \pm \sqrt{801}}{4}$$

$$x = \frac{1 \pm 3\sqrt{89}}{4}$$

Since the answer to this problem is a measurement, only the positive value is a solution.

$$x = \frac{1 + 3\sqrt{89}}{4} \approx 7.3$$

$$2x - 1 = 2\left(\frac{1 + 3\sqrt{89}}{4}\right) - 1$$

$$= \frac{1 + 3\sqrt{89}}{2} - \frac{2}{2}$$

$$= \frac{-1 + 3\sqrt{89}}{2} \approx 13.7$$

The width of the rectangle is 7.3 m. The length of the rectangle is 13.7 m.

59. Let x represent the width of the rectangular storage area. The length of the rectangular storage area is represented by $x + 2$.

$$6x(x + 2) = 240$$

$$6x^2 + 12x = 240$$

$$6x^2 + 12x - 240 = 0$$

$$x^2 + 2x - 40 = 0$$

$$x = \frac{-2 \pm \sqrt{2^2 - 4(1)(-40)}}{2(1)}$$

$$x = \frac{-2 \pm \sqrt{4 + 160}}{2}$$

$$x = \frac{-2 \pm \sqrt{164}}{2}$$

$$x = \frac{-2 \pm 2\sqrt{41}}{2}$$

$$x = -1 \pm \sqrt{41}$$

Since the answer to this problem is a measurement, only the positive value is a solution.

$$x = -1 + \sqrt{41} \approx 5.4$$

$$x + 2 = -1 + \sqrt{41} + 2 = 1 + \sqrt{41} \approx 7.4$$

The width of the rectangular storage area is 5.4 ft. The length is 7.4 ft. The height is 6 ft.

61. Let x represent the width of the rectangle. The length is represented by $x + 4$.

$$x(x + 4) = 72$$

$$x^2 + 4x = 72$$

$$x^2 + 4x - 72 = 0$$

$$x = \frac{-4 \pm \sqrt{4^2 - 4(1)(-72)}}{2(1)}$$

$$x = \frac{-4 \pm \sqrt{16 + 288}}{2}$$

$$x = \frac{-4 \pm \sqrt{304}}{2}$$

$$x = \frac{-4 \pm 4\sqrt{19}}{2}$$

$$x = -2 \pm 2\sqrt{19}$$

Since the answer to this problem is a measurement, only the positive answer is appropriate.

$$x = -2 + 2\sqrt{19} \approx 6.7$$

$$x + 4 = \left(-2 + 2\sqrt{19}\right) + 4$$

$$= 2 + 2\sqrt{19} \approx 10.7$$

The width is 6.7 ft. The length is 10.7 ft.

63. Let x represent the length of the first leg of the right triangle. The length of the second leg of the right triangle is represented by $x + 3$.

$$x^2 + (x + 3)^2 = 13^2$$

$$x^2 + x^2 + 6x + 9 = 169$$

$$2x^2 + 6x = 160$$

$$x^2 + 3x = 80$$

$$x^2 + 3x - 80 = 0$$

$$x = \frac{-3 \pm \sqrt{3^2 - 4(1)(-80)}}{2(1)}$$

$$x = \frac{-3 \pm \sqrt{9 + 320}}{2}$$

$$x = \frac{-3 \pm \sqrt{329}}{2}$$

Since the answer to this problem is a measurement, only the positive answer is a solution.

$$x = \frac{-3 + \sqrt{329}}{2} \approx 10.6$$

$$x + 3 = \frac{-3 + \sqrt{329}}{2} + 3 = \frac{3 + \sqrt{329}}{2} \approx 7.6$$

The lengths of the legs are 10.6 m and 7.6 m.

65. Let $h = 1$ in the equation $h = -16t^2 + 16t$ and solve for t using the quadratic formula.

$$1 = -16t^2 + 16t$$

$$0 = -16t^2 + 16t - 1$$

$$t = \frac{-16 \pm \sqrt{16^2 - 4(-16)(-1)}}{2(-16)}$$

$$t = \frac{-16 \pm \sqrt{256 - 64}}{-32}$$

$$t = \frac{-16 \pm \sqrt{192}}{-32}$$

$$t = \frac{-16 + \sqrt{192}}{-32} \quad \text{or} \quad t = \frac{-16 - \sqrt{192}}{-32}$$

$$t \approx 0.07 \qquad\qquad t \approx 0.93$$

Michael will be 1 ft off the ground 0.07 sec after leaving the ground (on the way up) and after 0.93 sec (on the way back down).

Problem Recognition Exercises

1. $6x^2 + 7x - 3 = 0$

(a) $(2x + 3)(3x - 1) = 0$

$$2x + 3 = 0 \quad \text{or} \quad 3x - 1 = 0$$

$$x = -\frac{3}{2} \quad \text{or} \quad x = \frac{1}{3}$$

(b) $\dfrac{6x^2}{6} + \dfrac{7x}{6} - \dfrac{3}{6} = 0$

$$x^2 + \frac{7}{6}x = \frac{1}{2}$$

$$x^2 + \frac{7}{6}x + \frac{49}{144} = \frac{1}{2} + \frac{49}{144}$$

$$\left(x + \frac{7}{12}\right)^2 = \frac{121}{144}$$

$$x + \frac{7}{12} = \pm\sqrt{\frac{121}{144}}$$

$$x = -\frac{7}{12} \pm \frac{11}{12}$$

$$x = -\frac{7}{12} + \frac{11}{12} \quad \text{or} \quad x = -\frac{7}{12} - \frac{11}{12}$$

$$= \frac{1}{3} \qquad\qquad = -\frac{18}{12}$$

$$\qquad\qquad\qquad = -\frac{3}{2}$$

(c) $x = \dfrac{-7 \pm \sqrt{(7)^2 - 4(6)(-3)}}{2(6)}$

$= \dfrac{-7 \pm \sqrt{49 + 72}}{12}$

$= \dfrac{-7 \pm \sqrt{121}}{12}$

$= \dfrac{-7 \pm 11}{12}$

$= \dfrac{4}{12}$ or $\dfrac{-18}{12}$

$= \dfrac{1}{3}$ or $-\dfrac{3}{2}$

3. (a) Quadratic

(b) $x(x - 8) = 6$

$x^2 - 8x - 6 = 0$

$x = \dfrac{-(-8) \pm \sqrt{(-8)^2 - 4(1)(-6)}}{2(1)}$

$= \dfrac{8 \pm \sqrt{64 + 24}}{2}$

$= \dfrac{8 \pm \sqrt{88}}{2}$

$= \dfrac{8 \pm \sqrt{4 \cdot 22}}{2}$

$= \dfrac{8 \pm 2\sqrt{22}}{2}$

$= 4 \pm \sqrt{22}$

5. (a) Linear

(b) $3(k - 6) = 2k - 5$

$3k - 18 = 2k - 5$

$k - 18 = -5$

$k = 13$

7. (a) Quadratic

(b) $8x^2 - 22x + 5 = 0$

$(2x - 5)(4x - 1) = 0$

$2x - 5 = 0$ **or** $4x - 1 = 0$

$x = \dfrac{5}{2}$ $\qquad$ $x = \dfrac{1}{4}$

9. (a) Rational

(b)

$\dfrac{2}{x-1} - \dfrac{5}{4} = -\dfrac{1}{x+1}$

$\left(4(x-1)(x+1)\right)\left(\dfrac{2}{x-1} - \dfrac{5}{4}\right) = \left(-\dfrac{1}{x+1}\right)\left(4(x-1)(x+1)\right)$

$4(2)(x+1) - 5(x-1)(x+1) = (-1)(4)(x-1)$

$8(x+1) - 5(x^2 - 1) = -4(x-1)$

$8x + 8 - 5x^2 + 5 = -4x + 4$

$0 = 5x^2 - 12x - 9$

$0 = (5x + 3)(x - 3)$

$5x + 3 = 0$ $\quad$ or $x - 3 = 0$

$x = -\dfrac{3}{5}$ $\qquad$ $x = 3$

11. (a) Radical

(b) $\sqrt{2y - 2} = y - 1$

$\left(\sqrt{2y - 2}\right)^2 = (y - 1)^2$

$2y - 2 = y^2 - 2y + 1$

$0 = y^2 - 4y + 3$

$0 = (y - 3)(y - 1)$

$y - 3 = 0$ or $y - 1 = 0$

$y = 3$ $\qquad$ $y = 1$

13. (a) Quadratic

(b) $(w + 1)^2 = 100$

$\sqrt{(w+1)^2} = \sqrt{100}$

$w + 1 = \pm 10$

$w = -1 \pm 10$

$w = -1 + 10 = 9$ or $w = -1 - 10 = -11$

15. **(a)** Rational

(b) $\dfrac{2}{x+1} = \dfrac{5}{4}$

$$2 \cdot 4 = 5(x+1)$$
$$8 = 5x + 5$$
$$3 = 5x$$
$$\frac{3}{5} = x$$

Calculator Exercises

1. Minimum at (-2, 3)

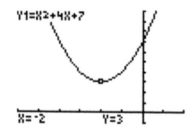

3. Maximum at (-1.5, -2.6)

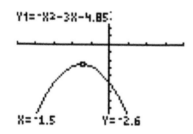

5. Minimum at $\left(\dfrac{5}{2}, 0\right)$

Section 9.4 Practice Exercises

1. **(a)** parabola

(b) $>$; $<$

(c) lowest; highest

(d) symmetry

(e) vertex

3. $3 + a(a+2) = 18$

$$3 + a^2 + 2a = 18$$
$$a^2 + 2a - 15 = 0$$
$$(a-3)(a+5) = 0$$
$$a - 3 = 0 \quad \text{or} \quad a + 5 = 0$$
$$a = 3 \qquad\qquad a = -5$$

5. $2z^2 + 4z - 10 = 0$

$$z^2 + 2z - 5 = 0$$
$$z = \frac{-2 \pm \sqrt{2^2 - 4(1)(-5)}}{2(1)}$$
$$= \frac{-2 \pm \sqrt{4 + 20}}{2}$$
$$= \frac{-2 \pm \sqrt{24}}{2}$$
$$= \frac{-2 \pm 2\sqrt{6}}{2}$$
$$= -1 \pm \sqrt{6}$$

7. $(x-2)^2 = 8$

$$x - 2 = \pm\sqrt{8}$$
$$x - 2 = \pm 2\sqrt{2}$$
$$x = 2 \pm 2\sqrt{2}$$

9. Linear

11. Quadratic

13. Neither

15. Linear

17. Quadratic

19. Neither

21. If $a > 0$ the parabola opens upward; if $a < 0$ the opens downward.

23. $a = 2$; upward

25. $a = -10$; downward

27. $x = \dfrac{-b}{2a} = \dfrac{-4}{2(2)} = \dfrac{-4}{4} = -1$

$y = 2(-1)^2 + 4(-1) - 6 = 2 - 4 - 6 = -8$

Vertex: $(-1, -8)$

29. $x = \dfrac{-b}{2a} = \dfrac{-2}{2(-1)} = \dfrac{-2}{-2} = 1$

$y = -(1)^2 + 2(1) - 5 = -1 + 2 - 5 = -4$

Vertex: $(1, -4)$

31. $x = \dfrac{-b}{2a} = \dfrac{-(-2)}{2(1)} = \dfrac{2}{2} = 1$

$y = (1)^2 - 2(1) + 3 = 1 - 2 + 3 = 2$

Vertex: $(1, 2)$

33. $x = \dfrac{-b}{2a} = \dfrac{-0}{2(1)} = 0$

$y = 0^2 - 4 = -4$

Vertex: $(0, -4)$

35. To find the x-intercept(s), substitute 0 for y and solve for x. To find the y-intercept, substitute 0 for x and solve for y.

$y = x^2 - 7$

x-intercepts: $x^2 - 7 = 0$

$\qquad\qquad x^2 = 7$

$\qquad\qquad x = \pm\sqrt{7}$

$\qquad \left(\sqrt{7}, 0\right) \ \left(-\sqrt{7}, 0\right)$

y-intercept: $\begin{aligned} y &= 0^2 - 7 \\ y &= -7 \end{aligned} \qquad (0, -7)$

This function is graph c.

37. $y = x^2 + 5x + 6$

x-intercepts: $\quad x^2 + 5x + 6 = 0$

$\qquad\qquad\qquad (x + 5)(x + 1) = 0$

$x + 5 = 0 \quad \textbf{or} \quad x + 1 = 0$

$x = -5 \qquad\qquad\quad x = -1$

$(-5, 0) \ (-1, 0)$

y-intercept: $y = 0^2 + 6(0) + 5$

$\qquad\qquad\qquad y = 5$

$\qquad\qquad\qquad (0, 5)$

This function is graph a.

39. $y = x^2 - 9$

(a) $a = 1$ \qquad\qquad\qquad Upward

(b) $x = \dfrac{-b}{2a} = \dfrac{-0}{2(1)} = 0$

$y = (0)^2 - 9 = -9 \qquad$ vertex: $(0, -9)$

(c) x-intercepts:

$x^2 - 9 = 0$

$x^2 = 9$

$x = \pm 3 \qquad (3, 0) \ (-3, 0)$

(d) y-intercept: $y = 0^2 - 9$

$\qquad\qquad\qquad y = -9 \qquad\qquad (0, -9)$

(e)

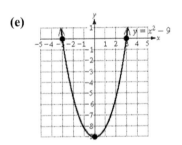

41. $y = x^2 - 2x - 8$

(a) $a = 1$ \qquad\qquad\qquad Upward

(b) $x = \dfrac{-b}{2a} = \dfrac{-(-2)}{2(1)} = \dfrac{2}{2} = 1$

$y = (1)^2 - 2(1) - 8 = 1 - 2 - 8 = -9$

Vertex: $(1, -9)$

(c) x-intercepts: $\quad x^2 - 2x - 8 = 0$

$\qquad\qquad\qquad\qquad (x + 2)(x - 4) = 0$

$x + 2 = 0$ or $x - 4 = 0$

$x = -2$ $x = 4$

$(-2, 0)$ $(4, 0)$

(d) y-intercept: $y = 0^2 - 2(0) - 8 = -8$

$(0, -8)$

(e)

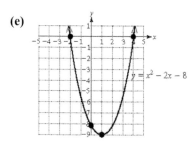

43. $y = -x^2 + 6x - 9$

(a) $a = -1$ Downward

(b) $x = \dfrac{-b}{2a} = \dfrac{-6}{2(-1)} = \dfrac{-6}{-2} = 3$

$y = -(3)^2 + 6(3) - 9 = -9 + 18 - 9 = 0$

Vertex: $(3, 0)$

(c) x-intercept:

$-x^2 + 6x - 9 = 0$

$x^2 - 6x + 9 = 0$

$(x - 3)^2 = 0$

$x - 3 = 0$

$x = 3$ $(3, 0)$

(d) y-intercept:

$y = -(0)^2 + 6(0) - 9$

$y = -9$ $(0, -9)$

(e)

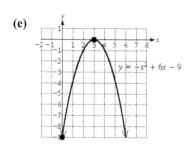

45. $y = -x^2 + 8x - 15$

(a) $a = -1$ Downward

(b) $x = \dfrac{-b}{2a} = \dfrac{-8}{2(-1)} = \dfrac{-8}{-2} = 4$

$y = -(4)^2 + 8(4) - 15$

$= -16 + 32 - 15$

$= 1$

Vertex: $(4, 1)$

(c) x-intercepts: $-x^2 + 8x - 15 = 0$

$x^2 - 8x + 15 = 0$

$(x - 3)(x - 5) = 0$

$x - 3 = 0$ or

$x = 3$

$x - 5 = 0$

$x = 5$

$(3, 0)$ $(5, 0)$

(d) y-intercept: $y = -(0)^2 + 8(0) - 15$

$y = -15$ $(0, -15)$

(e)

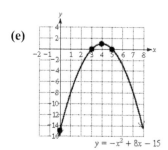

47. $y = x^2 + 6x + 10$

(a) $a = 1$ Upward

(b) $x = \dfrac{-b}{2a} = \dfrac{-6}{2(1)} = \dfrac{-6}{2} = -3$

$y = (-3)^2 + 6(-3) + 10$

$= 9 - 18 + 10$

$= 1$

Vertex: $(-3, 1)$

269

(c) x-intercepts: $x^2 + 6x + 10 = 0$

$$x = \frac{-6 \pm \sqrt{6^2 - 4(1)(10)}}{2(1)}$$

$$x = \frac{-6 \pm \sqrt{36 - 40}}{2}$$

$$x = \frac{-6 \pm \sqrt{-4}}{2} \qquad \text{None}$$

(d) y-intercept:

$$y = (0)^2 + 6(0) + 10$$

$$y = 10 \qquad (0, 10)$$

(e)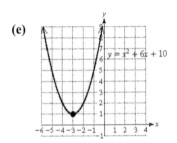

49. $y = -2x^2 - 2$

(a) $a = -2$ \qquad Downward

(b) $x = \dfrac{-b}{2a} = \dfrac{-0}{2(-2)} = 0$

$y = -2(0)^2 - 2 = 0 - 2 = -2$

Vertex: $(0, -2)$

(c) x-intercepts: $-2x^2 - 2 = 0$

$$-2x^2 = 2$$

$$x^2 = -1 \qquad \text{None}$$

(d) y-intercept:

$$y = -2(0)^2 - 2$$

$$y = -2 \qquad (0, -2)$$

(e)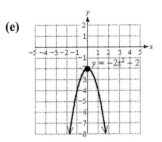

51. Because the graph opens downward, this is a true statement.

53. $a = 1.5$ which indicates the graph opens upward. Therefore, the statement is false.

55. $y = -16t^2 + 40t + 3$

(a) The maximum height of the ball is the y value of the vertex. (See part b for the x value of the vertex.)

$$y = -16(1.25)^2 + 40(1.25) + 3$$

$$= -25 + 50 + 3$$

$$= 28$$

The maximum height of the ball is 28 ft.

(b) The length of time needed for the ball to reach its maximum height is the x value of the vertex.

$$x = \frac{-b}{2a} = \frac{-40}{2(-16)}$$

$$= \frac{-40}{-32}$$

$$= 1.25$$

It will take the ball 1.25 seconds to reach its maximum height.

57. $y = -\dfrac{1}{40}x^2 + 10x - 500$

(a) The number of calendars that should be produced to maximize profit is the x value of the vertex.

270

$$x = \frac{-b}{2a} = \frac{-10}{2\left(-\frac{1}{40}\right)}$$

$$= \frac{-10}{\left(-\frac{1}{20}\right)}$$

$$= -10 \div \left(-\frac{1}{20}\right)$$

$$= -10 \cdot (-20)$$

$$= 200$$

200 calendars will need to be produced to maximize profit.

(b) Maximum profit is the y value of the vertex.

$$y = -\frac{1}{40}(200)^2 + 10(200) - 500$$

$$= -1000 + 2000 - 500$$

$$= 500$$

Maximum profit is $500.

59. (a) Let $t = 0.5$ in the equation $y = -16t^2 + 32t$ and solve for y.

$$y = -16(0.5)^2 + 32(0.5)$$

$$y = 12$$

Josh will be 12 feet high in 0.5 seconds.

(b) Josh will land when the height is 0. Let $y = 0$ in the equation $y = -16t^2 + 32t$ and solve for t.

$$0 = -16t^2 + 32t$$

$$0 = -16t(t - 2)$$

$$0 = -16t \quad \text{or} \quad 0 = t - 2$$
$$0 = t \qquad\qquad 2 = t$$

Josh will land in 2 seconds.

(c) Josh's maximum height is the y value of the vertex. First, find the x value of the vertex. Use it to then find the y value.

$$x = \frac{-b}{2a} = \frac{-32}{2(-16)}$$

$$= \frac{-32}{-32}$$

$$= 1$$

$$y = -16(1)^2 + 32(1)$$

$$y = 16$$

The maximum height is 16 feet.

Calculator Exercises

1.

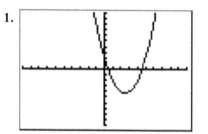

3.

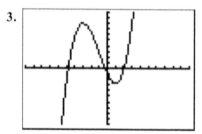

Section 9.5 Practice Exercises

1. (a) relation

(b) domain

(c) range

(d) function

(e) vertical

(f) $7x - 4$.

$$\left(\frac{1}{3}, \frac{7}{3}\right)$$

271

3. $x = \dfrac{-b}{2a} = \dfrac{-(-2)}{2(4)} = \dfrac{2}{8} = \dfrac{1}{4}$

$y = 4\left(\dfrac{1}{4}\right)^2 - 2\left(\dfrac{1}{4}\right) + 3 = \dfrac{11}{4}$

$\left(\dfrac{1}{4}, \dfrac{11}{4}\right)$

$\left(\dfrac{5}{2}, -\dfrac{17}{4}\right)$

5. Domain: $\{4, 3, 0\}$; Range: $\{2, 7, 1, 6\}$

7. Domain: $\left\{\dfrac{1}{2}, 0, 1\right\}$; Range: $\{3\}$

9. Domain: $\{0, 5, -8, 8\}$; Range: $\{0, 2, 5\}$

11. Domain: {Atlanta, Macon, Pittsburgh};

Range: {GA, PA}

13. Domain: {New York, California};

Range: {Albany, Los Angeles, Buffalo}

15. The relation is a function if each element in the domain has exactly one corresponding element in the range.

17. A relation is a function if for each element in the domain, there is exactly one element in the range. The relations in Exercises 7, 9, and 11 are functions.

19. Yes

21. No

23. No

25. Yes

27. Yes

29. **(a)** Substitute 0 for x and simplify.

$f(0) = 2(0) - 5 = -5$

(b) Substitute 2 for x and simplify.

$f(2) = 2(2) - 5 = -1$

(c) Substitute -3 for x and simplify.

$f(-3) = 2(-3) - 5 = -11$

31. **(a)** Substitute 1 for x and simplify.

$h(1) = \dfrac{1}{1+4} = \dfrac{1}{5}$

(b) Substitute 0 for x and simplify.

$h(0) = \dfrac{1}{0+4} = \dfrac{1}{4}$

(c) Substitute -2 for x and simplify.

$h(-2) = \dfrac{1}{-2+4} = \dfrac{1}{2}$

33. **(a)** Substitute 0 for x and simplify.

$m(0) = |5(0) - 7| = |0 - 7| = 7$

(b) Substitute 1 for x and simplify.

$m(1) = |5(1) - 7| = |5 - 7| = 2$

(c) Substitute 2 for x and simplify.

$m(2) = |5(2) - 7| = |10 - 7| = 3$

35. **(a)** Substitute 2 for x and simplify.

$n(2) = \sqrt{2 - 2} = \sqrt{0} = 0$

(b) Substitute 3 for x and simplify.

$n(3) = \sqrt{3 - 2} = \sqrt{1} = 1$

(c) Substitute 6 for x and simplify.

$n(6) = \sqrt{6 - 2} = \sqrt{4} = 2$

37. The domain is the set of all $x-$coordinates.

The range is the set of all $y-$coordinates.

Domain: $\{-3, 1, 2, 4\}$; Range: $\{-5, 0, 1, 2\}$

39. The domain is the set of all x-coordinates.

The range is the set of all y-coordinates.

Domain: $\{-4, -2, 0, 1, 5\}$; Range: $\{-3, 3, 4\}$

41. Graph b.

43. Graph c.

45. Domain: $(-\infty, \infty)$; Range: $[-2, \infty)$

47. Domain: $[-1, 1]$; Range: $[-4, 4]$

49. The function value at $x = 6$ is 2.

51. The function value at $x = \dfrac{1}{2}$ is $\dfrac{1}{4}$.

53. $f(2) = 7$ corresponds to the point $(2, 7)$.

55. **(a)** $s(t) = 32t$

$s(1) = 32(1)$

$s(1) = 32$

The speed of an object 1 second after being dropped is 32 ft/sec.

(b) $s(t) = 32t$

$s(2) = 32(2)$

$s(2) = 64$

The speed of an object 2 seconds after being dropped is 64 ft/sec.

(c) $s(t) = 32t$

$s(10) = 32(10)$

$s(10) = 320$

The speed of an object 10 seconds after being dropped is 320 ft/sec.

(d) $s(t) = 32t$

$s(9.2) = 32(9.2)$

$s(9.2) = 294.4$

The speed of the ball is 294.4 ft/sec just before hitting the ground.

57. **(a)** $h(t) = -16t^2 + 64t + 3$

$h(0) = -16(0)^2 + 64(0) + 3$

$h(0) = 0 + 0 + 3 = 3$

The initial height of the ball is 3 ft.

(b) $h(t) = -16t^2 + 64t + 3$

$h(1) = -16(1)^2 + 64(1) + 3$

$h(1) = -16 + 64 + 3 = 51$

The height of the ball 1 second after being kicked is 51 ft.

(c) $h(t) = -16t^2 + 64t + 3$

$h(2) = -16(2)^2 + 64(2) + 3$

$h(2) = -64 + 128 + 3 = 67$

The height of the ball 2 seconds after being kicked is 67 ft.

(d) $h(t) = -16t^2 + 64t + 3$

$h(4) = -16(4)^2 + 64(4) + 3$

$h(4) = -256 + 256 + 3 = 3$

The height of the ball 4 seconds after being kicked is 3 ft.

59. **(a)** To determine the cost for 3 hr of labor let $x = 3$ in the function $C(x) = 75 + 50x$ and simplify.

$C(3) = 75 + 50(3)$

$C(3) = 225$

The cost for 3 hr of labor is $225.

(b) To determine the number of hours of labor if Helena is charged $200, let $C(x) = 200$ and solve for x.

$200 = 75 + 50x$

$125 = 50x$

$2.5 = x$

Helena was charged for 2.5 hours.

(c) The domain is the set of all possible x values for the function. Since x

represents time, the domain must be greater than or equal to zero or $[0, \infty)$.

(d) The y–intercept occurs when $x = 0$ so it represents the cost for 0 hours of labor or the cost of the estimate.

Group Activity

1. The measurements and volume of the gutter are as follows:

Height, x	Base	Length	Volume
0.5 in.	7.5 in.	11 in.	41.25 in.^{3}
1.0 in.	6.5 in.	11 in.	71.5 in.^{3}
1.5 in.	5.5 in.	11 in.	90.75 in.^{3}
2.0 in.	4.5 in.	11 in.	99 in.^{3}
2.5 in.	3.5 in.	11 in.	96.25 in.^{3}
3.0 in.	2.5 in.	11 in.	82.5 in.^{3}
3.5 in.	1.5 in.	11 in.	57.75 in.^{3}

3. Let x represent the height of the gutter. The base is $8.5 - 2x$. The volume of gutter is

$$V = (8.5 - 2x)(x)(72) = -144x^2 + 612x$$

Chapter 9 Review Exercises

Section 9.1

1. Linear

3. Quadratic

5. $x^2 = 25$

$$x = \pm\sqrt{25}$$

$$x = \pm 5$$

7. $x^2 + 49 = 0$

$$x^2 = -49$$

$$x = \pm\sqrt{-49}$$

The equation has no real-valued solutions.

9. $(x+1)^2 = 14$

$$x + 1 = \pm\sqrt{14}$$

$$x = -1 \pm \sqrt{14}$$

11. $\left(x - \dfrac{1}{8}\right)^2 = \dfrac{3}{64}$

$$x - \frac{1}{8} = \pm\sqrt{\frac{3}{64}}$$

$$x - \frac{1}{8} = \pm\frac{\sqrt{3}}{8}$$

$$x = \frac{1}{8} \pm \frac{\sqrt{3}}{8}$$

Section 9.2

13. $n = \frac{1}{2}$ of 12, squared $\left[\dfrac{1}{2}(12)\right]^2 = 6^2 = 36$

15. $n = \frac{1}{2}$ of -5, squared

$$\left[\frac{1}{2}(-5)\right]^2 = \left(-\frac{5}{2}\right)^2 = \frac{25}{4}$$

17. $x^2 + 8x + 3 = 0$

$$x^2 + 8x = -3$$

$$x^2 + 8x + 16 = -3 + 16$$

$$(x+4)^2 = 13$$

$$x + 4 = \pm\sqrt{13}$$

$$x = -4 \pm \sqrt{13}$$

19. $2x^2 - 6x - 6 = 0$

$x^2 - 3x - 3 = 0$

$x^2 - 3x = 3$

$x^2 - 3x + \dfrac{9}{4} = 3 + \dfrac{9}{4}$

$\left(x - \dfrac{3}{2}\right)^2 = \dfrac{21}{4}$

$x - \dfrac{3}{2} = \pm\sqrt{\dfrac{21}{4}}$

$x - \dfrac{3}{2} = \pm\dfrac{\sqrt{21}}{2}$

$x = \dfrac{3}{2} \pm \dfrac{\sqrt{21}}{2}$

21. Let x represent the length of each of the legs of the triangle.

$x^2 + x^2 = 15^2$

$2x^2 = 225$

$x^2 = 112.5$

$x \approx \pm 10.6$

Since the answer to this problem is a measurement, only the positive value is a solution. The length of the legs is about 10.6 ft.

Section 9.3

23. For $ax^2 + bx + c = 0$,

$x = \dfrac{-b \pm \sqrt{b^2 - 4ac}}{2a}$

25. $x^2 + 4x + 4 = 0$

$x = \dfrac{-4 \pm \sqrt{4^2 - 4(1)(4)}}{2(1)}$

$x = \dfrac{-4 \pm \sqrt{16 - 16}}{2}$

$x = -\dfrac{4}{2}$

$x = -2$

27. $2x^2 - x - 3 = 0$

$x = \dfrac{-(-1) \pm \sqrt{(-1)^2 - 4(2)(-3)}}{2(2)}$

$x = \dfrac{1 \pm \sqrt{1 + 24}}{4}$

$x = \dfrac{1 \pm \sqrt{25}}{4}$

$x = \dfrac{1 \pm 5}{4}$

$x = \dfrac{1 + 5}{4}$ or $x = \dfrac{1 - 5}{4}$

$x = \dfrac{6}{4} \qquad\qquad x = \dfrac{-4}{4}$

$x = \dfrac{3}{2} \qquad\qquad x = -1$

29. $\dfrac{1}{6}x^2 + x + \dfrac{1}{3} = 0$

$x^2 + 6x + 2 = 0$

$x = \dfrac{-6 \pm \sqrt{6^2 - 4(1)(2)}}{2(1)}$

$x = \dfrac{-6 \pm \sqrt{36 - 8}}{2}$

$x = \dfrac{-6 \pm \sqrt{28}}{2}$

$x = \dfrac{-6 \pm 2\sqrt{7}}{2}$

$x = -3 \pm \sqrt{7}$

31. $0.01x^2 - 0.02x - 0.04 = 0$

$$x^2 - 2x - 4 = 0$$

$$x = \frac{-(-2) \pm \sqrt{(-2)^2 - 4(1)(-4)}}{2(1)}$$

$$x = \frac{2 \pm \sqrt{4+16}}{2}$$

$$x = \frac{2 \pm \sqrt{20}}{2}$$

$$x = \frac{2 \pm \sqrt{4 \cdot 5}}{2}$$

$$x = \frac{2 \pm 2\sqrt{5}}{2}$$

$$x = 1 \pm \sqrt{5}$$

33. $(x-1)(x-7) = -18$

$$x^2 - 7x - x + 7 = -18$$

$$x^2 - 8x + 25 = 0$$

$$x = \frac{-(-8) \pm \sqrt{(-8)^2 - 4(1)(25)}}{2(1)}$$

$$x = \frac{8 \pm \sqrt{64-100}}{2}$$

$$x = \frac{8 \pm \sqrt{-36}}{2}$$

The equation has no real-valued solutions.

35. Let x represent the height of the parallelogram. The base of the parallelogram is represented by $x + 1$.

$$x(x+1) = 24$$

$$x^2 + x = 24$$

$$x^2 + x - 24 = 0$$

$$x = \frac{-1 \pm \sqrt{1^2 - 4(1)(-24)}}{2(1)}$$

$$x = \frac{-1 \pm \sqrt{1+96}}{2}$$

$$x = \frac{-1 \pm \sqrt{97}}{2}$$

Since the answer to this problem is a measurement, only the positive value is a solution.

$$x = \frac{-1 + \sqrt{97}}{2} \approx 4.4$$

$$x + 1 = \frac{-1 + \sqrt{97}}{2} + 1 = \frac{1 + \sqrt{97}}{2} \approx 5.4$$

The height of the parallelogram is 4.4 cm.
The base of the parallelogram is 5.4 cm.

Section 9.4

37. $a = 1$; upward

39. $a = -2$; downward

41. $y = 3x^2 + 6x + 4$

$$x = \frac{-b}{2a} = \frac{-6}{2(3)} = \frac{-6}{6} = -1$$

$$y = 3(-1)^2 + 6(-1) + 4$$

$$= 3 - 6 + 4$$

$$= 1$$

Vertex: $(-1, 1)$

43. $y = -2x^2 + 12x - 5$

$$x = \frac{-b}{2a} = \frac{-12}{2(-2)} = \frac{-12}{-4} = 3$$

$$y = -2(3)^2 + 12(3) - 5$$

$$= -18 + 36 - 5$$

$$= 13$$

Vertex: $(3, 13)$

45. $y = x^2 + 2x - 3$

(a) $a = 1$ Upward

(b) $x = \frac{-b}{2a} = \frac{-2}{2(1)} = -1$

$$y = (-1)^2 + 2(-1) - 3$$

$$= 1 - 2 - 3$$

$$= -4$$

Vertex: $(-1, -4)$

(c) x-intercepts: $x^2 + 2x - 3 = 0$

$$(x+3)(x-1) = 0$$

$$x + 3 = 0 \quad \text{or}$$

$$x = -3$$

$$x - 1 = 0$$

$$x = 1$$

$$(-3, 0) \quad (1, 0)$$

(d) y-intercept:

$$y = (0)^2 + 2(0) - 3$$
$$y = -3 \qquad\qquad (0, -3)$$

(e)

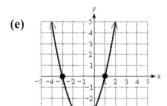

$$y = x^2 + 2x - 3$$

47. $y = -3x^2 + 12x - 9$

(a) $a = -3$ Downward

(b) $x = \dfrac{-b}{2a} = \dfrac{-(12)}{2(-3)} = \dfrac{-12}{-6} = 2$

$$y = -3(2)^2 + 12(2) - 9$$
$$y = -12 + 24 - 9$$
$$y = 3$$

Vertex: $(2, 3)$

(c) x – intercepts: $-3x^2 + 12x - 9 = 0$

$$-3(x^2 - 4x + 3) = 0$$
$$-3(x - 3)(x - 1) = 0$$

$$x - 3 = 0 \quad \text{or} \quad x - 1 = 0$$
$$x = 3 \qquad\qquad x = 1$$

$(3, 0)$ and $(1, 0)$

(d) y-intercept: $y = -3(0)^2 + 12(0) - 9$

$$y = 0 + 12 - 9$$
$$y = -9$$

$(0, -9)$

(e) Note to production: show graph of

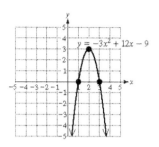

$$y = -3x^2 + 12x - 9$$

49. $y = -16t^2 + 256t$

(a) The maximum height is the y value of the vertex. (See part b for the t value.)

$$y = -16(8)^2 + 256(8)$$
$$y = -1024 + 2048$$
$$y = 1024$$

The maximum height reached is 1024 ft.

(b) $t = \dfrac{-b}{2a} = \dfrac{-256}{2(-16)} = 8$

The time required for the object to reach its maximum height is 8 sec.

51. Domain: $\{2\}$; Range: $\{0, 1, -5, 2\}$; The relation is not a function because for the domain value 2, there is more than one range value.

53. Domain: $(-\infty, \infty)$; Range: $[-2, \infty)$; The relation is a function because a vertical line will intersect the relation only once.

55. Domain: $\{3, -4, 0, 2\}$; Range: $\left\{0, \dfrac{1}{2}, 3, -12\right\}$; The relation is a function because for every domain value there is only one range value.

57. **(a)** $g(0) = \dfrac{0}{5 - 0} = \dfrac{0}{5} = 0$

(b) $g(4) = \dfrac{4}{5 - 4} = \dfrac{4}{1} = 4$

(c) $g(-1) = \dfrac{-1}{5 - (-1)} = -\dfrac{1}{6}$

(d) $g(3) = \dfrac{3}{5-3} = \dfrac{3}{2}$

(e) $g(-5) = \dfrac{-5}{5-(-5)} = \dfrac{-5}{10} = -\dfrac{1}{2}$

Chapter 9 Test

1. $(x+1)^2 = 14$

$\qquad x+1 = \pm\sqrt{14}$

$\qquad x = -1 \pm \sqrt{14}$

3. $\qquad 3x^2 - 5x = -1$

$3x^2 - 5x + 1 = 0$

$x = \dfrac{-(-5) \pm \sqrt{(-5)^2 - 4(3)(1)}}{2(3)}$

$x = \dfrac{5 \pm \sqrt{25 - 12}}{6}$

$x = \dfrac{5 \pm \sqrt{13}}{6}$

5. $(c-12)^2 = 12$

$\quad c - 12 = \pm\sqrt{12}$

$\qquad c = 12 \pm \sqrt{12}$

$\qquad c = 12 \pm \sqrt{4 \cdot 3}$

$\qquad c = 12 \pm 2\sqrt{3}$

7. $3t^2 = 30$

$\quad t^2 = 10$

$\quad t = \pm\sqrt{10}$

9. $6p^2 - 11p = 0$

$p(6p - 11) = 0$

$p = 0 \quad \text{or} \quad 6p - 11 = 0$

$\qquad\qquad\qquad p = \dfrac{11}{6}$

11. $4\pi r^2 = 201$

$r^2 = \dfrac{201}{4\pi}$

$r = \pm\sqrt{\dfrac{201}{4\pi}}$

Since the answer to this problem is a measurement, only the positive value is a solution.

$r = \sqrt{\dfrac{201}{4\pi}} \approx 4.0 \text{ in.}$

13. For $y = ax^2 + bx + c$ if $a > 0$, the parabola opens upward; if $a < 0$, the parabola opens downward.

15. $y = 3x^2 - 6x + 8$

$x = \dfrac{-b}{2a} = \dfrac{-(-6)}{2(3)} = 1$

$y = 3(1)^2 - 6(1) + 8$

$\quad = 3 - 6 + 8$

$\quad = 5$

$(1, 5)$

17. The parabola has no x-intercepts.

19. $y = -x^2 + 25$

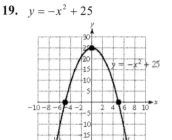

Vertex: $(0, 25)$
x-intercepts: $(-5, 0)$ and $(5, 0)$
y-intercept: $(0, 25)$

21. (a) Domain: $(-\infty, 0]$; Range: $(-\infty, \infty)$;
The relation is not a function because a vertical line will intersect the relation more than once.

(b) Domain: $(-\infty, \infty)$; Range: $(-\infty, 4]$;
The relation is a function because a vertical line will intersect the relation only once.

23. $D(x) = \frac{1}{2}x(x-3)$

(a) $D(5) = \frac{1}{2}(5)(5-3) = \frac{1}{2}(5)(2) = 5$

A five-sided polygon has five diagonals.

(b) $D(10) = \frac{1}{2}(10)(10-3) = \frac{1}{2}(10)(7) = 35$

A 10-sided polygon has 35 diagonals.

(c) $20 = \frac{1}{2}x(x-3)$

$40 = x(x-3)$

$0 = x^2 - 3x - 40$

$0 = (x-8)(x+5)$

$x - 8 = 0$ or $x + 5 = 0$

$x = 8 \qquad x = -5$

Since x represents the number of sides of a polygon, it must be a positive value. So, a polygon with 20 diagonals has 8 sides.

Cumulative Review Exercises
Chapters 1–9

1.
$3x - 5 = 2(x-2)$

$3x - 5 = 2x - 4$

$3x - 2x - 5 = 2x - 2x - 4$

$x - 5 = -4$

$x - 5 + 5 = -4 + 5$

$x = 1$

3.
$\frac{1}{2}y - \frac{5}{6} = \frac{1}{4}y + 2$

$6y - 10 = 3y + 24$

$6y - 3y - 10 = 3y - 3y + 24$

$3y - 10 = 24$

$3y - 10 + 10 = 24 + 10$

$3y = 34$

$y = \frac{34}{3}$

5. $y = -37.6x + 1353$ where $8 \le x \le 13$

(a) Decreases

(b) $m = -37.6$
For each additional increase in education level, the death rate decreases by approximately 38 deaths per 100,000 people.

(c) $y = -37.6(12) + 1353$
$y = -451.2 + 1353$
$y = 901.8$
The expected death rate would be 901.8 per 100,000.

(d) $977 = -37.6x + 1353$
$-376 = -37.6x$
$10 = x$
The approximate median education level for a city with 977 deaths per 100,000 is about 10th grade.

7. $(5.2 \times 10^7)(365) = 1898 \times 10^7$
$= 1.898 \times 10^{10}$ diapers

9. $(2x-3)^2 - 4(x-1)$
$= (2x)^2 - 2(2x)(3) + 3^2 - 4x + 4$
$= 4x^2 - 12x + 9 - 4x + 4$
$= 4x^2 - 16x + 13$

11. $2x^2 - 9x - 35 = (2x+5)(x-7)$

13. Let x represent the height of a triangle. The length of the base of the triangle is represented by $x + 1$.

279

$$\frac{1}{2}x(x+1)=36$$

$$x(x+1)=72$$

$$x^2+x-72=0$$

$$(x-8)(x+9)=0$$

$$x-8=0 \quad \text{or} \quad x+9=0$$

$$x=8 \qquad\qquad x=-9$$

Since the answer to this problem is a measurement, only the positive value is a solution.

$$x=8$$

$$x+1=9$$

The base is 9 m and the height is 8 m.

15.
$$\frac{x^2+10x+9}{x^2-81}\cdot\frac{18-2x}{x^2+2x+1}$$

$$=\frac{(x+1)(x+9)}{(x-9)(x+9)}\cdot\frac{-2(x-9)}{(x+1)^2}$$

$$=-\frac{2}{x+1}$$

17.
$$\frac{\frac{1}{x+1}-\frac{1}{x-1}}{\frac{x}{x^2-1}}=\frac{\frac{1}{x+1}-\frac{1}{x-1}}{\frac{x}{(x-1)(x+1)}}$$

$$=\frac{\frac{1}{x+1}-\frac{1}{x-1}}{\frac{x}{(x-1)(x+1)}}\cdot\frac{(x-1)(x+1)}{(x-1)(x+1)}$$

$$=\frac{(x-1)-(x+1)}{x}$$

$$=\frac{x-1-x-1}{x}$$

$$=-\frac{2}{x}$$

19
$$y-3=\frac{1}{2}(x-(-2))$$

$$y-3=\frac{1}{2}(x+2)$$

$$y-3=\frac{1}{2}x+1$$

$$y=\frac{1}{2}x+4$$

21. $4x+12=0$

(a) x-intercept: $4x+12=0$
$$4x=-12$$
$$x=-3 \quad (-3,0)$$

(b) y-intercept: $4(0)+12=0$
$$12=0 \quad \text{None}$$

(c) It has been shown in part b that this equation can be written in the form $x=$ a constant. This indicates the line is a vertical line whose slope is undefined.

(d)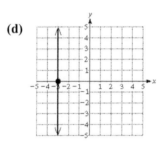

23. $2x-y=8$

$$4x-4y=3x-3$$

Since the coefficient on y in the first equation is -1, solve this equation for y.

$$2x-y=8$$

$$-y=-2x+8$$

$$y=2x-8$$

Substitute this value for y into the second equation and solve for x.

$$4x-4(2x-8)=3x-3$$

$$4x-8x+32=3x-3$$

$$-4x+32=3x-3$$

$$-7x+32=-3$$

$$-7x=-35$$

$$x=5$$

Substitute the value found for x into the first equation and solve for y.

$$2(5)-y=8$$

$$10-y=8$$

$$-y=-2$$

$$y=2 \qquad (5,2)$$

25.

	Number	Value	Total value
Dimes	x	0.10	$0.10x$
Quarters	y	0.25	$0.25y$
	27		4.80

$$x + y = 27$$
$$0.10x + 0.25y = 4.80$$

Solve this system of equations by the substitution method. Solve the first equation for x.

$$x + y = 27$$
$$x = 27 - y$$

Multiply the second equation by 100 to eliminate the decimals.

$$0.10x + 0.25y = 4.80$$
$$10x + 25y = 480$$

Substitute the value found for x into the above equation.

$$10(27 - y) + 25y = 480$$
$$270 - 10y + 25y = 480$$
$$270 + 15y = 480$$
$$15y = 210$$
$$y = 14$$

Substitute the value found for y into the first equation and solve for x.

$$x + 14 = 27$$
$$x = 13$$

There are 13 dimes and 14 quarters.

27. π, $\sqrt{7}$

29.
$$\frac{\sqrt{16x^4}}{\sqrt{2x}} = \sqrt{\frac{16x^4}{2x}}$$
$$= \sqrt{8x^3}$$
$$= \sqrt{4 \cdot 2 \cdot x^2 \cdot x}$$
$$= 2x\sqrt{2x}$$

31. $-3\sqrt{2x} + \sqrt{50x} = -3\sqrt{2x} + \sqrt{25 \cdot 2x}$
$$= -3\sqrt{2x} + 5\sqrt{2x}$$
$$= 2\sqrt{2x}$$

33.
$$\sqrt{x + 11} = x + 5$$
$$\left(\sqrt{x + 11}\right)^2 = (x + 5)^2$$
$$x + 11 = x^2 + 10x + 25$$
$$0 = x^2 + 9x + 14$$
$$0 = (x + 2)(x + 7)$$
$$x + 2 = 0 \quad \text{or} \quad x + 7 = 0$$
$$x = -2 \qquad\qquad x = -7$$

When checking these values in the original equation, $x = -7$ does not check.
Solution: $x = -2$

35. b

37. Domain: $\{2, -1, 9, -6\}$
Range: $\{4, 3, 2, 8\}$

39.
$$-4x - 5y = 10$$
$$-5y = 4x + 10$$
$$y = -\frac{4}{5}x - 2$$
$$m = -\frac{4}{5}$$

41.
$$2x^2 + 12x + 6 = 0$$
$$x^2 + 6x + 3 = 0$$
$$x^2 + 6x = -3$$
$$x^2 + 6x + 9 = -3 + 9$$
$$(x + 3)^2 = 6$$
$$x + 3 = \pm\sqrt{6}$$
$$x = -3 \pm \sqrt{6}$$

43.

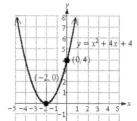

Vertex: $(-2, 0)$
x-intercept: $(-2, 0)$
y-intercept: $(0, 4)$